SO-AYT-704

Advances in

INORGANIC CHEMISTRY

AND

RADIOCHEMISTRY

Volume 30

Advances in

INORGANIC CHEMISTRY

AND

RADIOCHEMISTRY

EDITORS

H. J. EMELÉUS

A. G. SHARPE

University Chemical Laboratory
Cambridge, England

VOLUME 30

1986

ACADEMIC PRESS, INC.
Harcourt Brace Jovanovich, Publishers

Orlando   San Diego   New York   Austin
Boston   London   Sydney   Tokyo   Toronto

COPYRIGHT © 1986 BY ACADEMIC PRESS, INC.
ALL RIGHTS RESERVED.
NO PART OF THIS PUBLICATION MAY BE REPRODUCED OR
TRANSMITTED IN ANY FORM OR BY ANY MEANS, ELECTRONIC
OR MECHANICAL, INCLUDING PHOTOCOPY, RECORDING, OR
ANY INFORMATION STORAGE AND RETRIEVAL SYSTEM, WITHOUT
PERMISSION IN WRITING FROM THE PUBLISHER.

ACADEMIC PRESS, INC.
Orlando, Florida 32887

United Kingdom Edition published by
ACADEMIC PRESS INC. (LONDON) LTD.
24–28 Oval Road, London NW1 7DX

LIBRARY OF CONGRESS CATALOG CARD NUMBER: 59-7692

ISBN 0–12–023630–3 (alk. paper)

PRINTED IN THE UNITED STATES OF AMERICA

86 87 88 89   9 8 7 6 5 4 3 2 1

CONTENTS

Catenated Nitrogen Ligands Part I. Transition Metal Derivatives of Triazenes, Tetrazenes, Tetrazadienes, and Pentazadienes

DAVID S. MOORE AND STEPHEN D. ROBINSON

The Coordination Chemistry of 2,2′:6′,2″-Terpyridine and Higher Oligopyridines

E. C. CONSTABLE

High-Nuclearity Carbonyl Clusters: Their Synthesis and Reactivity

MARIA D. VARGAS AND J. NICOLA NICHOLLS

Inorganic Chemistry of Hexafluoroacetone

M. WITT, K. S. DHATHATHREYAN, AND H. W. ROESKY

CATENATED NITROGEN LIGANDS PART I.[1] TRANSITION METAL DERIVATIVES OF TRIAZENES, TETRAZENES, TETRAZADIENES, AND PENTAZADIENES

DAVID S. MOORE*[2] and STEPHEN D. ROBINSON**

* Department of Chemistry, Dover College, Dover, Kent CT17 9RL, England, and
** Department of Chemistry, King's College London, Strand, London WC2R 2LS, England

I. Introduction

Although the great strength of the nitrogen–nitrogen triple bond relative to the corresponding double and single bonds (946 vs. 418 and 160 kJ, respectively) strongly militates against the stability of molecules containing catenated nitrogen systems, a substantial number of such structures are in fact known. These range from 3-nitrogen systems such as the azide anion (N_3^-) and 1,3-disubstituted triazenes ($RN=N-NHR$) through 4-, 5-, 6-, and 7- to 8- and, in one instance, 10-nitrogen chains (*10*). Whereas the parent "hydronitrogens" are generally unknown or, at best, dangerously explosive, their substituted analogues frequently enjoy much greater stability, particularly when aryl substituents are present. The very extensive chemistry of these catenated nitrogen systems, which originated with Peter Griess' discovery of 1,3-diaryltriazenes in 1859 (*91*), has recently been comprehensively reported in F. A. Benson's book *The High Nitrogen Compounds* (*10*).

The coordination of catenated nitrogen ligands to transition metals also dates back to the early work of Griess (*89, 90*), which included references to copper and silver derivatives of 1,3-diphenyltriazene. Around the turn of the century Meldola and Streatfeild (*146–148*), Meunier (*150–152*), Niementowski and Roszkowski (*159*), Cuisa and Pestalozza (*55, 56*), and others reported extensively on triazene complexes of copper, silver, and mercury, and in the late 1930s and early 1940s Dwyer and colleagues (*69–74*) extended this work to include derivatives of nickel and palladium. However, most work on the coordination chemistry of triazenes and other catenated

---

[1] Part II. Transition metal derivatives of *vic*-triazoles, tetrazoles, and pentazoles. *Adv. Inorg. Chem. Radiochem.* **31**, in preparation.

[2] Present address: Department of Chemistry, St. Edwards School, Oxford OX2 7NN, England.

1

Copyright © 1986 by Academic Press, Inc.
All rights of reproduction in any form reserved.

nitrogen ligands has been reported in the past 25 years and the field appears ripe for continued rapid growth.

The observation that, with a few notable exceptions, coordination to transition metals imparts stability to catenated nitrogen systems promises exciting developments in the future. Although organic molecules containing up to 10 linked nitrogen atoms have been synthesized, metal complexes are known to date only for systems containing 2-, 3-, 4-, or 5-nitrogen chains. The chemistry of two-nitrogen ligands—notably hydrazines, diazenes, and, of course, dinitrogen itself—has received an enormous boost in recent years from the work on nitrogen fixation. However, the exciting developments in this field have recently been extensively reviewed elsewhere (*101*) and are therefore omitted from the present article. The three-nitrogen systems include, in addition to the azide anion, triazenes, and their *N*-oxides. The chemistry of azide complexes, a large field meriting a review in its own right, is specifically excluded from the present article. Complexes of triazene *N*-oxides are also excluded since they were recently covered in an exhaustive review containing over 200 references (*68*). However complexes containing ArN=N—N(Ar)C(O) or "phosphazide" (ArN=N—N=PR$_3$) ligands are covered. The chemistry of four-nitrogen ligands is dominated by complexes of tetrazadienes, RN=N—N=NR, which are covered in the present article together with the few isolated examples of tetrazene derivatives. The latter include a binuclear tungsten derivative of the hypothetical *iso*-tetrazene molecule (H$_2$N)$_2$N=N. Finally, we include the first transition metal derivatives of pentazadienes RN=N—N(H)—N=NR.

Cyclic catenated nitrogen ligands will be covered in Section II.

II. Triazenide Complexes

1,3-Diaryltriazenes, the first organic compounds containing three or more nitrogen atoms in sequence, were prepared by Griess in 1859 (*91*). The less stable alkylaryl and dialkyl triazenes were first obtained by Dimroth in 1903 (*60*) and 1906 (*63*), respectively. The extremely unstable parent molecule, HN=N—NH$_2$, is formed during electrode-less discharge in a high-speed stream of hydrazine (*83*), and is also considered to be present in aqueous hydrazine solution after pulse radiolysis (*100*). However, the conjugate base, HN=N—NH$^-$, has been stabilized as a ligand in the osmium cluster [Os$_3$(μ-H)(μ-HNNNH)(CO)$_{10}$] (*113*). Reviews of triazene chemistry include articles on triazene structures and stability by Süling (*204*) and Smith (*202*), diaryltriazenes by Campbell and Day (*28*), and monoalkylmonoaryltriazenes by Vaughan and Stevens (*221*).

Although the ability of triazenes to complex transition metal ions, notably copper, silver, and mercury, was recognized at a very early stage, and was further explored by Dwyer in the 1940s, most of the work in this field has been reported since 1965. Papers published between 1965 and 1974 by Corbett and Hoskins (47–50), Knoth (120), Robinson and Uttley (190, 191), and Brinckman et al. (17) on triazenide complexes of the group VIII metals aroused new interest in the field. Subsequent reports by many authors describe triazenide complexes of these and most other d-block transition metals. Numerous examples of metal complexes containing monodentate, chelate, or bridging triazenide ligands are now known. In contrast no complexes containing the fully saturated triazanes R_2N—$N(R)$—NR_2 appear to have been reported.

A. SYNTHESIS AND PROPERTIES

Diaryl-, alkylaryl-, and, to a lesser degree, dialkyltriazenes are readily available. Consequently most triazenide complexes are synthesized from the free triazene or one of its salts. The salient methods, together with illustrative examples, are given below.

1. From metal halides and free triazenes in the presence of base (71).

$$CuCl + ArNNNHAr + NaOH + 2py \longrightarrow$$
$$Cu(ArNNNAr)(py)_2 + NaCl + H_2O(py, \text{ pyridine}) \quad (1)$$

2. From metal carboxylates and free triazenes (71, 111).

$$Cu(O_2CMe)_2 + 2ArNNNHAr \longrightarrow Cu(ArNNNAr)_2 + 2MeCO_2H \quad (2)$$

$$[Pd(allyl)(O_2CMe)]_2 + 2ArNNNHAr \longrightarrow$$
$$[Pd(allyl)(ArNNNAr)]_2 + 2MeCO_2H \quad (3)$$

3. From metal halides and lithium (120), sodium (42), or magnesium (17) triazenides.

$$RhCl(PPh_3)_3 + LiBu^n + ArNNNHAr \longrightarrow$$
$$Rh(ArNNNAr)(PPh_3)_2 + LiCl + C_4H_{10} + PPh_3 \quad (4)$$

$$[RhCl(CO)_2]_2 + 2Na(ArNNNAr) \longrightarrow$$
$$[Rh(ArNNNAr)(CO)_2]_2 + 2NaCl \quad (5)$$

$$TiCl_4 + 4(MeNNNMe)MgI \longrightarrow Ti(MeNNNMe)_4 + 4MgICl \quad (6)$$

4. From metal halides and organosilicon or tin triazenides (1).

$$Me_3Si(ArNNNAr) + MnBr(CO)_5 \longrightarrow$$
$$Mn(ArNNNAr)(CO)_4 + CO + SiBrMe_3 \quad (7)$$

$$Me_3Sn(ArNNNAr) + MoCl(C_5H_5)(CO)_3 \longrightarrow$$
$$CO + Mo(ArNNNAr)(C_5H_5)(CO)_2 + SnClMe_3 \quad (8)$$

5. By oxidative addition of triazenes to low oxidation state metal complexes (*135*).

$$Ru(CO)_3(PPh_3)_2 + ArNNNHAr \longrightarrow$$

$$RuH(ArNNNAr)(CO)(PPh_3)_2 + 2CO \quad (9)$$

6. From metal hydrides (*133*) or alkyls (*53*) and free triazenes.

$$OsH_4(PPh_3)_3 + ArNNNHAr \longrightarrow$$

$$OsH_3(ArNNNAr)(PPh_3)_2 + H_2 + PPh_3 \quad (10)$$

$$Li_4[Cr_2Me_8] + 4ArNNNHAr \longrightarrow$$

$$Cr_2(ArNNNAr)_4 + 4CH_4 + 4LiMe \quad (11)$$

7. From metal hydrides (*113*) or alkyls (*40*) and aryl or silyl azides.

$$Os_3H_2(CO)_{10} + Me_3SiN_3 \longrightarrow Os_3H(HNNNH)(CO)_{10} + SiHMe_3 \quad (12)$$

$$ZrMe_2(C_5H_5)_2 + PhN_3 \longrightarrow ZrMe(PhNNNMe)(C_5H_5)_2 \quad (13)$$

8. From arylamines, isoamyl nitrite, and metal acetates or nitrates (*225*).

$$4ArNH_2 + 2RONO + Hg(O_2CMe)_2 \longrightarrow$$

$$Hg(ArNNNAr)_2 + 2MeCO_2H + 2ROH + 2H_2O \quad (14)$$

9. From metal nitrites and aromatic amines (*159*).

$$AgNO_2 + 2PhNH_2 \longrightarrow Ag(PhNNNPh) + 2H_2O \quad (15)$$

10. From metallic mercury and free triazenes in the presence of oxygen (*231*).

$$2Hg + 4ArNNNHAr + O_2 \longrightarrow 2Hg(ArNNNAr)_2 + 2H_2O \quad (16)$$

Finally, copper(I), silver(I), and mercury(II) triazenides react with selected rhodium(I), iridium(I), and platinum(II) halide complexes to afford metal-metal bonded binuclear triazenide-bridged species (*129*).

$$IrCl(CO)(PPh_3)_2 + Ag(ArNNNAr) \longrightarrow$$

$$(PPh_3)_2(CO)Ir(\mu\text{-}ArNNNAr)AgCl \quad (17)$$

Most triazenide complexes of the transition metals are air-stable, crystalline solids, soluble in many common nonpolar organic solvents. Most are deeply colored, ranging from bright yellow to deep red or brown. Dialkyl triazenide complexes tend to be rather less stable and those of titanium and zirconium are reported to be sensitive to air and moisture (*17*).

B. STRUCTURAL PROPERTIES

The free triazenes, $RN{=}N{-}NHR$, are generally believed to adopt a trans configuration (1) and this arrangement has been confirmed for several diaryltriazenes by X-ray diffraction studies (*122, 123, 163*). No examples of salts containing noncoordinated triazenide anions appear to have been characterized by X-ray diffraction methods. The triazenide anions, $RN{=}N{-}NR^-$, are formally analogous to the nitrite anion, $O{=}N{-}O^-$, and might be expected to display a similar variety of coordination modes. The most feasible of these are illustrated in formulas **2-8**. Numerous examples of complexes containing monodentate (**2**), chelate (**3**), or bridging (**4**) triazenide ligands have been reported. The N-1,N-1- and N-1,N-2-bridging arrangements (**5** and **6**) have not been observed, and coordination through N-2 (**7**)—analogous to that found in the nitro form of NO_2^- coordination—is also unknown. Finally the η^3-structure (**8**), which is formally related to that found in η^3-allyl complexes, is still only a hypothetical bonding option.

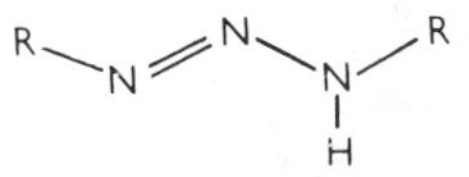

(1)

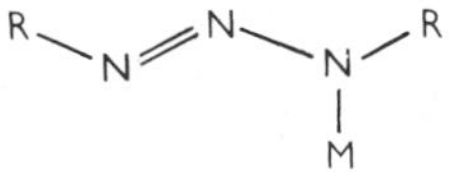

(2)

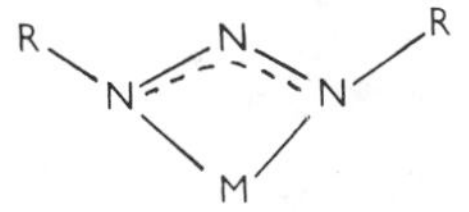

(3)

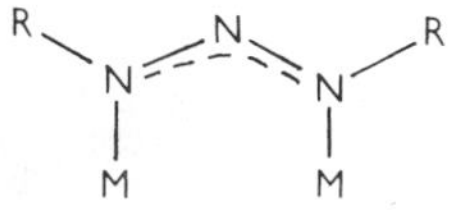

(4)

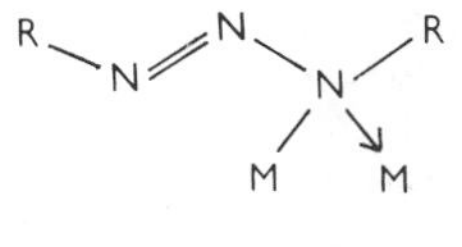

(5)

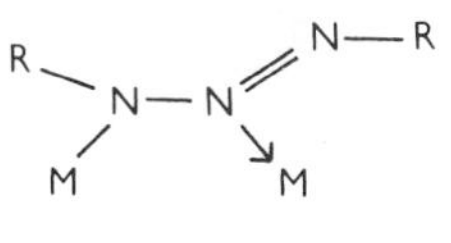

(6)

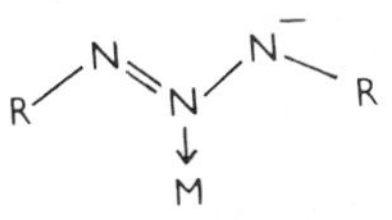

(7)

(8)

1. Monodentate Triazenide Structures

Although monodentate triazenide ligands were first proposed (erroneously) by Meunier in 1900 (*150, 152*) for the copper(I) complex "Cu(PhNNNPh)," this mode of coordination was not finally confirmed until 1976 when Brown and Ibers (*21, 22*) reported the X-ray crystal structure of *cis*-Pt(PhNNNPh)$_2$(PPh$_3$)$_2$·C$_6$H$_6$ (Fig. 1). Monodentate triazenide ligands remain relatively rare and are mainly confined at present to the four-coordinate d^8 metal ions, Rh(I), Ir(I), Pd(II), and Pt(II). To date only six examples have been characterized by diffraction methods (see Table I). In some of these, notably *trans*-PtH(*p*-tol-NNN-tol-*p*)(PPh$_3$)$_2$, there is a short nonbonded contact (2.91 Å) between the metal and the terminal (N-3) nitrogen atom (*109*). Variable-temperature proton NMR studies (see Section II,C,3) have established that η^1-triazenide ligands, like their η^1-allyl counterparts, display fluxional behavior.

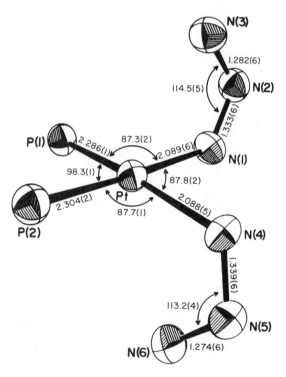

Fig. 1. View of the environment of the platinum atom in *cis*-Pt(PhNNNPh)$_2$(PPh$_3$)$_2$ showing salient bond lengths and angles.

2. Chelate Triazenide Structures

Chelate structures in which the triazenide anion coordinates to the metal through nitrogens N-1 and N-3 were first advanced by Dwyer in 1939 for silver(I) and mercury(II) triazenide complexes (69, 70), and subsequently for nickel(II) and palladium(II) derivatives (71, 74). This proposal was at first criticized by Sidgwick (199) and others (98), who reasoned that closure of four-membered chelate rings in triazenide or carboxylate complexes would impose excessive steric strain. Although Dwyer's ideas were later vindicated, some of his original structures were incorrect, and it was not until 1967 that Corbett and Hoskins' X-ray crystal structure of the cobalt complex Co(PhNNNPh)$_3$ (Fig. 2) finally provided unequivocal proof that triazenide anions could serve as chelate ligands (49). It is interesting to note that this structure was the first in which the presence of three four-membered chelate rings within the same octahedral coordination sphere was established. Fourteen X-ray structures involving chelate triazenide ligands are listed in Table I together with salient bond lengths and angles for each. The twofold symmetry of the triazenide ligands in the tris(chelates) M(PhNNNPh)$_3$ (M = Cr, Co) clearly establishes the electron delocalization in the chelate triazenide ligand. The high degree of strain in the four-membered chelate rings is amply demonstrated by the angles subtended at the metal center ($\angle$ N-1—M—N-3 $\sim$ 56–65°) and at the central nitrogen ($\angle$ N-1—N-2—N-3 $\sim$ 99–107°).

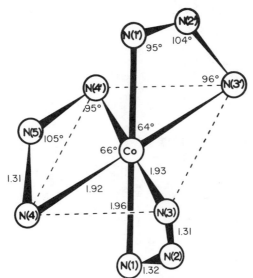

FIG. 2. View of the environment of the cobalt atom in Co(PhNNNPh)$_3$ showing salient bond lengths (in Å) and angles (in degrees).

TABLE I

SELECTED BOND LENGTHS[a,b] AND ANGLES[a,b] FOR TRIAZENIDE COMPLEXES

Generic triazenide skeleton labeling shown in header diagrams: R1—N1—N2=N3—R3 with metal atoms M1 (and M3).

Molecular formula	R1	R3	M1—M3	M1—N1	M1—N3	M3—N3	N1—N2	N2—N3	∠N1M1N3	∠N1N2N3	∠M1M3N3	∠M3M1N1	References
Monodentate triazenide													
trans-Pd(RNNNR)Cl(PPh$_3$)$_2$	p-tol	p-tol	—	2.033	(2.836)	—	1.336	1.286	—	113.0	—	—	12, 13, 209
trans-Pt(RNNNR)Cl(PPh$_3$)$_2$	p-tol	p-tol	—	2.114	(3.008)	—	1.266	1.257	—	116.4	—	—	12, 209
trans-Pt(RNNNR)H(PPh$_3$)$_2$	p-tol	p-tol	—	2.090	(2.908)	—	1.310	1.264	—	111.6	—	—	109, 209
cis-Pt(RNNNR)$_2$(PPh$_3$)$_2$	Ph	Ph	—	2.088	(2.974)	—	1.336	1.278	—	113.9	—	—	21, 22
trans-Ir(RNNNR)(CO)(PPh$_3$)$_2$	p-tol	p-tol	—	2.16	(2.58)	—	1.31	1.28	—	109.7	—	—	110
PhHg(RNNNR)	o-ClC$_6$H$_4$	Ph	—	2.14	(2.46)	—	1.36	1.33	—	99	—	—	133
Chelate triazenide													
Cr(RNNNR)$_3$	Ph	Ph	—	2.01	2.01	—	1.31	1.31	62	105	—	—	53
Mo$_2$(RNNNR)$_4$	p-tol	p-tol	2.212	2.271	2.231	—	1.310	1.316	55.8	106.7	—	—	37
Mo(RNNNR)(C$_5$H$_5$)(CO)$_2$	C$_6$H$_3$(CF$_3$)$_2$	C$_6$H$_3$(CF$_3$)$_2$	—	2.12		—	1.31		56	101	—	—	179
Mo(RNNNR)(C$_5$H$_5$)(CO)$_2$	Ph	Ph	—	2.114	2.126	—	1.326	1.292	56.8	100.8	—	—	182
W$_2$(RNNNR)$_2$(NMe)Et$_2$	Ph	Ph	2.304	2.232	2.160	—	1.308	1.325	56.68	104.8	—	—	35a
W$_2$(RNNNR)$_2$(NMe$_2$)$_4$	Ph	Ph	2.314	2.186	2.211	—	1.32		56.6	104.5	—	—	39
Tc(RNNNR)(CO)$_2$(PMe$_2$Ph)$_2$	p-tol	Ph	—	2.186		—	1.316		57.2	105.4	—	—	142a
Re(RNNNR)(CO)$_2$(PPh$_3$)$_2$	Ph	Ph	—	2.21	2.18	—	1.33	1.31	57	105	—	—	87
Re(RNNNR)Cl$_2$(PPh$_3$)$_2$	p-tol	p-tol	—	2.08	1.99	—	1.31	1.30	58	99	—	—	192
Ru(RNNNR)H(CO)(PPh$_3$)$_2$	p-tol	p-tol	—	2.149	2.179	—	1.318	1.310	57.7	105.2	—	—	20, 22
Co(RNNNR)$_3$ (monoclinic)	Ph	Ph	—	1.94		—	1.31	1.31	65	105	—	—	49
Co(RNNNR)$_3$ (trigonal)	Ph	Ph	—	1.92		—	1.31	1.31	64.8	103.2	—	—	127
(OC)$_5$MnHg(RNNNR)	p-ClC$_6$H$_4$	p-ClC$_6$H$_4$	2.557	2.314	2.435	—	1.332	1.262	52.82	109.2	—	—	112
W(PhNNNPPh$_3$)Br$_2$(CO)$_3$[c]	Ph	PPh$_3$	—	2.163	2.220	—	1.279	1.364	56.7	103.8	—	—	105
[(C$_8$H$_{12}$)Ir(RNNNR)$_2$HgCl]$_2$[d]	p-tol	p-tol	—	2.19	2.06	—	1.30	1.29	58.2	106	—	—	217

Bridging triazenide

Complex	R													Ref
$Cr_2(RNNNR)_4$	Ph	1.858	2.052	—	2.037	—	1.297	1.302	—	—	112.8	93.2	94.2	53
$Mo_2(RNNNR)_4$	Ph	2.083	2.145	—	2.13	—	1.33	1.30	—	—	113	91.4	91.4	53
$Mo_2(RNNNR)_2(NMe_2)_2Me_2$	p-tol	2.174	2.283	—	2.157	—	1.290	1.134	—	—	113.1	86.1	91.7	38
$Mo_2(RNNNR)_2(NMe_2)_2Et_2$	p-tol	2.171	2.205	—	2.212	—	1.283	1.315	—	—	113.7	89.3	89.3	34
$W_2(RNNNR)_2(NMe_2)_2Et_2$	p-tol	2.267	2.156	—	2.254	—	1.305	1.316	—	—	—	—	—	35a
$W_2(RNNNR)_2(dmhp)_2$	Ph	2.169	2.108	—	2.097	—	1.33	1.33	—	—	111.9	90.6	90.1	52
$Os_3(RNNNR)H(CO)_{10}$	H	2.923	2.12	—	2.13	—	1.32	1.32	—	—	118	—	—	113
$Os_3(RNNNR)H(CO)_{10}$	H	2.960	2.137	—	2.110	—	1.32	1.29	—	—	117.5	—	—	26
$Rh_2(RNNNR)_2(CO)_2(PPh_3)_2$	p-tol	2.900	2.135	—	2.140	—	—	—	—	—	—	—	—	43
$[Rh_2(RNNNR)_2(CO)_2(PPh_3)_2]PF_6$	p-tol	2.698	2.095	—	2.098	—	—	—	—	—	—	—	—	43
$Ni_2(RNNNR)_4$	Ph	2.395	1.924	—	1.907	—	—	—	1.312	—	116.5	86.8	86.4	45, 46, 50
$Pd_2(RNNNR)_4$	Ph	2.5626	2.049	—	2.033	—	—	—	1.313	—	117.6	85.0	84.9	45, 46
$Pd_2(RNNNR)_2(methallyl)_2$	Me	(2.97)	2.115	—	2.128	—	1.32	1.30	—	—	118.3	—	—	103
$Pd_2(RNNNR)_2(methallyl)_2$	p-tol	(2.86)	2.098	—	2.120	—	1.298	1.301	—	—	116.7	—	—	29
$Cu_2(RNNNR)_4$	Ph	2.441	2.032	—	2.007	—	—	—	1.296	—	117.1	85.7	85.4	45, 46
$Cu_2(RNNNR)_2$	Me	2.45	1.90	—	1.94	—	1.27	1.29	—	—	115.8	86.2	86.0	19
$Cu_4(RNNNR)_4$	Ph	2.66	1.87	—	1.87	—	—	—	1.29	—	116.0	—	—	161
$Au_4(RNNNR)_4$	Ph	2.85	2.04	—	2.04	—	1.27	1.30	—	—	118.6	—	—	6a
$Zn_4O(RNNNR)_6$	Ph	(3.11)	2.04	—	2.04	—	—	—	1.31	—	117.0	—	—	48
$(Ph_3P)_2(CO)Rh(RNNNR)CuCl$	Me	2.730	2.14	—	1.91	—	1.25	1.27	—	—	119	85.0	79.1	124
$(Me_2PhP)_2(CO)Ir(RNNNR)CuCl$	Me	2.686	2.08	—	1.89	—	1.29	1.29	—	—	116	86.9	78.8	125
$(Ph_3P)_2(CO)Ir(RNNNR)AgO_2CPr^i$	p-tol	2.874	2.086	—	2.111	—	1.298	1.339	—	—	115.1	80.6	81.3	131
$[(C_8H_{12})Ir(RNNNR)_2HgCl]_2$ [d]	p-tol	2.618	2.10	—	2.42	—	1.31	1.27	—	—	117	80.7	82.3	217

[a] Bond lengths in angstroms (Å), angles in degrees (°). Values in parentheses refer to nonbonded distances.

[b] For complexes of high symmetry, bond lengths and angles quoted are averaged values for equivalent bonds.

[c] P–N–3 = 1.672 Å.

[d] This complex contains chelate and bridging triazenide ligands; see entries under both headings.

3. Bridging Triazenide Structures

A binuclear structure with chelating and bridging triazenide ligands was proposed by Dwyer and Mellor in 1941 (74) for the nickel(II) complex $Ni_2(PhNNNPh)_4$. However, this structure was incorrect and it was not until 1958 that Harris *et al.* (98) suggested that this complex and related copper(II) and palladium(II) derivatives all possess the now familiar copper(II) acetate or "lantern" structure (Fig. 3; M = Ni, Cu, or Pd). X-Ray diffraction studies on the nickel complex $Ni_2(PhNNNPh)_4$ (46, 50), the copper complex $Cu_2(PhNNNPh)_4$ (46), and the palladium complex $Pd_2(PhNNNPh)_4$ (46) subsequently established the correctness of Harris' suggestion. More recently, the same "lantern" structure has been established for $Cr_2(PhNNNPh)_4$ and $Mo_2(PhNNNPh)_4$ (53) by diffraction methods, and it is now recognized that the geometry of triazenide ligands is particularly suited to bridge formation in bi- or polynuclear metal complexes. Other interesting structures involving triazenide ligands include the copper(I) derivatives $Cu_2(PhNNNPh)_2$ (Fig. 4)

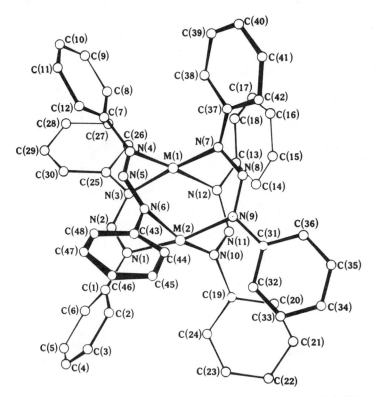

FIG. 3. View of the $M_2(PhNNNPh)_4$ dimers (M = Cu, Ni, Pd), showing the bridging of the metal atoms by the four triazenide ligands.

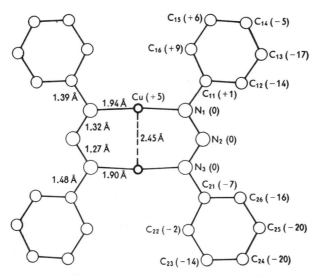

FIG. 4. Molecular structure of [Cu(PhNNNPh)]₂ dimers.

(19) and $Cu_4(MeNNNMe)_4$ (Fig. 5) (161), the allylpalladium complexes $Pd_2(RNNNR)_2(allyl)_2$ (29, 103), and the tetranuclear zinc complex $Zn_4O(PhNNNPh)_6$ (48), which has a "basic beryllium acetate" structure. A number of heterobinuclear complexes have been shown to contain bridging triazenide ligands supported by metal–metal bonds (see Sections II,E,5 and II,E,6). A full list of diffraction studies on triazenide complexes is given in Table I.

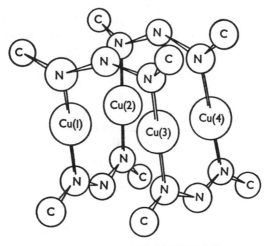

FIG. 5. Molecular structure of [Cu(MeNNNMe)]₄ tetramers.

C. Spectroscopic Studies

1. Vibrational Spectra

1,3-Diaryltriazenes display infrared absorptions in the range of $\sim 1100\text{-}1600 \text{ cm}^{-1}$, attributable to vibrations of the $-N{=}N{-}N{<}$ skeleton (*128*), which are perturbed on complex formation. Several authors have attempted to use these perturbations to determine the mode of coordination of the triazenide ligand. Robinson and Uttley (*190, 191*) reported that monodentate triazenide ligands absorbed at ~ 1150, 1190–1210, 1260–1300, and 1580–1600 cm^{-1}, whereas their chelate analogues absorbed only at 1260–1300 and 1580–1600 cm^{-1}. Knoth (*120*), in a later paper, reported that bridging triazenide ligands, in particular those in $Ni_2(PhNNNPh)_4$, show infrared spectral characteristics similar to those observed for monodentate triazenide ligands. Knoth further claimed that the complexes M(PhNNNPh)(CO)(PPh$_3$)$_2$ (M = Rh or Ir), which according to the criteria of Robinson and Uttley contain monodentate triazenide ligands, were really five-coordinate chelate complexes. However, this assertion was based on an erroneous interpretation of NMR data (*120*) and has subsequently been shown to be incorrect by an X-ray diffraction study of Ir(*p*-tol-NNN-tol-*p*)(CO)(PPh$_3$)$_2$, which has established the monodentate nature of the triazenide ligand (*110*). Kuyper *et al.* (*129, 130*) suggested that the absorptions at 1580–1600 cm^{-1} arose from the aryl groups rather than the triazenide skeleton, and further reported (*102, 129, 130*) that bridging triazenide ligands display a characteristic skeletal vibration at 1350–1375 cm^{-1}. This observation, taken with those of Robinson and Uttley, suggests that with care triazenide coordination modes can be assigned on the basis of vibrational spectra in most simple structures. However, in complex structures the situation is more ambiguous. This point is highlighted by the observation (*215*) that the triazenide skeletal vibrations for the binuclear complexes [2,6-(Me$_2$NCH$_2$)C$_6$H$_3$](*p*-tol-NNNR)PtAgBr (R = Me, Et) do not correspond to any of the categories given above.

2. Electronic Spectra

Studies of electronic spectra are confined almost exclusively to derivatives of d^{10} silver(I) (*236, 237*) and mercury(II) (*79, 194, 234*), both of which have been investigated extensively. Data have also been recorded for copper(II) triazenide complexes (*79, 236*) and nickel(II) (*84*).

3. Nuclear Magnetic Resonance

Proton NMR has been used extensively in the routine assignment of stereochemistry, particularly for triazenide ligands containing alkyl, *p*-tolyl,

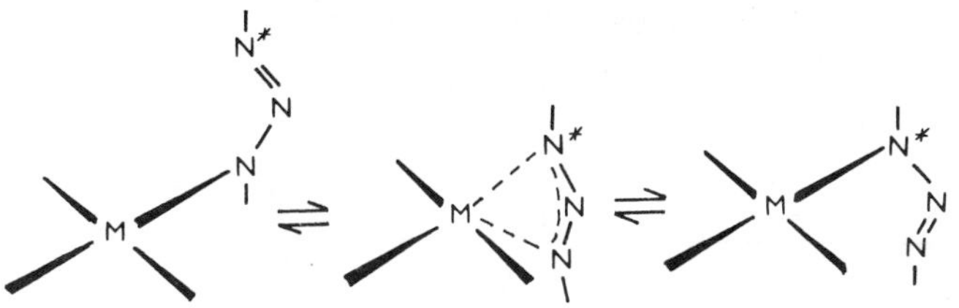

SCHEME 1. 1,3-Metallotropic equilibria in η^1-triazenide complexes.

or p-methoxyphenyl groups (1, 17, 102, 120, 135, 190). Variable-temperature proton NMR has been employed to study 1,3-metallotropic equilibria in η^1-triazenide complexes (Scheme 1) (54, 180, 191, 208, 209). Other processes investigated by variable-temperature proton NMR include dynamic interchange of chelate and bridging triazenide ligands (217), and the exchange of free and coordinated triazenes (158). Variable-temperature $^{13}C\{^1H\}$ NMR has also been used to examine fluxional processes in triazenide complexes (126, 158, 179, 183, 217). Intramolecular 1,3-metallotropic-exchange reactions in phenylmercuric triazenide complexes have been studied by ^{19}F, ^{15}N, and ^{199}Hg NMR (126, 157, 172, 174, 175).

4. Electron Spin Resonance

Few triazenide complexes reported to date are paramagnetic. Examples studied by ESR are restricted to the low-spin d^5 osmium(III) complexes $OsX_2(PhNNNPh)(PPh_3)_2$ (3), the binuclear mixed-valence rhodium salts $[\{Rh(\mu\text{-}ArNNNAr)(CO)(PPh_3)\}_2][PF_6]$ (43), and the low-spin d^7 cobalt(II) species $Co(ArNNNAr)(C_5H_5)L$ (178).

5. Mass Spectroscopy

Most triazene complexes are thermally stable and, where mass spectra have been recorded, molecular ions are usually observed. Fragmentation initially involves loss of ancillary ligands, notably CO; cleavage of the diaryltriazenide ligands generates aryl, aryldiazo, and arylazo ions together with nitrene species $[L_nM{=}NAr]^+$ (1, 42, 174, 179).

D. CHEMICAL REACTIVITY

Triazenide ligands when bound to transition metals are very stable entities and the $-N{=}N{=}N-$ system does not fragment except under the most vigorous conditions. Reactions undergone by coordinated triazenide ligands include protonation (*44*),

$$Rh(ArNNNAr)(CO)(PPh_3)_2 + HBF_4 \longrightarrow$$

$$[Rh(ArNNNHAr)(CO)(PPh_3)_2][BF_4] \quad (18)$$

reductive elimination (*207, 208*),

$$PtH(ArNNNAr)(PPh_3)_2 + 2PPh_3 \longrightarrow Pt(PPh_3)_4 + ArNNNHAr \quad (19)$$

and carbonyl (*177*) or isocyanide (*180*) "insertion."

$$CoI_2(C_5H_5)(CO) + Ag(ArNNNAr) \longrightarrow$$

$$\overline{CoI\{ArNNN(Ar)C(O)\}}(C_5H_5) + AgI \quad (20)$$

$$Ni(ArNNNAr)(C_5H_5)(CNR) \longrightarrow \overline{Ni\{ArNNN(Ar)C(NR)\}}(C_5H_5) \quad (21)$$

Electrochemical redox reactions have been reported for triazenide complexes of iron (*214*), cobalt (*214*), and rhodium (*43*).

E. GROUP SURVEY

To date no triazenide complexes appear to have been reported for the scandium, yttrium, and lanthanum group of metals.

1. Titanium, Zirconium, and Hafnium

Titanium and zirconium tetrahalides react with dimethyltriazenide-magnesium iodide, prepared from MeMgI and MeN$_3$, to afford the volatile, moisture-sensitive dark red (M = Ti) or orange (M = Zr) complexes M(MeNNNMe)$_4$. Proton NMR data (δ_{Me} single sharp peak) are consistent with symmetrical tetrakis(chelate) structures for both complexes (*17*). Treatment of the same tetrahalides with Ag(PhNNNPh) under anhydrous conditions yields TiCl(PhNNNPh)$_3$ and Zr(PhNNNPh)$_4$ (*18*). The organozirconium complex ZrMe$_2$(C$_5$H$_5$)$_2$ reacts with Me$_3$SiN$_3$ to produce the azide complex Zr(N$_3$)Me(C$_5$H$_5$)$_2$ but with phenyl azide affords a yellow air-sensitive triazenide derivative Zr(MeNNNPh)Me(C$_5$H$_5$)$_2$, for which a chelate structure (**9**) has been proposed (*40*). The analogous complex Zr(PhNNNPh)Ph(C$_5$H$_5$)$_2$ was prepared in a similar fashion (*40*). Triazenide complexes have been implicated as reactive intermediates in the formation of hafnium amides from azides and hafnium hydrides (Scheme 2) (*106*). Nitrogen-15 NMR-labeling experiments confirm that the terminal nitrogen of the azide formally inserts into the Hf—H bond. Isolation of diamides Hf(NHR)$_2$(C$_5$Me$_5$)$_2$ as the sole products when excess azide, RN$_3$, is used has tentatively been attributed to the formation and subsequent decomposition of the

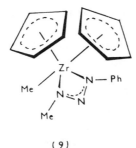

(9)

unstable η^1-triazenides $Hf(HNNNR)_2(C_5Me_5)_2$ and $Hf(NHR)(HNNNR)$ $(C_5Me_5)_2$ (106). Species of this type, notably $HfH(HNNNAr)(C_5Me_5)_2$ and $HfH(ArNNNAr)(C_5Me_5)_2$ have been obtained from reactions of $HfH_2(C_5Me_5)_2$ with azides (ArN_3) and triazenes (ArNNNHAr), respectively (106).

$$HfH_2(\eta^5-C_5Me_5)_2 \ + \ RN_3 \ \longrightarrow \ (\eta^5-C_5Me_5)_2Hf$$

$$HfH(NHR)(\eta^5-C_5Me_5)_2 \ \xleftarrow{-N_2} \ (\eta^5-C_5Me_5)_2Hf$$

SCHEME 2. Pathways for the formation of the hafnium amides from azides and hafnium hydrides. Reprinted with permission from *Organometallics* 1, 1025. Copyright (1982) American Chemical Society.

To date no triazenide complexes have been reported for the vanadium, niobium, and tantalum triad of elements.

2. Chromium, Molybdenum, and Tungsten

The complex anion $[Cr(PhNNNPh)(CO)_4]^-$, prepared from $Cr(CO)_6$ and $Na[PhNNNPh]$, has been isolated as an orange tetramethylammonium salt (118). Stoichiometric reactions of $CrPh_3(THF)_3$ with diphenyltriazene in tetrahydrofuran (THF) solution afford the triazenide

chelates $CrPh_2(PhNNNPh)(THF)_2$, $CrPh(PhNNNPh)_2(THF)$,   and
$Cr(PhNNNPh)_3$. Thermal and hydrolytic stability increases with increasing
triazene content (195). The tris(chelate) $Cr(PhNNNPh)_3$ has also been
obtained as the major product from the reaction of $[Cr_2Me_8]^{4-}$ with
diphenyltriazene, and has been shown to possess a severely strained octahedral structure (53). The minor product of this reaction, $Cr_2(PhNNNPh)_4$,
has been found to possess the expected "lantern" structure (Fig. 3, M = Cr),
analogous to that found for chromium(II) acetate (53). However, there is
considerable torsional rotation away from a perfectly eclipsed structure and
the chromium–chromium distance [1.858(1) Å] is extremely short even for a
quadruple bond [cf. 2.362(1) Å for chromous acetate (51)].

The analogous molybdenum complex, obtained from $[Mo_2Me_8]^{4-}$ and
diphenyltriazene, has a similar "lantern" structure [Mo≡Mo 2.083(2) Å]
(53). Binuclear molybdenum(II) triazenide complexes $Mo_2(ArNNNAr)_4$
(Ar = Ph, p-tol) have also been obtained from the organomolybdenum(II)
species $Mo_2R_2(NMe_2)_4$ (R = Et, Pr^i, Bu^n, Bu^s, Bu^t, or Bz) (34, 35), and a
"lantern" structure has been confirmed for $Mo_2(p\text{-}tol\text{-}NNN\text{-}tol\text{-}p)_4$ (34).
However, with closely related precursors $Mo_2R_2(NMe_2)_4$ (R = Me or
CH_2SiMe_3) substitution is incomplete and the dialkyl products
$Mo_2R_2(NMe_2)_2(ArNNNAr)_2$ are obtained (34). The dimethyl complex
$Mo_2Me_2(NMe_2)_2(p\text{-}tol\text{-}NNN\text{-}tol\text{-}p)_2$ (38) and its diethyl analogue
$Mo_2Et_2(NMe_2)_2(p\text{-}tol\text{-}NNN\text{-}tol\text{-}p)_2$, isolated as an intermediate from the
reaction of $Mo_2Et_2(NMe_2)_4$ with di-p-tolyltriazene (34), have been shown to
possess a pair of molybdenum atoms [Mo≡Mo 2.174(1) and 2.171(4) Å,
respectively] bridged by a *cis* pair of triazenide ligands (Fig. 6). A similar
structure (10) has been proposed on the basis of ^{1}H NMR data for the
complex $Mo_2(OPr^i)_4(PhNNNPh)_2$ isolated from the reaction of $Mo_2(OPr^i)_6$
and diphenyltriazene (36). However, the closely related complex
$Mo_2(NMe_2)_4(p\text{-}tol\text{-}NNN\text{-}tol\text{-}p)_2$, obtained from $Mo_2(NMe_2)_6$ and the free
triazene, adopts a rather different structure (Fig. 7), with chelate triazenide
ligands and an unsupported molybdenum–molybdenum triple bond
[2.212(1) Å] (37).

A range of mononuclear molybdenum complexes containing chelate
triazenide ligands has also been reported. The orange anion

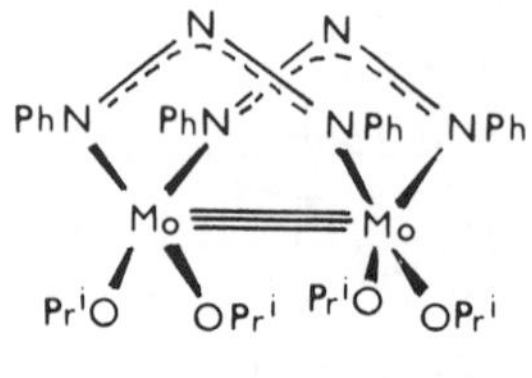

(10)

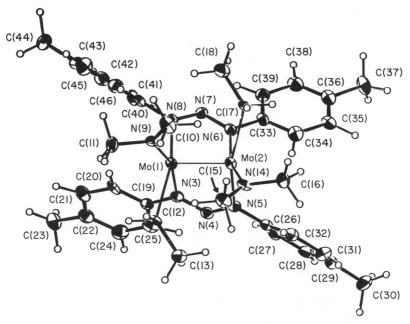

FIG. 6. An ORTEP view of the Mo$_2$Et$_2$(NMe$_2$)$_2$(p-tol-NNN-tol-p)$_2$ molecule.

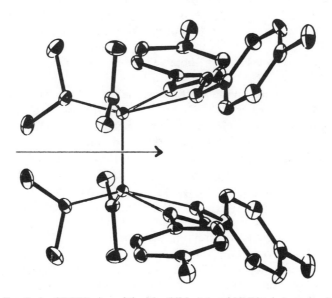

FIG. 7. An ORTEP view of the Mo$_2$(NMe$_2$)$_4$(p-tol-NNN-tol-p)$_2$ molecule.

[Mo(PhNNNPh)(CO)$_4$]$^-$, prepared from Mo(CO)$_6$ and Na[PhNNNPh] in boiling 1,2-dimethoxyethane, can be isolated as its tetramethylammonium salt (*118*). The complexes Mo(ArNNNAr)(C$_5$H$_5$)(CO)$_2$ have been obtained from MoCl(C$_5$H$_5$)(CO)$_3$ and Na- or K[PhNNNPh] (*58, 118*), ArNNNAr·SnMe$_3$ (*1*), Ag(ArNNNAr) (*179*), or ArNNNHAr/py (*117*). A carbonyl "insertion" product (**11**) has also been isolated from the reaction of

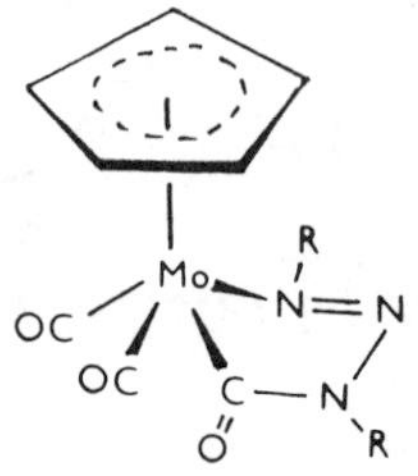

(11)

MoCl(C$_5$H$_5$)(CO)$_3$ with K[PhNNNPh] (*58*). An X-ray diffraction study has established a "piano stool" structure with a symmetrically bound triazenide ligand for the complex Mo(ArNNNAr)(C$_5$H$_5$)(CO)$_2$ [Ar = 3,5-C$_6$H$_3$(CF$_3$)$_2$] (*182*). However, variable-temperature ^{13}C NMR spectra reveal that for derivatives of nonsymmetric diaryltriazenes, Mo(ArNNNR)(C$_5$H$_5$)(CO)$_2$, a fluxional process involving interchange of the two carbonyl ligands via a Berry-type pseudo-rotation is operative when R = aryl but not when R = alkyl (*179*). In contrast, the nitrosyl complexes MoX(NO)(ArNNNAr)(C$_5$H$_5$), obtained from the dimers [MoX$_2$(NO)(C$_5$H$_5$)]$_2$ and Ag(ArNNNAr) (X = Cl, Br, I) (*181*) or PhNNNHPh (X = I) (*118*), are nonfluxional on the NMR timescale (*181*). Attempts to replace the halides X by neutral ligands (CO, PR$_3$, etc.) in the presence of AgPF$_6$ or TlPF$_6$ were unsuccessful (*181*). A "phosphazide" structure similar to that established for the tungsten analogue (see below) has been assigned to the complexes MoBr$_2$(CO)$_3$(ArNNNPPh$_3$), obtained from the aryl azides ArN$_3$ (Ar = Ph, *p*-tol) and MoBr$_2$(CO)$_3$(PPh$_3$)$_2$ (*105*).

The binuclear tungsten complexes W$_2$(NMe$_2$)$_4$(PhNNNPh)$_2$ (*39*) and W$_2$Bz$_2$(NMe$_2$)$_2$(PhNNNPh)$_2$ (*35*) have been obtained from the reaction of diphenyltriazene with W$_2$(NMe$_2$)$_6$ and W$_2$Bz$_2$(NMe$_2$)$_4$, respectively. The first of these has a binuclear structure with chelate triazenide ligands and an unsupported W≡W bond [2.314(1) Å] (*39*), similar to that found for the molybdenum analogue (*37*). The second is assigned a triazenide-bridged structure (*35*) similar to that found for the molybdenum complex Mo$_2$Me$_2$(NMe$_2$)$_2$(*p*-tol-NNN-tol-*p*)$_2$ (Fig. 6). *Anti* and *gauche* forms of W$_2$Et$_2$(NMe$_2$)$_4$ react with diaryltriazenes ArNNNHAr (Ar = Ph or *p*-tol) to

isomer A   isomer B   isomer C

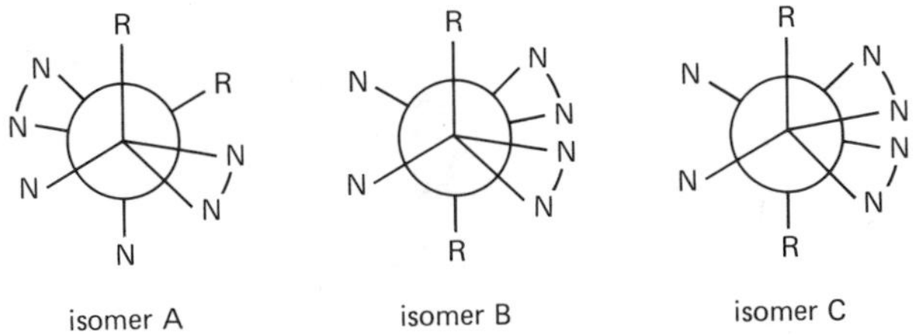

(12)

give three isomers of $W_2Et_2(NMe_2)_2(ArNNNAr)_2$ (**12a,b,c**). With time, isomer **12a** is converted to **12b** which reacts further to give an equilibrium mixture of **12b** and **12c**. Triazenide chelated and bridged structures have been confirmed for **12b** (Ar = Ph) and **12c** (Ar = p-tol) by diffraction methods (35a). The mixed ligand complex $W_2(PhNNNPh)_2(dmhp)_2$ (dmhpH=2,4-dimethyl-6-hydroxypyrimidine), obtained from $W_2(dmhp)_4$ and Li(PhNNNPh), has a "lantern" structure with a transoid arrangement of bridging triazenide ligands and a tungsten–tungsten distance of 2.169(1) Å (52). Several mononuclear tungsten triazenide complexes have been described. Tungsten hexacarbonyl and sodium diphenyltriazenide afford the orange anion $[W(PhNNNPh)(CO)_4]^-$, which is isolable as an air-stable orange tetramethylammonium salt (118). The complexes $W(ArNNNAr)(C_5H_5)(CO)_2$, obtained from $WCl(C_5H_5)(CO)_3$ and potassium (58) or silver triazenides, display fluxional behavior similar to that of their molybdenum analogues (179). Irradiation of $WCl(C_5H_5)$ $(CO)_3$/PhNNNHPh/py mixtures affords the same product (117). The tungsten monoaryl triazenide $W(HNNN\text{-}tol\text{-}p)(CO)(NO)(PPh_3)_2$ has been obtained from $WH(CO)_2(NO)(PPh_3)_2$ and p-tolyl azide (107). Finally, infrared and NMR studies [$\delta_{PPh} = 42.0$ ppm, no $J(PW)$ coupling] support a novel "phosphazide" structure for the product obtained from p-tolyl azide and $WBr_2(CO)_3(PPh_3)_2$ (107), and this assignment has subsequently been confirmed by X-ray diffraction (Fig. 8) (105).

3. Manganese, Technetium, and Rhenium

Manganese pentacarbonyl bromide reacts with Na[PhNNNPh] (118), ArNNNAr·SiMe₃ (1), or ArNNNAr·SnMe₃ (1) to generate the chelate complexes $Mn(ArNNNAr)(CO)_4$, which in turn readily lose a carbonyl ligand to form monosubstituted products $Mn(ArNNNAr)(CO)_3L$ (L = PPh₃, PMe₂Ph, and AsPh₃) (1).

The first technetium triazenide complexes $Tc(ArNNNAr)(CO)_2(PR_2Ph)_2$ (R = Me, Ph) have been prepared from Li(ArNNNAr) and $TcCl(CO)_2(PMe_2Ph)_3$ or $TcCl(CO)_3(PPh_3)_2$ in boiling benzene. The di-p-tolyltriazenide derivative $Tc(p\text{-tol-NNN-tol-}p)(CO)_2(PMe_2Ph)_2$ has been

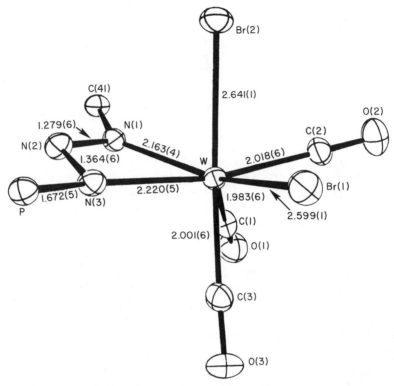

FIG. 8. View of the environment of the tungsten atom in WBr$_2$(CO)$_3$(p-tol-NNN-PPh$_3$).

shown to possess a distorted octahedral structure with chelate triazenide and *trans* phosphine ligands (*142a*).

Rhenium(I) triazenide complexes Re(ArNNNAr)(CO)$_2$(PPh$_3$)$_2$ prepared by a parallel method have been shown by X-ray diffraction methods (Ar = *p*-tolyl) to possess a similar structure (*87*). The reaction of ReH(CO)$_2$(PPh$_3$)$_3$ with *p*-tolyl azide affords Re(HNNN-tol-*p*)(CO)$_2$(PPh$_3$)$_2$ (*107*). Reduction of ReOCl$_2$(PPh$_3$)$_2$ with excess triphenylphosphine in the presence of lithium triazenides yields paramagnetic rhenium(III) complexes Re(ArNNNAr)Cl$_2$(PPh$_3$)$_2$, one of which (Ar = *p*-tol) has been shown to possess a distorted octahedral structure with a chelate triazenide ligand and a trans pair of phosphines (*192*). On heating under reflux in CCl$_4$ solution this complex decomposes to form the nitrene species ReCl$_3$(N tol-*p*)(PPh$_3$)$_2$ in good yield (*193a*). A symmetrical trinuclear structure with chelate triazenide ligands has been proposed on the basis of molecular weight and spectroscopic data for the brown product Re$_3$Cl$_3$(CH$_2$SiMe$_3$)$_3$(PhNNNPh)$_3$, obtained by treatment of Re$_3$Cl$_3$(CH$_2$SiMe$_3$)$_6$ with diphenyltriazene (*75*).

4. Iron, Ruthenium, and Osmium

High yields of the moisture-sensitive tris(chelates) $Fe(ArNNNAr)_3$ have been obtained by shaking anhydrous iron(III) chloride with silver(I) triazenides in dry ether at room temperature (18). Silver triazenides also react with the iron complexes $FeX(C_5H_5)L_2$ [L = PPh_3, $P(OMe)_3$, $P(OPh)_3$, or CO] via unstable binuclear species "$(C_5H_5)LFe(ArNNNAr)AgX$" to give the chelate triazenide complexes $Fe(ArNNNAr)(C_5H_5)L$ (13). The complexes

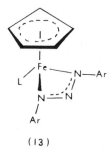

(13)

have been shown by ^{31}P and ^{13}C NMR to be rigid on the NMR timescale (183). The reversible electrochemical oxidations shown in Eq. (22) occur in

$$Fe(ArNNNAr)(C_5H_5)L \rightleftharpoons [Fe(ArNNNAr)(C_5H_5)L]^+ \qquad (22)$$

the potential range 0.25–0.65 V using an Ag–AgI electrode in CH_2Cl_2 solution. However, attempted chemical oxidation with $NO \cdot PF_6$ affords iron(II) nitrosyl derivatives $[Fe(ArNNNAr)(C_5H_5)(NO)(CO)][PF_6]$, which are thought to contain monodentate triazenide ligands (214).

Several series of chelate diaryl triazenide complexes of ruthenium have been reported. The hydrides $RuH_2(PPh_3)_4$, $RuH_2(CO)(PPh_3)_2$, and $RuHCl(CO)(PPh_3)_3$ react with free diaryltriazenes to afford the brightly colored, air-stable complexes $Ru(ArNNNAr)_2(PPh_3)_2$, $RuH(ArNNNAr)(CO)(PPh_3)_2$, and $RuCl(ArNNNAr)(CO)(PPh_3)_2$, respectively (135, 190). Lithium triazenides react with $RuCl_2(PPh_3)_3$ and $RuHCl(PPh_3)_3$ to yield $Ru(ArNNNAr)_2(PPh_3)_2$ and $RuH(ArNNNAr)$ $(PPh_3)_3$, respectively (120). Finally, free triazenes react with $RuCl_2(PPh_3)_3$ and $RuCl_2(CO)(PPh_3)_3$ in the presence of base (NEt_3) to form $Ru(ArNNNAr)_2(PPh_3)_2$ and $RuH(ArNNNAr)(CO)(PPh_3)_2$, respectively (54), and with $RuCl_2(PPh_3)_3$ in the presence of air to form $RuCl_2(ArNNNAr)(PPh_3)_2$ (41). A distorted octahedral structure [$\angle N$—Ru—$N = 57.7(1)°$] with trans phosphine ligands has been reported for $RuH(p\text{-tol-NNN-tol-}p)(CO)(PPh_3)_2$; the symmmetrical nature of the chelate triazenide ligands is confirmed by the N—N bond lengths [1.318(4) and 1.310(4) Å] (20).

Nonrigid seven-coordinate osmium trihydrides, $OsH_3(ArNNNAr)$ $(PPh_3)_2$, have been obtained by treatment of $OsH_4(PPh_3)_3$ with diaryl-triazenes in refluxing benzene/2-methoxyethanol (*135, 190*). Similar reactions involving $OsHCl(CO)(PPh_3)_3$ or $OsH_2(CO)(PPh_3)_3$ afford the complexes $OsH(ArNNNAr)(CO)(PPh_3)_2$ (*135, 190*). A single-pot reaction between Na_2OsCl_6, diaryltriazene, and base (KOH) in refluxing 2-methoxyethanol affords the products *cis*-$Os(ArNNNAr)_2(PPh_3)_2$ (*135*). Osmium(III) triazenides, $OsX_2(PhNNNPh)(PPh_3)_2$ (X = Cl or Br) have been obtained by treatment of the osmium complexes *trans*-$Os(O)_2X_2(PPh_3)_2$ or $OsCl_4(PPh_3)_2$ with Li[PhNNNPh]. The green paramagnetic products exhibit X-band ESR spectra (CH_2Cl_2 solution, $-160°C$, *g* values ~ 2.7, 2.1, and 1.0) consistent with d^5 low-spin osmium(III) in a low-symmetry environment (*3*). Reactions between osmium carbonyl clusters and various organic azides have yielded complexes in which triazenide ($HN{=}N{-}NH^-$) and monosubstituted triazenide ($RN{=}N{-}NH^-$) anions are stabilized as coordinated ligands. The parent triazenide complex $Os_3(\mu\text{-}H)(\mu\text{-}HNNNH)(CO)_{10}$, obtained from Me_3SiN_3 and $Os_3H_2(CO)_{10}$ after heating under reflux in hexane for 24 hours, has been shown to possess a triazenide-bridged structure (Fig. 9) (*113*). Corresponding reactions using alkyl or aryl azides RN_3 (R = Bu^n, Cy, Ph, Bz, or C{Ph}$={}$CH$_2$) afford mono-substituted triazenide derivatives $Os_3(\mu\text{-}H)(\mu\text{-}HNNNR)(CO)_{10}$, one of which (R = Ph) has been shown to possess a similar structure (**14**) (*26, 43*).

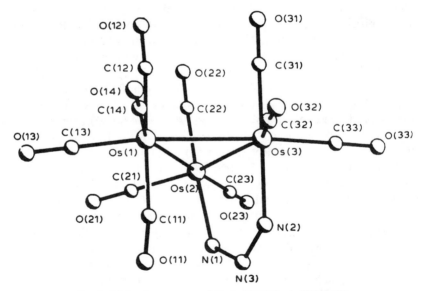

FIG. 9. Molecular structure of $Os_3(\mu\text{-}H)(CO)_{10}(\mu\text{-}HNNNH)$.

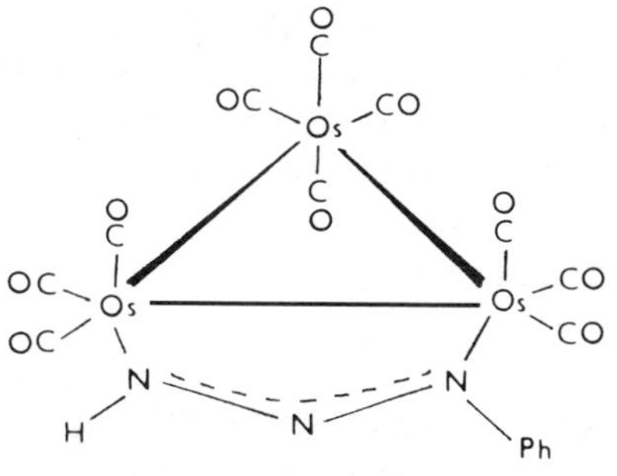

(14)

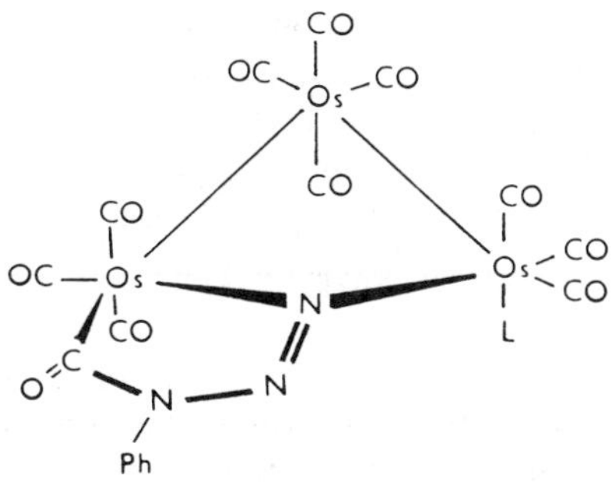

(15)

Thermolysis of these products generates the face-capped nitrenes $Os_3(\mu\text{-}H)_2(\mu_3\text{-}NR)(CO)_9$ (24). The substituted clusters $Os_3(CO)_{11}L$ react with phenyl azide to give the products $Os_3(CON_3Ph)(CO)_{10}L$, whose structures (15, L = py or MeCN) have been established by diffraction methods (25).

5. Cobalt, Rhodium, and Iridium

The first cobalt triazenide complex $Co(ArNNNAr)_2$ (Ar = $p\text{-}C_6H_4NO_2$) was obtained by Meldola and Streatfeild in 1887 from an ammoniacal solution of $Co(NO_3)_2$ and $NH_4[ArNNNAr]$ (146). A cobalt(III) triazenide $Co(p\text{-}tol\text{-}NNN\text{-}tol\text{-}p)_2(NO_2)(H_2O)$ was reported in 1908 (108). X-Ray crystal structure determinations have been published for the monoclinic (47, 49) and trigonal (127) forms of $Co(PhNNNPh)_3$. Both confirm the tris(chelate) structure and the strain within the chelate rings ($\angle N\text{—}Co\text{—}N = \sim 65°$, $\angle N\text{—}N\text{—}N = \sim 105°$). The original synthesis of tris(triazenide)cobalt(III) complexes does not appear to have been published (67), but a more recent paper (137) mentions a route based on the thermal decomposition of the bis(pyridine) adducts $Co(ArNNNAr)_2py_2$. A brief note (225) mentions the synthesis of cobalt(II) and cobalt(III) triazenides from the corresponding acetates or nitrates and arylamines in the presence of isoamyl nitrite. Magnetic data and electronic spectra have been reported for the cobalt(III) complexes $Co(ArNNNAr)_3$ ($\mu_{eff} = \sim 0.44\text{-}0.72$ BM) and for the cobalt(II) adducts $Co(ArNNNAr)_2py_2$ ($\mu_{eff} = 5.08\text{-}5.43$ BM) (137). The compound $mer\text{-}CoMe_3(PMe_3)_3$ reacts with Me_3SiN_3 to form the azide $Co(N_3)Me_2(PMe_3)_3$; however, with p-tolyl azide, a formal "insertion" reaction generates the green air-sensitive triazenide complex $CoMe_2(p\text{-}tol\text{-}NNNMe)(PMe_3)_2$ ($trans\text{-}PMe_3$ isomer) (40). Treatment of the complexes $CoI_2(C_5H_5)L$ [L = PEt_3, PPh_3, $P(OMe)_3$, or $P(OPh)_3$] with silver(I) triazenides, followed by anion exchange with $TlPF_6$, affords triazenide chelate salts $[Co(ArNNNAr)(C_5H_5)L][PF_6]$ (177). Electrochemical reduction of these salts $(-0.2-+0.1\ \text{V}$ versus an Ag/AgCl electrode in acetone solution) affords the cobalt(II) products $\text{Co(ArNNNAr)(C}_5\text{H}_5)\text{L}$ (214). The cobalt(II) derivatives $(\text{L} = \text{PPh}_3)$ have also been obtained by treatment of $\text{CoCl(C}_5\text{H}_5)(\text{PPh}_3)$ with copper(I) or silver(I) triazenides, and can be reoxidized using AgPF_6 (177).

$$\text{Co(ArNNNAr)(C}_5\text{H}_5)(\text{PPh}_3) + \text{AgPF}_6 \longrightarrow$$

$$[\text{Co(ArNNNAr)(C}_5\text{H}_5)(\text{PPh}_3)][\text{PF}_6] + \text{Ag}^0 \quad (23)$$

Poorly resolved ESR spectra for the complexes $\text{Co(ArNNNAr)(C}_5\text{H}_5)\text{L}$ are consistent with low-spin d^7 cobalt(II) in a low-symmetry environment $(g_x \neq g_y \neq g_z, g_{\text{iso}} = \sim 2.1)$ (178). The reaction between $\text{CoCl}_2(\text{PPh}_3)_2$ and silver triazenides affords cobalt(II) species Co(ArNNNAr)_2, which can be isolated as the previously known (137) pyridine adducts $\text{Co(ArNNNAr)}_2\text{py}_2$ (178). Magnetic data and electronic spectra have been analyzed for the complexes Co(PhNNNPh)_2 and $\text{Co(PhNNNPh)}_2\text{py}_2$ (84, 149). The cobalt(III) triazenide $\text{Co(C}_3\text{F}_7)(\text{PhNNNPh})(\text{C}_5\text{H}_5)$ has been obtained from $\text{CoI(C}_3\text{F}_7)(\text{C}_5\text{H}_5)(\text{CO})$ and Na[PhNNNPh] (118). However, reaction of $\text{CoI}_2(\text{C}_5\text{H}_5)(\text{CO})$ with silver(I) triazenides is accompanied by a carbonyl "insertion" step leading to formation of the products $\overline{\text{Co\{RNNN(R)C(O)\}}}(\text{C}_5\text{H}_5)(\text{CO})$ (16) (177). The first stable metal "phosphazide" complex $\text{Co\{Cy}_3\text{P=N—N=NC(O)C}_4\text{H}_3\text{O\}Br}_2$ has been obtained by treatment of $\text{CoBr}_2(\text{PCy}_3)_2$ with furoyl azide at 0°C (9).

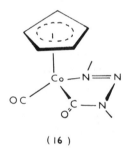

(16)

Diaryltriazenes react with the rhodium precursors $\text{RhCl(PPh}_3)_3$, $\text{RhH(CO)(PPh}_3)_3$, $\text{RhH(PPh}_3)_4$, and $\text{Rh(NO)(PPh}_3)_3$ to afford the 1,3-diaryl triazenide products $\text{RhCl(ArNNNAr)}_2(\text{PPh}_3)$, $\text{Rh(ArNNNAr)(CO)(PPh}_3)_2$, $\text{RhH}_2(\text{ArNNNAr})(\text{PPh}_3)_2$, and $\text{Rh(ArNNNAr)}_2(\text{NO})(\text{PPh}_3)$, respectively (135, 190). Treatment of $\text{RhCl(PPh}_3)_3$ and $[\text{RhCl(C}_8\text{H}_{12})]_2$ with Li[ArNNNAr] yields $\text{Rh(ArNNNAr)(PPh}_3)_2$ and binuclear $[\text{Rh(ArNNNAr)(C}_8\text{H}_{12})]_2$, respectively (120). Diaryltriazenes in the presence of base (NEt_3) react with $\text{RhCl(C}_8\text{H}_{12})(\text{PPh}_3)$, $\text{RhCl(CO)(PPh}_3)_2$, and $[\text{RhCl(CO)}_2]_2$ to afford $\text{Rh(ArNNNAr)(C}_8\text{H}_{12})(\text{PPh}_3)$, $\text{Rh(ArNNNAr)(CO)(PPh}_3)_2$, and binuclear

[Rh(ArNNNAr)(CO)$_2$]$_2$, respectively (54). The last-named complexes have also been obtained by treatment of [RhCl(CO)$_2$]$_2$ with Na[ArNNNAr] (42) or Me$_3$Sn(ArNNNAr) (1), and by carbonylation (40 psi, 25°C) of the complexes Rh(ArNNNAr)(C$_8$H$_{12}$) (120). Heterobimetallic complexes (see below), formed between RhCl(CO)(PPh$_3$)$_2$ and Ag[ArNNNAr], break apart to yield rhodium(I) triazenides Rh(ArNNNAr)(CO)(PPh$_3$)$_2$ (129). The complexes Rh(ArNNNAr)(C$_8$H$_{12}$) form adducts Rh(ArNNNAr)(C$_8$H$_{12}$)(NH$_3$) with ammonia (22 psig, 25°C) and react with dihydrogen to afford {RhH$_2$(PhNNNPh)}$_3$(C$_8$H$_{12}$), {RhH$_2$(PhNNNPh)}$_2$(C$_8$H$_{14}$), and {Rh(p-F-C$_6$H$_4$-NNN-C$_6$H$_4$-F-p)}$_3$(C$_8$H$_{12}$) (120). Proton NMR data for {RhH$_2$(PhNNNPh)}$_3$(C$_8$H$_{12}$), which show all hydride ligands to be equivalent and to couple to all three [103]Rh nuclei, have been taken to indicate the novel macrocyclic structure 17 (120). The complexes Rh(ArNNNAr)(PPh$_3$)$_2$

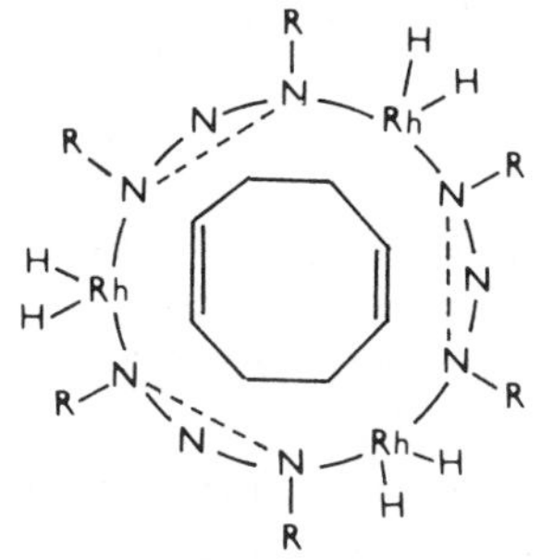

(17)

serve as olefin hydrogenation catalysts (THF, 46 psig H$_2$) for ethylene and hexene (120). They also form adducts Rh(ArNNNAr)(PPh$_3$)$_2$(O$_2$), Rh(ArNNNAr)(PPh$_3$)$_2$(CO), and Rh(ArNNNAr)(PPh$_3$)$_2$(NH$_3$), and undergo oxidative addition with MeI and H$_2$ to form Rh(ArNNNAr)(Me)I(PPh$_3$)$_2$ and RhH$_2$(ArNNNAr)(PPh$_3$)$_2$, respectively (120). The triazenide bridged rhodium(I) carbonyl dimers [Rh(ArNNNAr)(CO)$_2$]$_2$ undergo oxidative addition with I$_2$ to form rhodium(II) species [RhI(ArNNNAr)(CO)$_2$]$_2$ (18) and are substituted by PPh$_3$ and dienes (cyclooctadiene, norbornadiene) to yield

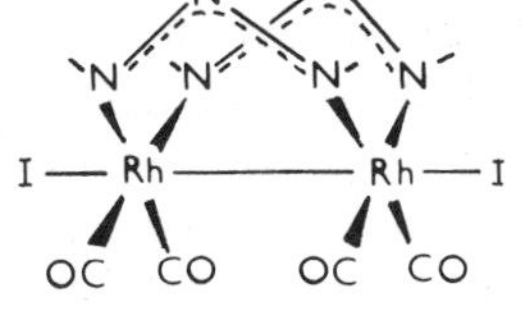

(18)

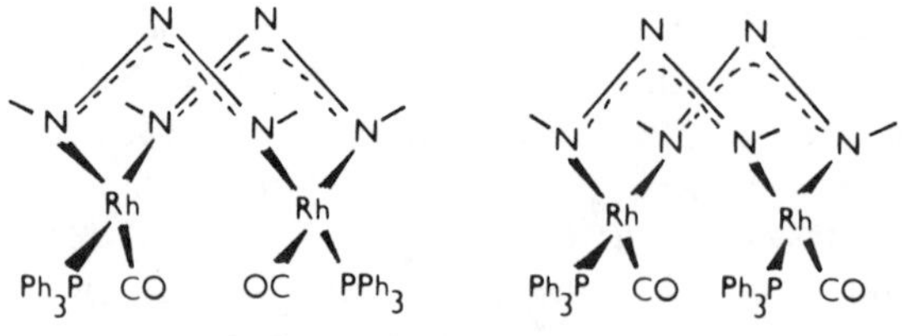

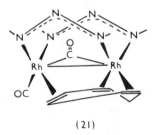

[Rh(ArNNNAr)(CO)(PPh₃)]₂ (**19a/b**) and [Rh(ArNNNAr)(diene)]ₙ (**20**), respectively (*42*). Ambiguous molecular weight and mass spectroscopic data fail to distinguish between *n* values of 1 and 2 for [Rh(ArNNNAr)(diene)]ₙ, but NMR spectra favor the triazenide-bridged binuclear structure shown. The blue-black products obtained by treatment of [Rh(ArNNNAr)(CO)₂]₂ with cycloheptatriene were tentatively assigned structure **21** (*42*). Cycloocta-

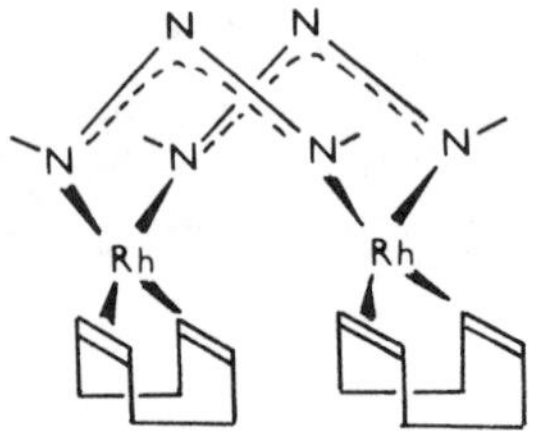

1,5-diene also affords Rh₂(ArNNNAr)₂(CO)₂(C₈H₁₂) (*42*). One-electron electrochemical oxidation of [Rh(ArNNNAr)(CO)(PPh₃)]₂ generates the rhodium(I/II) cation [Rh(ArNNNAr)(CO)(PPh₃)]₂⁺ and is accompanied by a decrease in the Rh–Rh distance from 2.96 to 2.67 Å, which is taken to indicate that the electron is lost from an orbital which is anti-bonding with respect to the Rh—Rh bond (*43*). Electron spin resonance spectra for the cation lack fine structure and therefore do not distinguish between delocalized and trapped valence (Rh^I/Rh^II) electronic structures. Protonation (HBF₄) of rhodium(I) triazenides Rh(ArNNNAr)(CO)L₂

[L = PMePh$_2$ or P(C$_6$H$_4$-F-p)$_3$] affords the cationic diaryltriazene complexes [Rh(ArNNNHAr)(CO)L$_2$][BF$_4$] (**44**), one of which, [Rh(PhNNNHPh)(CO)(PPh$_3$)$_2$][BF$_4$], has been shown to possess an N-3-coordinated structure (**22**) (*201*). Treatment of the same complexes (L = PPh$_3$) with [Ar'N$_2$][BF$_4$] leads to formation of the salts [Rh{ArNNN(Ar)C̄(O)}(N$_2$Ar')(PPh$_3$)$_2$][BF$_4$] (*212*). The binuclear complex [RhCl$_2$(C$_5$Me$_5$)]$_2$ reacts with free triazenes and base or with silver(I) triazenides to afford the chelate triazenide products RhCl(ArNNNAr)(C$_5$Me$_5$), which, on treatment with AgPF$_6$ in the presence of donor ligands [L = MeCN or P(OMe)$_3$], form the salts [Rh(ArNNNAr)(C$_5$Me$_5$)L][PF$_6$] (*189*). The azide complex Rh$_2$(N$_3$)$_4$(C$_5$Me$_5$)$_2$ forms an adduct Rh$_2$(N$_3$)$_4$(C$_5$Me$_5$)$_2$(ArNNNHAr), which has been assigned structure **23** with a labile triazene NH proton, on the basis of infrared and variable-temperature NMR studies (*189*).

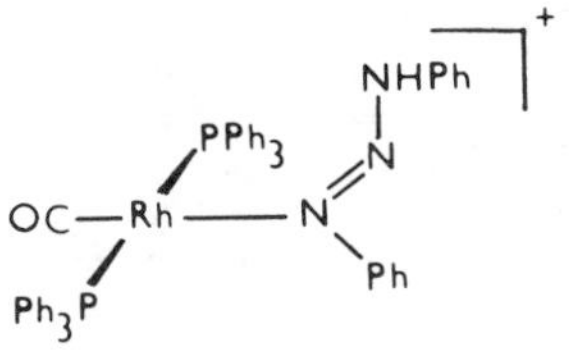

(22 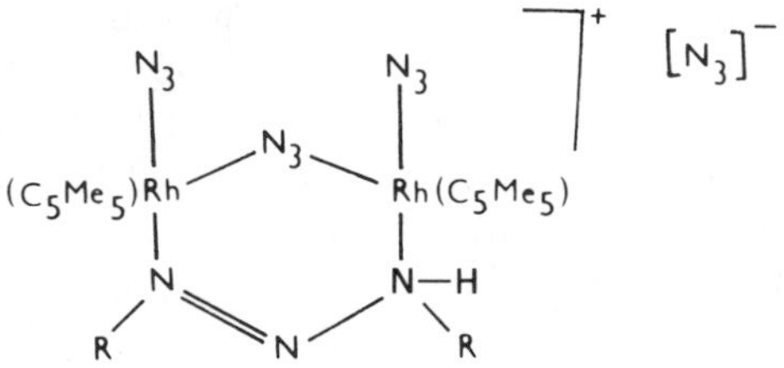)   (23)

Iridium precursors *mer*-IrH$_3$(PPh$_3$)$_3$ and IrHCl$_2$(PPh$_3$)$_3$ react with free diaryltriazenes to form IrH$_2$(ArNNNAr)(PPh$_3$)$_2$ and IrHCl(ArNNNAr)(PPh$_3$)$_2$, respectively. The former products have also been obtained from IrH(CO)(PPh$_3$)$_3$/ArNNNHAr and by a single-pot synthesis involving Na$_2$IrCl$_6$/PPh$_3$/ArNNNHAr and KOH in boiling 2-methoxyethanol (*135, 190*). Vaska's complex reacts with lithium (*120*) or silver (*129*) triazenides to yield monodentate triazenide complexes Ir(ArNNNAr)(CO)(PPh$_3$)$_2$. These in turn add aryldiazonium salts [Ar'N$_2$][BF$_4$] to form salts [Ir(ArNNNAr)(N$_2$Ar')(CO)(PPh$_3$)$_2$][BF$_4$] (*212*), and they carbonylate reversibly to yield the carbonyl insertion products Ir{ArNNN(Ar)C̄(O)}(CO)$_2$(PPh$_3$) (*129*). The same precursors undergo protonation by HBF$_4$ at the metal to yield salts [IrH(ArNNNAr)(CO)(PPh$_3$)$_2$][BF$_4$] (contrast the behavior of corresponding rhodium precursors) (*44*). However, triazene complexes [Ir(ArNNNHAr)(CO)(PPh$_3$)$_2$][BF$_4$] have been obtained from [Ir(CO)(OCMe$_2$)(PPh$_3$)$_2$][BF$_4$] by substitution using free triazenes and by solid-state isomerization of the hydrides [IrH(ArNNNAr)(CO)(PPh$_3$)$_2$][BF$_4$] (*44*).

$$\begin{array}{c} X \\ \diagdown \\ M' \longleftarrow N \diagup R \\ \uparrow \quad\; \diagdown \\ \mid \swarrow PPh_3 \quad \vdots N \\ OC - M \underline{\qquad} N \diagdown \\ Ph_3P \diagup \qquad\qquad R \end{array}$$

(24)

Vrieze and co-workers have reported the formation of novel triazenide-bridged heterobimetallic complexes (24, M = Rh^I, Ir^I; M' = Cu^I, Ag^I; X = Cl, Br, I, or O_2CR; L = P- or As-donor ligand) by addition of copper(I) or silver(I) triazenides to square-planar d^8 rhodium and iridium complexes $MX(CO)L_2$ (129, 130). Thus treatment of $MX(CO)L_2$ with copper(I) triazenides $[Cu(RNNNR')]_n$ (n = 4, R = R' = Me; n = 2, R = R' = p-tol or R = Me, R' = p-tol) in refluxing THF affords the air-stable products $L_2(CO)M(\mu\text{-}RNNNR')CuX$, in which the bridging triazenide ligand supports a dative $M^I \rightarrow Cu^I$ bond (130). Directly analogous but somewhat less stable products containing dative ($M^I \rightarrow Ag^I$) bonds are obtained from silver(I) triazenides $Ag(RNNNR')$ (R = Me, Et, Bu^t; R' = Me or p-tol) and the precursors $MCl(CO)(PPh_3)_2$ (M = Rh, X = Cl; M = Ir, X = Cl, Br, I, or O_2CCF_3) (129, 132, 219). These products afford the first examples of compounds containing metal-to-copper(I) (130) or -silver(I) (129) donor bonds. For both series of reactions, formation of the observed products involves a rather unusual migration of halide or carboxylate ligands from the group VIII metal to the coinage metal. Two possible mechanisms proposed for this process are shown in Scheme 3 (130, 219). The ease of formation and the stability of the products are dependent upon the identity of the metals M and M', the triazenide bridges, and the auxiliary ligands L (130). Generally the rhodium products are less stable than their iridium analogues, possibly because of the greater electron donor capacity of the latter metal (130). The five-membered rings formed by the bridging triazenide ligands are thought to possess some delocalized electron density and to play an important role in stabilizing the M → M' bonds (130). Crystal structures reported for $(Ph_3P)_2(CO)M(MeNNNMe)CuCl$ [M = Rh (124) or Ir (125)] and $(Ph_3P)_2(CO)Ir(MeNNN\text{-tol-}p)Ag(O_2CPr)$ (131) confirm the proposed general structure. Triazene-exchange reactions between $(Ph_3P)_2(CO)Rh(MeNNNMe)AgCl$ and $RNNNHR'$ (R = Me, p-tol; R' = p-tol) have been reported; similar reactions with the corresponding iridium complexes are much slower and lead to decomposition (129, 130). The iridium complexes $(Ph_3P)_2(CO)Ir(RNNNR')AgCl$ (R = Me, p-tol; R' = Me) carbonylate readily to form acyl triazenide products $Ir\{R'NNN(R)C(O)\}(CO)_2(PPh_3)$ (129), identical to those obtained from the

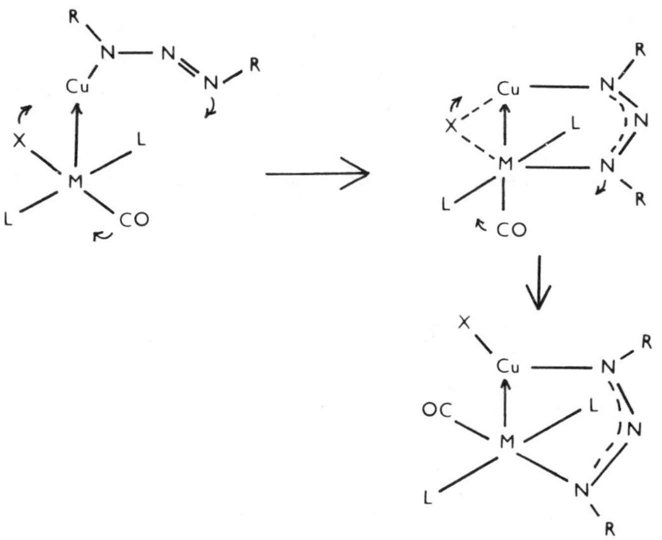

(A)

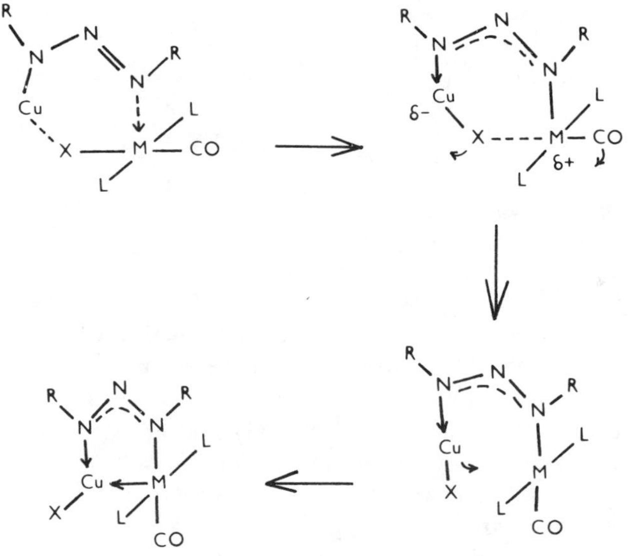

(B)

SCHEME 3. Mechanisms for the formation of heterobimetallic triazenide complexes. Adapted with permission from *J. Organomet. Chem.* **96**, 289 (1975).

compounds Ir(R'NNNR)(CO)(PPh$_3$)$_2$ (see above). Vrieze and co-workers have also examined reactions between the complexes MX(CO)(PPh$_3$)$_2$ and mercury triazenides HgX(RNNNR) or Hg(RNNNR)$_2$, from which they have harvested a rich variety of products (218). The rhodium precursor RhCl(CO)(PPh$_3$)$_2$ reacts with HgCl(MeNNNMe) and HgI(MeNNNMe) to form the acyl derivatives 25 and 26, respectively. A closely related product (27) is obtained from Rh(O$_2$CCF$_3$)(CO)(PPh$_3$)$_2$ and HgI(MeNNNMe).

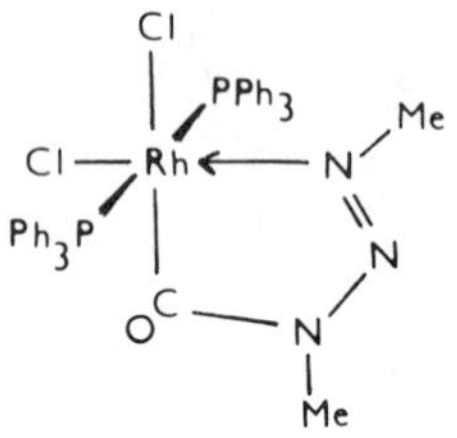

(25)

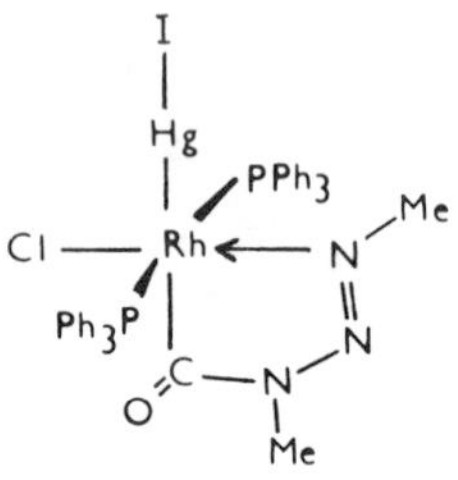

(26)

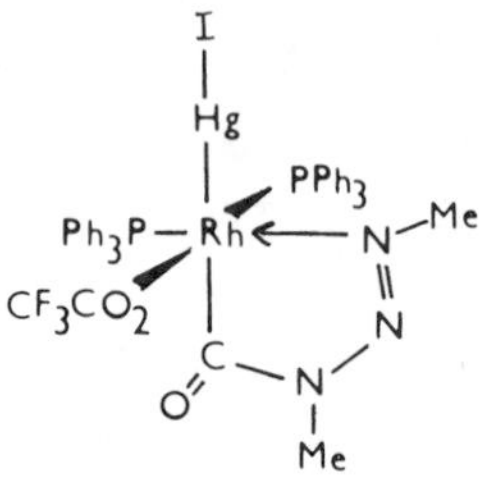

(27)

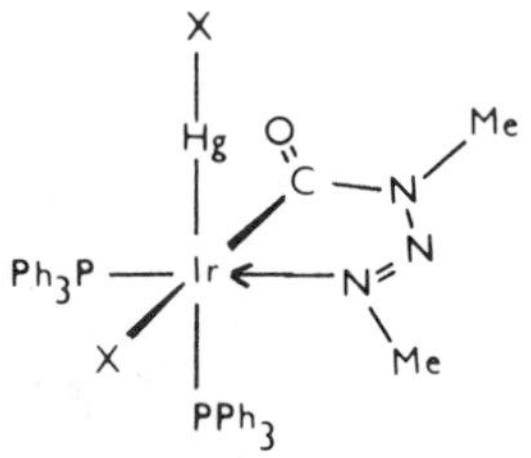

(28)

Acyl triazenide complexes 28 are also isolated from reactions between IrX(CO)(PPh$_3$)$_2$ and HgX(MeNNNMe) (X = Cl, Br, I, or O$_2$CCF$_3$), but corresponding reactions involving HgX(p-tol-NNNMe) afford triazenide-bridged products 29. Finally, treatment of the rhodium and iridium complexes MX(CO)(PPh$_3$)$_2$ with Hg(MeNNNMe)$_2$ generated 30 (218). Formation of the M—Hg bonds is relatively easy and less dependent upon the presence of stabilizing bridging ligands (218). However, it is not clear why small changes in X, R, and M give rise to such variations in product structure (218). The tetranuclear products [(diene)(RNNNR')$_2$MHgCl]$_2$ (M = Rh, Ir; diene = C$_8$H$_{12}$ or C$_7$H$_8$; R = Me, Et, p-tol; R' = p-tol) have been obtained from the reactions of [MCl(diene)]$_2$ with Hg(RNNNR')$_2$, and of [(diene)MCl·HgCl$_2$]$_2$ with [Ag(RNNNR')]$_n$ (217). An X-ray diffraction study reveals the chloride-bridged dimer structure 31. Variable-temperature ^{1}H- and ^{13}C-NMR establish that the Rh/Hg complexes, but not their Ir/Hg

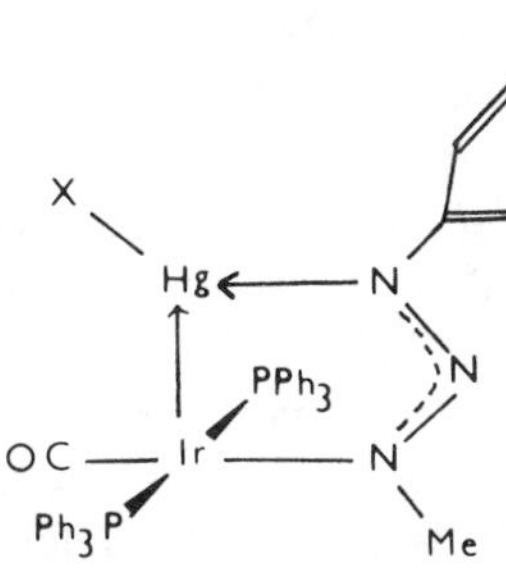

(29)

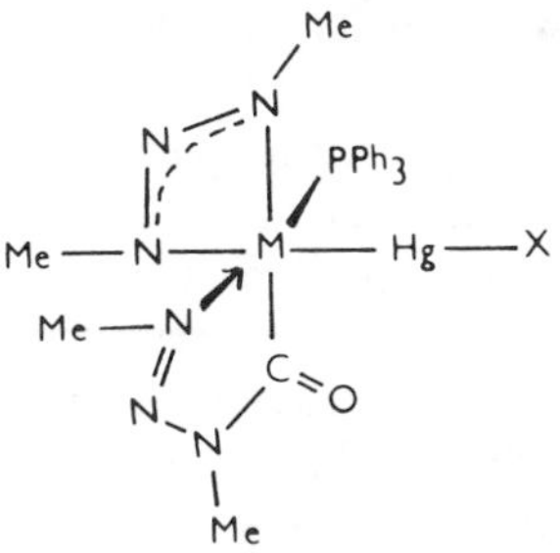

(30)

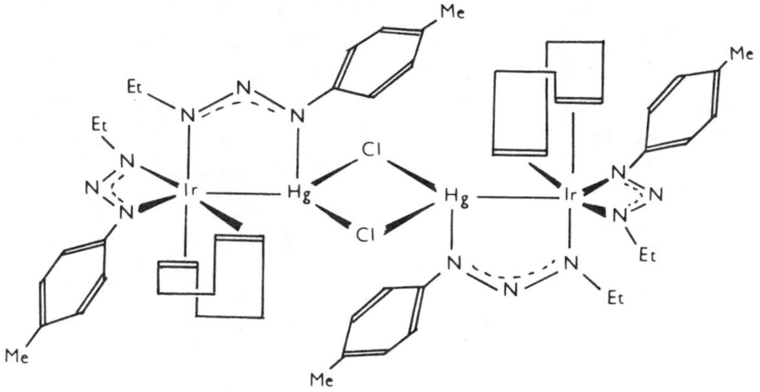

(31)

analogues, are fluxional with chelate and bridging triazenide ligands exchanging via monodentate intermediates (217). Attempts to prepare complexes containing Rh/Ir—Tl bonds were frustrated by the instability of thallium(III) triazenides $Tl(RNNNR)_nCl_{3-n}$ ($n = 1-3$) (220).

6. Nickel, Palladium, and Platinum

An early report by Meldola and Streatfeild (146) described an impure sample of $Ni(ArNNNAr)_2$ (Ar = p-C_6H_4-NO_2). Formation of nickel(II) diaryltriazenide complexes $Ni(ArNNNAr)_2py_2$ by treatment of nickel acetate with triazenes and pyridine followed by base (Na_2CO_3), and the subsequent thermal decomposition of these products to leave the pyridine-free complexes "$Ni(ArNNNAr)_2$," was first reported by Dwyer and Mellor in 1941 (74). Prolonged, vigorous treatment of the latter products with pyridine, ethylenediamine (en), and o-phenanthroline (o-phen) gave the adducts

Ni(ArNNNAr)$_2$py$_2$, Ni(ArNNNAr)$_2$(en) (*74*), and Ni(ArNNNAr)$_2$(*o*-phen) (*98*), respectively. Nickel(II) diaryltriazenide complexes have also been obtained from the reactions of arylamines with isoamyl nitrite in the presence of nickel salts (*225*). The paramagnetic character of the pyridine and *o*-phenanthroline adducts (μ_{eff} = 3.38 BM) led to their formulation as poly-meric triazenide-bridged derivatives of octahedral nickel(II) (*98*). However, a monomeric structure with chelate triazenide ligands seems more probable. Binuclear chelate/bridging triazenide structures were originally proposed for the complexes [Ni(ArNNNAr)$_2$]$_n$ (**32**) (*74*). The binuclear character was subsequently confirmed (*98, 138*), and an alternative triazenide-bridged "lantern" structure (**33**) was advanced (*98, 99*). An X-ray diffraction study later showed the "lantern" structure to be correct (*46*). Magnetic data and electronic spectra have been recorded and analyzed for Ni$_2$(PhNNNPh)$_4$ and Ni(PhNNNPh)$_2$py$_2$ (*84*). Silver triazenides, Ag(RNNNR), react with NiCl(C$_5$H$_5$)(PPh$_3$) in the presence of ligands L [L = PPh$_3$, CO, CNR'] to yield the monodentate triazenide nickel(II) derivatives Ni(RNNNR)(C$_5$H$_5$)L. When L = PPh$_3$, the complexes are fluxional (Scheme 4). However, when L = CO or CNR, "insertion" occurs to give **34**

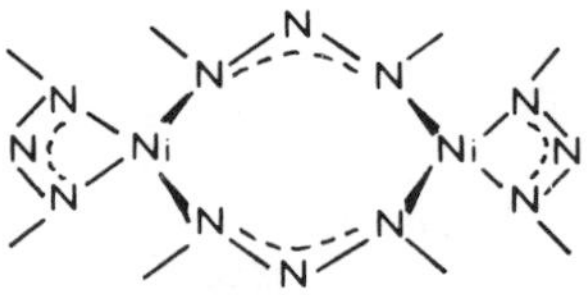

(32)

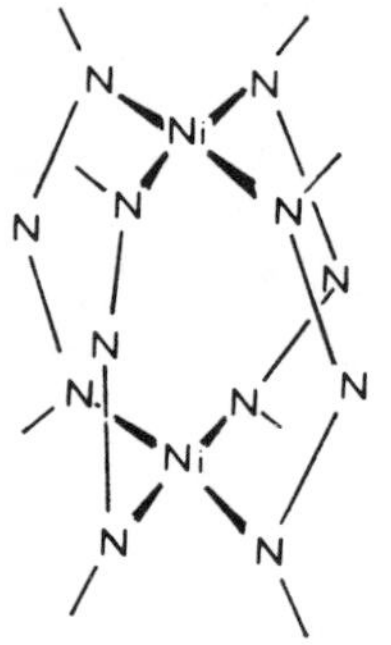

(33)

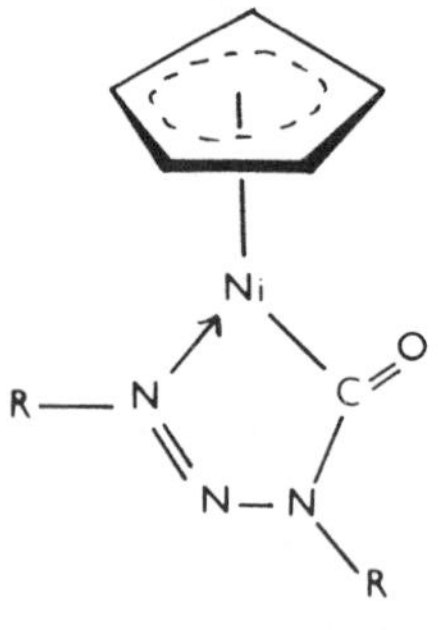

(34)

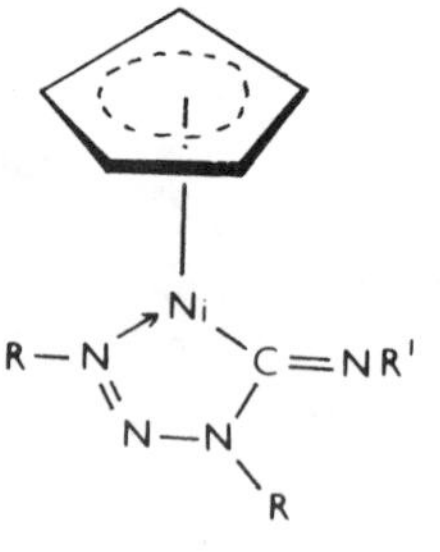

(35)

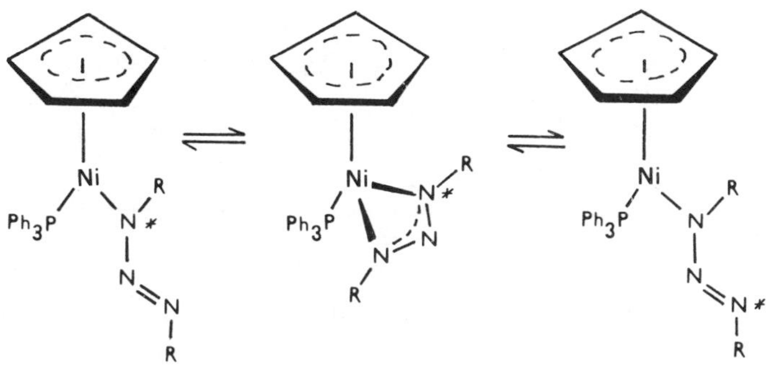

SCHEME 4. Fluxional behavior of the nickel (II) complexes $Ni(RNNNR)(C_5H_5)(PPh_3)$. Reproduced with permission from *Transition Met. Chem (Weinheim)* **2**, 240 (1977).

and **35**, respectively (*180*). The isocyanide insertion products are also fluxional (Scheme 5) (*180*).

Triazenide complexes of palladium were first described by Dwyer in 1941 (*71*). Addition of Na_2PdCl_4 and sodium acetate to cold aqueous solutions of diaryltriazenes afforded voluminous precipitates of "$Pd(ArNNNAr)_2(ArNNNHAr)$," which readily lost triazene to yield "$Pd(ArNNNAr)_2$" and formed adducts $Pd(ArNNNAr)_2py_2$

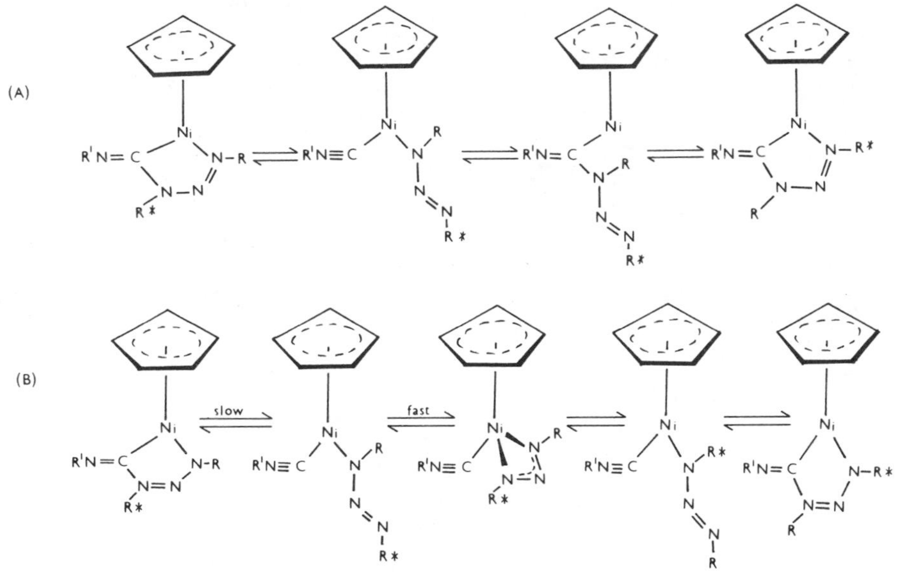

SCHEME 5. Fluxional behavior of $Ni\{RNNN(R)C(NR)\}(C_5H_5)$. Reproduced with permission from *Transition Met. Chem. (Weinheim)* **2**, 240 (1977).

and   Pd(ArNNNAr)$_2$en   (71).   The   triazene   adducts
Pd(ArNNNAr)$_2$(ArNNNHAr) were claimed as the first examples of octahedral palladium(II) (71). However, subsequent attempts to prepare these
compounds invariably led to bis-adducts which were formulated as square-
planar palladium(II) species (98).

The complexes Pd(ArNNNAr)$_2$, which were originally regarded
as square-planar bis-chelates (71), were subsequently shown by
X-ray diffraction methods to possess a binuclear triazenide-bridged "lantern"
structure similar to that of the corresponding nickel(II) complexes (46). More
recently, monodentate triazenide derivatives of square-planar palladium(II)
trans-PdCl(ArNNNAr)L$_2$ and trans- Pd(ArNNNAr)$_2$L$_2$ (L = phosphine or
arsine ligand) have been prepared by treatment of the precursors
trans-PdX$_2$L$_2$ with ArNNNHAr/NEt$_3$ (54) or Li[ArNNNAr] (13, 209). The
latter products (L = PPh$_3$) have also been obtained from
Pd(PPh$_3$)$_3$/ArNNNHAr mixtures in warm benzene (135, 190). Confirmation
of the square-planar coordination geometry and the monodentate triazenide
ligand has been provided by the results of an X-ray diffraction study on trans-
PdCl(p-tol-NNN-tol-p)(PPh$_3$)$_2$ (13). The fluxional behavior of these complexes, first reported by Robinson and Uttley in 1972 (191), has been
confirmed by more recent variable-temperature NMR studies, and a mechanism involving chelate triazenide intermediates (Scheme 1) has been proposed
(54, 209). Formation of a palladium triazenide intermediate is thought to be a
key step in the PdCl$_2$(PPh$_3$)$_2$-catalyzed carbonylation (CO atmosphere) of
diaryltriazenes to the corresponding arylamides ArHNC(O)Ar (Scheme 6)
(210).

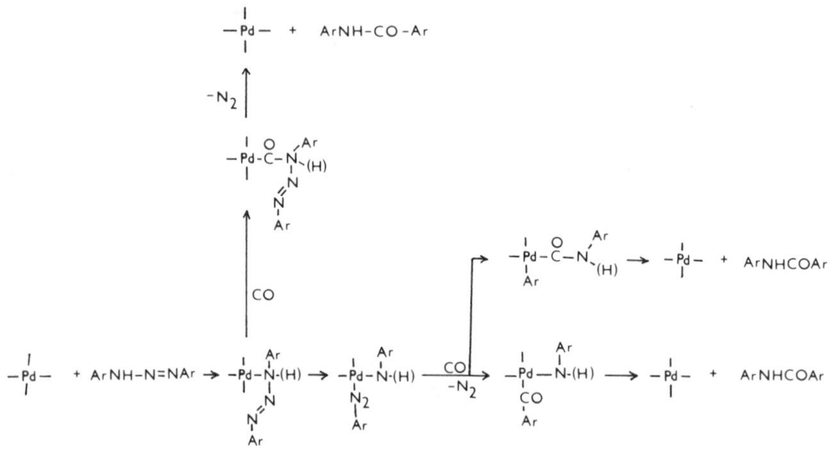

SCHEME 6. Pathway for the PdCl$_2$(PPh$_3$)$_2$-catalyzed carbonylation of diaryltriazenes to the
corresponding arylamides. Adapted with permission from Inorg. Chim. Acta 35, L367 (1979).

Alkylpalladium chloride dimers are split by aryltriazenes to yield the products PdCl(η^3-C$_3$H$_5$)(ArNNNHAr), which participate in two fluxional processes—triazene dissociation and triazene/chloride exchange (Scheme 7A/B). The latter process, which has the lower activation energy, is thought to proceed via a pentacoordinate chloro-bridged intermediate (*14*). Binuclear

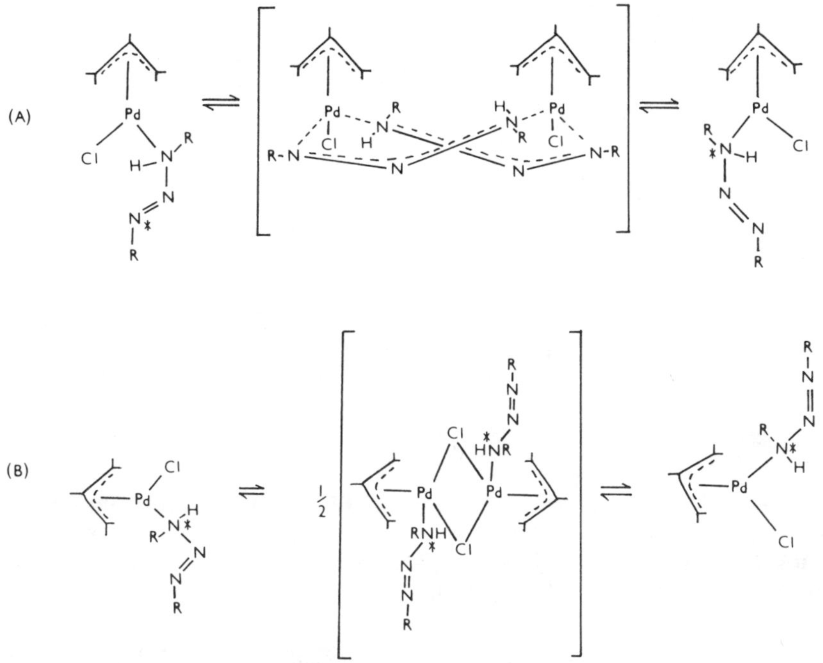

SCHEME 7. Fluxional processes for (A) triazene and (B) triazene/chloride exchange in PdCl(RN=N—NHR)(allyl) complexes. Adapted with permission from *Inorg. Chim. Acta* **34**, 37 (1979).

triazenide-bridged palladium allyls [Pd(RNNNR)(η^3-C$_3$H$_4$R')]$_2$ have been obtained by treatment of the corresponding chlorides [PdCl(η^3-C$_3$H$_4$R')]$_2$ [R' = H or Me] with KOBu^t/ArNNNHAr (*102*), LiBu^n/ArNNNHAr (*30*), or Ag(MeNNNMe) (*102, 103*); or by treatment of the acetates [Pd(O$_2$CMe)(η^3-C$_3$H$_4$R')]$_2$ with free triazenes (*111*). X-Ray diffraction studies on [Pd(p-tol-NNN-tol-p)(η^3-C$_3$H$_5$)]$_2$ (*29*) and [Pd(MeNNNMe)(η^3-C$_3$H$_5$)]$_2$ (*103*) reveal structures of the form **36a**. In solution, conformer **36a** co-exists with small amounts of **36b** (*30, 111*) and, according to more recent reports, **36c** (*102*). The ^{1}H NMR spectra of the triazenide-bridged complexes are temperature independent (*30, 111*), and there is no evidence for the operation of nondissociative, low-energy ring flip,

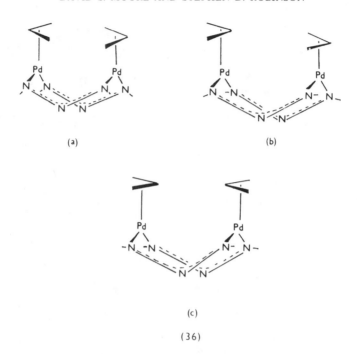

(a)   (b)

(c)

(36)

bridge inversion, or bimolecular exchange processes similar to those found for the analogous acetate-bridged complexes (*111*). However the complexes Pd(ArNNNAr)(C$_5$H$_5$)(PPh$_3$), prepared from PdBr(C$_5$H$_5$)(PPh$_3$) and Ag(ArNNNAr), show fluxional behavior (*180*) similar to that observed for the corresponding nickel species.

The early work of Griess on diaryltriazene synthesis included the preparation of complex salts of platinum(II) "ArNNNHAr·HCl·PtCl$_2$" (*89*). However, most work on platinum triazene complexes is of relatively recent origin. Platinum(II) triazene adducts PtX$_2$(ArNNNHMe)$_2$ (X = Cl, Br)—incorrectly formulated as PtX$_2$(ArNNNMe)$_2$ in the original paper—show some evidence of antitumor activity (*116*). The monodentate triazenide derivatives of the four-coordinate platinum(II) compound *cis*-PtCl(ArNNNAr)L$_2$ (L = phosphine, arsine; L$_2$ = cycloocta-1,5-diene) have been prepared from the corresponding dichlorides and Li[ArNNNAr] (*209*) or ArNNNHAr/NEt$_3$ (*54*), or from *trans*-PtHCl(PPh$_3$)$_2$ and free triazene (*135*). Treatment of *cis*-PtCl$_2$(PPh$_3$)$_2$ with free triazene in the presence of hydrazine affords *trans*-PtH(ArNNNAr)(PPh$_3$)$_2$ (*208*). X-Ray diffraction studies on PtCl(*p*-tol-NNN-tol-*p*)(PPh$_3$)$_2$ (*12, 209*) and PtH(*p*-tol-NNN-tol-*p*)(PPh$_3$)$_2$ (*109, 209*) confirm the monodentate nature of the triazenide ligands (Pt—N-3 = 2.908 and 3.008 Å, respectively). Oxidative addition of triazenes ArNNNHAr to Pt(PPh$_3$)$_3$ affords Pt(ArNNNAr)$_2$(PPh$_3$)$_2$ (*135, 190*). An X-ray diffraction study on *cis*-Pt(PhNNNPh)$_2$(PPh$_3$)$_2$ (Fig. 1) by Brown and Ibers in 1976 provided the first unambiguous structural evidence of monodentate triazenide ligands in a transition metal complex (*21, 22*). Variable-temperature NMR studies on these platinum(II) triazenide complexes reveal fluxional behavior (*54, 191, 208, 209*) similar to that discussed above for the corresponding palladium(II) complexes. Activation energies are low (≤ 50 kJ/mol) and decrease Pd > Pt. Proton NMR studies on a series of triazenide complexes, *trans*-PtH(ArNNNAr)(PPh$_3$)$_2$, indicate that the triazenide ligand is a good σ donor (*211*). The same complexes react with CO, ArNC, (NC)$_2$CC(CN)$_2$, and PhC≡CPh under mild conditions to yield platinum(0) derivatives by reductive elimination of triazenes (*207, 208*). The products [{2,6-(Me$_2$NCH$_2$)$_2$C$_6$H$_3$}Pt(μ-*p*-tol-NNN-R)AgBr] (R = Me, Et, Pr^i), obtained from {2,6-(Me$_2$NCH$_2$)$_2$C$_6$H$_3$}PtBr and [Ag(*p*-tol-NNN-R)]$_n$, have been assigned the Pt → Ag bonded structure **37** on the basis of NMR data. The asymmetry of the triazenide ligands *p*-tol-NNN-R permits two isomeric forms of these complexes, both of which are observed. It has been suggested that the basicity of the platinum atom, enhanced by the N-donor ligands, contributes to the stability of the Pt → Ag dative bond (*215*). An analogous reaction involving HgCl(*p*-tol-NNN-R) produces quantitative yields of [{2,6-(Me$_2$NCH$_2$)$_2$C$_6$H$_3$}Pt(*p*-tol-NNN-R)HgBrCl] (**38**). These products can also be prepared from the corresponding carboxylates [{2,6-(Me$_2$NCH$_2$)$_2$C$_6$H$_3$}Pt(O$_2$CR)Hg(O$_2$CR)Br and free triazenes, *p*-tol-NNNHR (*216*).

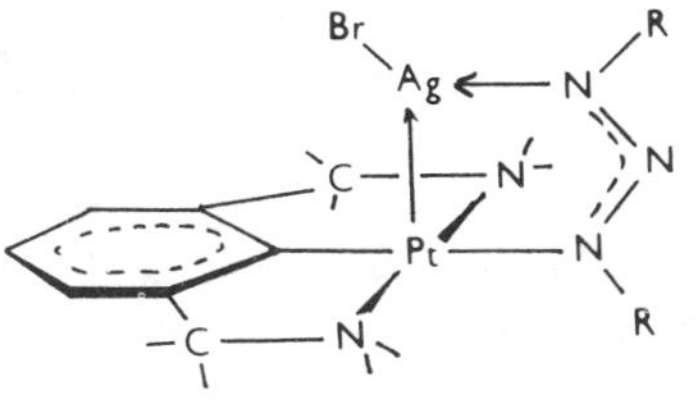

(37) 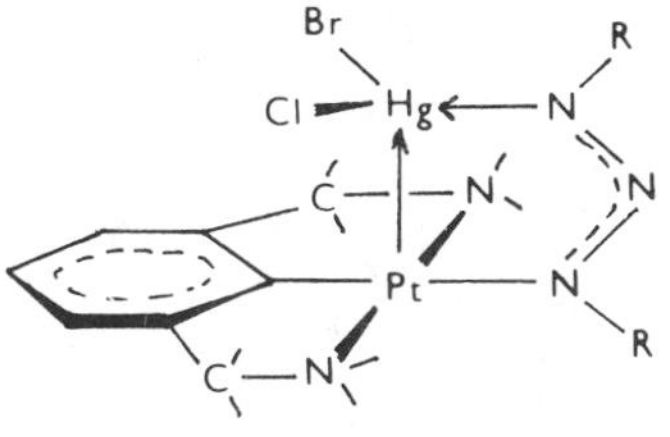   (38)

7. Copper, Silver, and Gold

The ability of copper(I) and silver(I) salts to react with triazenes was first noted by Peter Griess in 1866 (*90*), and triazenide complexes of these metals have attracted considerable attention during the intervening years. In marked contrast the first gold triazenide complexes have only recently been reported.

Copper(I) diphenyltriazenide was first properly characterized by Meunier and Rigot (*150, 152*), who obtained the complex by direct reaction of the free triazene with copper powder. Other copper(I) triazenide complexes, including Cu(PhNNNH) (64), Cu(PhNNNMe) (61), Cu(PhNNNEt) (62), and Cu(MeNNNMe) (63), were prepared by Dimroth around the turn of the century. The bis(pyridine) adduct Cu(PhNNNPh)py$_2$, first described by Dwyer in 1941 (71), loses pyridine at 100°C to form Cu(PhNNNPh) as a lemon yellow powder. The latter product was first formulated as a Cu^I—Cu^I-bonded dimer (150, 152), but was subsequently shown to possess a triazenide-bridged structure (Fig. 4) (19). In contrast, the corresponding dimethyl triazenide derivative, which was more recently obtained from CuCl and Al(MeNNNMe)$_3$ (17), has been found to possess a novel tetranuclear structure (Fig. 5) (161). The original suggestion that triazenes existed in cis and trans isomeric forms—"malenoid" and "fumaroid"—which gave rise to different colored copper(I) and silver(I) complexes (140–142), was later proved to be incorrect; color differences were attributed to impurities (69).

A diaryl triazenide complex of copper(II), Cu(ArNNNAr)$_2$ (Ar = p-C$_6$H$_4$-NO$_2$), was obtained by Meldola and Streatfeild in 1887 from an ammoniacal solution of CuSO$_4$·5H$_2$O and the free triazene (146). Similar complexes were mentioned by Meunier (150) and subsequently described in more detail by Mangini and Dejudicibus (142). Purer compounds were obtained by addition of cupric acetate to the free triazenes in methanol solution (71) and by treatment of cupric nitrate with triazene and base in ethanol (231). The adducts Cu(ArNNNAr)$_2$py$_2$ and Cu(ArNNNAr)$_2$(en) were also described (71). A bis-chelate structure was first proposed for the complexes Cu(ArNNNAr)$_2$ (71), but this was subsequently discarded in favor of a binuclear "lantern" structure (98, 99), which was later confirmed by an X-ray diffraction study on the diphenyl triazenide derivative (45, 46). However, molecular weight data for some copper(II) triazenides Cu(ArNNNAr)$_2$ (Ar = p-Cl-C$_6$H$_4$, p-Me-C$_6$H$_4$) favor a mononuclear ⇌ dimer equilibrium (138) and a combination of molecular weight and magnetic data indicates that copper(II) complexes of bis(trichlorophenyl)triazenes are paramagnetic monomers (235, 236). Infrared and electronic spectra have been discussed for Cu(p-tol-NNN-tol-p)$_2$ (237). Reactions of arylamines with copper(II) acetate in the presence of isoamyl nitrite provide a convenient route to copper(II) triazenide complexes (225). Thermal decomposition of copper(II) diaryltriazenides to form the corresponding copper(I) species has been described; side products included N$_2$, biaryls, and azoaryls (233). In the presence of dioxygen, intermediate paramagnetic copper-containing species are formed (145). A series of papers by Russian workers record the synthesis and spectroscopic properties of a wide range of copper(II) diaryl triazenide complexes (79, 194, 206, 235–237). Magnetic data and electronic spectra have been recorded for Cu$_2$(PhNNNPh)$_4$ (84), and polarographic reduction data have been reported for Cu^I and Cu^{II} diphenyl triazenide complexes (27).

Following the initial brief report by Griess (*90*), silver triazenide complexes were next mentioned by Meldola and Streatfeild (*146*), who, in 1887, described the bright red, thermally unstable compound "ArNNNAgAr" (Ar = p-C_6H_4-NO_2), obtained from the free triazene and silver nitrate in ammoniacal solution, and who later prepared other silver triazenides by this method (*147, 148*). This approach was subsequently employed by Dimroth to prepare Ag(MeNNNMe) (*63*), Ag(PhNNNH) (*64*), Ag(PhNNNMe) (*61*), and Ag(PhNNNEt) (*62*), and by Mangini (*140*) and Dwyer (*69–71*) in the 1930s to prepare a range of silver diaryl triazenide complexes. In 1897 Niementowski and Roszkowski (*159*) obtained Ag(PhNNNPh) from anilinium sulfate and silver nitrite; a similar reaction involving $AgNO_2$, $ArNH_2$, and aqueous CO_2 was successfully employed by Meunier (*151*) in 1903. More recently, Vernin and co-workers (*225*) prepared silver(I) triazenides from $AgNO_3$ or AgO_2CMe and arylamines in the presence of isoamyl nitrite. Syntheses of silver(I) diaryltriazenide complexes have also been reported by Mangini (*141*). A mononuclear chelate triazenide structure was originally proposed for the silver(I) salts on the basis of molecular weight data obtained for solutions in pyridine (*71*). However, pyridine adduct formation is likely to occur under these conditions, and it has subsequently been suggested that dimeric or polymeric triazenide-bridged structures are more probable for the nonsolvated complexes (*98*). A series of papers from Russian workers report synthesis of silver(I) triazenide complexes and record vibrational and electronic spectra (*194, 206, 236, 237*). A very extensive range of silver(I) triazenide complexes Ag(RNNNR'), in which R and/or R' are functionalized organic groups, have been reported (*5, 11, 16, 66, 88, 119, 162, 184–188, 200, 229*).

The first gold triazenide complex [Au(PhNNNPh)]$_4$, prepared from AuI and Na(PhNNNPh) in liquid NH_3 (*6a*), forms yellow, air stable crystals with a tetrameric structure similar to that found for [Cu(MeNNNMe)]$_4$ (Fig. 5).

8. Zinc, Cadmium, and Mercury

Diethylzinc reacts with dimethyltriazene to afford the moisture-sensitive complex Zn(MeNNNMe)$_2$ (*17*). However, a similar reaction involving diphenyltriazene gave a product which, after recrystallization from dry benzene, was found to be $Zn_4O(PhNNNPh)_6$ and to possess a "basic beryllium acetate" type of structure (*48*).

The first cadmium triazenide complex Cd(ArNNNAr)$_2$ (Ar = p-C_6H_4-NO_2) was obtained by Meldola and Streatfeild in 1887 on adding ammoniacal $CdCl_2$ to a hot alcoholic solution of the triazene (*146*). It was described as a steely blue crystalline material which turned red on drying and exploded on heating. The diphenyltriazene analogue, prepared more recently from CdI_2 and $Ag(PhNNNPh)$ in dry ether, forms moisture-sensitive yellow crystals (18).

Mercury forms at least three series of complexes with triazenes: the adducts $HgX_2(ArNNNHAr)_2$, and the triazenide derivatives $Hg(ArNNNAr)_2$ and $HgX(ArNNNAr)$ (X = halogen, acetate, aryl, etc.). The adduct $HgCl_2(PhNNNHPh)$ was reported as early as 1897 (97). A later report (139) that mercury salts HgX_2 form 1/4 adducts $HgX_2(PhNNNPh)_4$ was subsequently refuted (121). Attempts to repeat the work gave 1/2 adducts $HgX_2(PhNNNHPh)_2$ (X = Cl, Br) or complexes $Hg(PhNNNPh)_2$ (X = NO_3, $MeCO_2$) (98, 121). The first mercury(II) triazenide complex appears to have been $Hg(PhNNNPh)_2$, obtained by Cuisa and Pestalozza in 1911 from HgO and the free triazene (55, 56). This complex has also been obtained from mercuric acetate and phenylhydrazine (223) and from metallic mercury and free triazene in the presence of oxygen (231). Numerous papers report the synthesis of mercury(II) triazenides $Hg(ArNNNAr)_2$ (98, 140–142, 205) or $Hg(O_2CMe)(ArNNNAr)$ (222) from mercuric acetate and free triazenes. Two more recent notes describe isolation of $Hg(ArNNNAr)_2$ complexes from the reaction of arylamines with isoamyl nitrite in the presence of mercuric acetate (225, 226). Treatment of mercuric salts HgX_2 (X = Cl, I) with stoichiometric amounts of silver(I) triazenides has been used to generate the complexes $HgX(RNNNR')$ and $Hg(RNNNR')_2$ (R, R' = Me or p-tol) in solution (218).

Several papers report synthesis (205) and vibrational, electronic, and proton NMR spectral data (234, 235, 237) for a range of mercury(II) triazenides $Hg(ArNNNAr)_2$ (Ar = polyhalophenyl). Molecular weight and spectroscopic data indicate that these products are best formulated as monomers with chelate triazenide ligands (205, 235). Core binding energies [C $1s$, N $1s$, Hg $4f_{5/2}$, and Hg $4f_{7/2}$] have been measured for a range of mercury(II) triazenides $Hg(p\text{-}X\text{-}C_6H_4\text{-}NNN\text{-}C_6H_4\text{-}X\text{-}p)_2$ (X = H, Me, MeO, or Cl); within limits of error (± 0.2 eV) only a single N $1s$ signal could be detected for each sample. Insensitivity of core binding energies to the nature of the para substituents has been interpreted as evidence of ionic character in the Hg—N bonds (136). The high lattice energy, reflected in the low solubility, of the mercury(II) chelates $Hg(ArNNNAr)_2$ apparently causes the species $HgX(ArNNNAr)$ to undergo disproportionation. However, if the aryl groups bear ortho substituents the bis(triazenide) complexes are much more soluble and are thought to contain monodentate triazenide ligands bound to two-coordinate mercury. Using these ortho-substituted diphenyltriazenes it has proved possible to isolate the mixed complexes $HgX(ArNNNAr)$ (X = Cl, Br, I, or CN) by synproportionation of equimolar amounts of HgX_2 and $Hg(ArNNNAr)_2$ (172).

Phenylmercuric triazenide complexes $PhHg(ArNNNAr)$, prepared from phenylmercuric hydroxide or acetate and free triazene (174) or by synproportionation of $HgPh_2$ and $Hg(ArNNNAr)_2$ (*172*, *173*, *175*), have attracted considerable attention. An X-ray diffraction study on $PhHg(2\text{-}ClC_6H_4NNNPh)$ has established linear two-coordinate mercury(II) with a monodentate N-1-bonded triazenide ligand (*133*). Multinuclear (^1H, ^{13}C, ^{15}N, ^{19}F, and ^{199}Hg) NMR studies have shown that the 1,3-diaryl triazenide ligands undergo a rapid metallotropic rearrangement, which occurs mainly by an intramolecular mechanism (*126*, *157*, *174*, *175*).

$$\underset{\underset{\text{HgPh}}{|}}{\text{X-C}_6\text{H}_4\text{-NNN-C}_6\text{H}_4\text{-X}} \quad\rightleftharpoons\quad \underset{\underset{\text{HgPh}}{|}}{\text{X-C}_6\text{H}_4\text{-NNN-C}_6\text{H}_4\text{-X}} \qquad (24)$$

Exchange reactions between diaryltriazenes and their phenylmercury derivatives have been examined by proton NMR, and equilibrium constants have been determined (*158*).

III. Tetrazane, Tetrazene, and Tetrazadiene Complexes

Organic derivatives of tetrazane ("buzane") H_2N—NH—NH—NH_2 have been known since 1893 (*227*), but to date no transition metal tetrazane complexes have been fully characterized. However, iron tetrazane species have been proposed as intermediates in the iron(II)-catalyzed aerobic oxidation of hydrazines (*104, 230*) and are thought to account for the red coloration observed in some of these reactions (*230*).

The first tetrazene Et_2N—N=N—NEt_2 was reported by Fischer (*81*) in 1878 and the first complex, $HgCl_2(Et_2N$—N=N—$NEt_2)$, was described shortly thereafter (*82*). Despite the ready availability of these ligands, remarkably little further work has been reported and examples of characterized complexes are restricted to metals of group IIB and possibly molybdenum. However, a tungsten complex containing a bridging isotetrazenide (N_4^{4-}) anion, derived from the unknown isotetrazene $(H_2N)_2N$=N, has recently been reported (see Section III,E,1).

In contrast, tetrazadienes, RN=N—N=NR, which are unknown in the free state, are found in a growing number of transition metal complexes. The ligands can be generated *in situ* from organic azides or diazonium cations and, following the discovery of $Fe(MeNNNNMe)(CO)_3$ by Dekker and Knox in 1967 (*59*), an extensive range of transition metal tetrazadiene complexes has been synthesized. Finally, several papers report theoretical calculations on the stability and structure of N_4 ligands bound to transition metals (*196–198*).

A. SYNTHESIS AND PROPERTIES

Tetrazane complexes may form as unstable intermediates during the transition metal catalyzed oxidation of hydrazines. Tetrazene adducts are obtained from the free ligand and the appropriate metal halide. Tetrazadiene complexes are usually prepared by treatment of suitable transition metal complexes with organic azides (*164*) or diazonium salts (*86*).

$$Co(C_5H_5)(CO)_2 + 2PhN_3 \longrightarrow$$

$$Co(C_5H_5)(PhNNNNPh) + N_2 + 2CO \quad (25)$$

$$IrCl(CO)(PPh_3)_2 + 2ArN_2^+ + 2e \longrightarrow$$

$$[Ir(ArNNNNAr)(CO)(PPh_3)_2]^+ + Cl^- \quad (26)$$

Reactions involving transfer of tetrazadiene ligands between metal centers have also been reported (*170*).

$$Ni(ArNNNNAr)_2 + Ni(Bu^tNC)_4 \longrightarrow 2Ni(ArNNNNAr)(Bu^tNC)_2 \quad (27)$$

Tetrazadiene complexes are noted for their brilliant color and for their stability, which is much greater than that of analogous 1,4-diazabutadiene derivatives.

B. Structural Properties

There is no structural information available on tetrazane complexes. Tetrazenes can exist in cis or trans forms, with N-1,N-4-chelation of the cis form originally being advanced as the most probable mode of coordination (*160*). However the only X-ray crystal structure reported to date revealed a trans-tetrazene ligand coordinated through nitrogens N-1 and N-3 (*57*). Tetrazadienes are unknown. However, the parent tetrazabutadiene $HN=N-N=NH$ is isoelectronic with a range of well-known ligands, notably 1,4-diazabutadiene, butadiene, and dinitrogen tetroxide (bound as hyponitrite, $N_2O_4^{2-}$). Although there are numerous possible modes of attachment of a four-nitrogen chain to one or more metal centers, only three have assumed importance to date in discussion of metal tetrazadiene structures. There are the η^4-diene structure **39** analogous to that found in

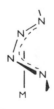

(39)

(a)   (b)

(40)

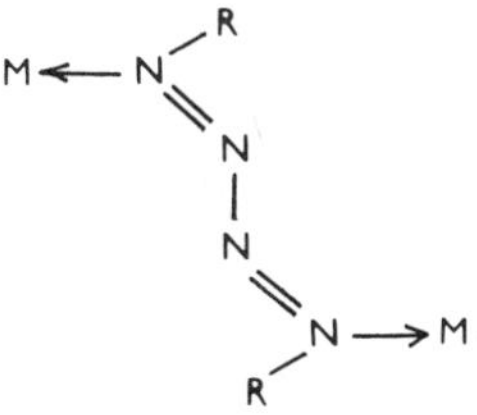

(41)

butadiene iron tricarbonyl, the N-1,N-4-metallocyclic structure **40a/b**, and the N-1,N-4-bridging structure **41**. Theoretical calculations indicate that the η^4-structure (**39**) is likely to be less stable than the N-1,N-4-metallocycle (**40**), and, in keeping with this conclusion, no examples of η^4-tetrazadiene complexes have been reported to date. By virtue of their delocalized π systems, the N-1,N-4-chelated tetrazadienes are "suspect" or "noninnocent" ligands (86) like the 1,2-diimines and 1,2-dithiolenes. Consequently, the electronic structures of their complexes may approximate to one of two canonical forms $M^{n+}/RN=N-N=NR$ and $M^{(n+2)+}/RN^--N=N-N^-R$, represented by **40a** and **40b**, respectively. In some instances the coordination geometry of the metal center fixes the formal oxidation state of the metal and hence the electronic structure of the tetrazadiene ligands. Thus octahedral $Pt(C\equiv CPh)_2(ArNNNNAr)(PEt_3)_2$ is clearly a platinum(IV) rather than a platinum(II) complex, and should be formulated as a tetrazenediyl $(ArN^--N=N-N^-Ar)$ derivative (85), whereas the pseudo-tetrahedral complexes $Ni(ArNNNNAr)_2$ are best regarded as nickel(0) tetrazadiene derivatives (167). In other cases electronic structures are assigned on the basis of theoretical calculations and/or ligand geometry. Structures containing short central N—N bonds are generally regarded as tetrazene-1,4-diyl[3]

---

[3] Some authors incorrectly refer to complexes formulated in this manner as tetrazene complexes. In this article we use the term *tetrazadiene* to describe all ligands of stoichiometry RNNNNR, irrespective of the formal bonding mode, and retain the name tetrazene for ligands of the form $R_2NN=NNR_2$ (R = H, alkyl, or aryl).

complexes (40b), whereas those in which N—N bonds are of essentially equal length are classified as tetrazadiene derivatives (40a). Finally, in the case of diaryltetrazadiene complexes, further evidence is provided by the orientation of the aryl groups relative to the plane of the metallocycle. Structures of the type 40a have aryl groups conjugated to and fully coplanar with the MN_4 metallocycle, whereas those of type 40b have aryl groups tilted at angles of ~ 45–65 and 130° to the plane (167).

Several binuclear structures thought to contain bridging tetrazadiene ligands have been described and others have been postulated as intermediates in tetrazadiene-transfer reactions but none has proved suitable for X-ray diffraction study. Although N-1,N-4 bridging has been cited (169, 171) as the most probable form of linkage, N-1,N-2 or N-1,N-3 bridges cannot be excluded. Some other possible bridging arrangements have also been discussed (171). X-Ray diffraction studies reported to date for tetrazadiene complexes are listed in Table II.

C. SPECTROSCOPIC STUDIES

Salient examples of the applications of spectroscopic techniques to the study of tetrazadiene complexes are given below. In most instances, further details are given when the complexes concerned are discussed in Section III,E.

1. Vibrational Spectra

In the absence of spectra for the free ligands relatively little use has been made of infrared and Raman spectroscopy in the study of tetrazadiene complexes. However *prima facie* evidence that the π-acceptor ability of 1,4-dimethyltetrazadiene is comparable to that of a pair of carbonyl ligands is provided by the observation that the degeneracy weighted average $v(CO)$ frequency for $Fe(MeNNNNMe)(CO)_3$ is the same (2028.6 cm^{-1}) as that of $Fe(CO)_5$ (213). Infrared data [$v(NC)$ 2168 and 2146 cm^{-1}] for $Ni(ArNNNNAr)(BuNC)_2$ complexes point to a similar conclusion (167).

2. Electronic Spectra

Intense electronic transitions (470–520 and 350–390 nm) exhibited by the complexes $Fe(MeNNNNMe)(CO)_nL_{3-n}$ (L = P-donor, n = 1–3) have been attributed to the presence of a low-lying unoccupied metallocycle π* orbital (115, 213). Electronic spectra have also been reported and analyzed for $[Co(ArNNNNAr)(C_5H_5)]^-$ (143), $Co(ArNNNNAr)(C_5H_5)$ (164), and $Ni(ArNNNNAr)_2$ (167).

TABLE II

Selected Bond Lengths[a] and Angles[b] for Tetrazadiene Complexes

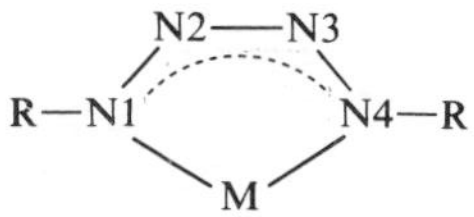

Molecular formula	R	M—N1/N4	N1—N2/N3—N4	N2—N3	∠α[c]	∠β[c]	Reference
Fe(RNNNNR)(CO)₃	Me	1.83	1.32		—	—	65
Co(RNNNNR)(C₅H₅)	C₆F₅	1.802/1.819	1.360/1.355	1.270	61.45	90.24	95
[Ir(RNNNNR)(CO)(PPh₃)₂][BF₄]	4-FC₆H₄	1.941/1.971	1.400/1.350	1.270	66	131	78
Ni(RNNNNR)₂	3,5-Me₂C₆H₃	1.851	1.328	1.309	~0	~0	167
Ni(RNNNNR)(C₅H₅)	4-MeC₆H₄	1.853/1.843	1.344/1.346	1.278	45.0	44.8	166
Pt(RNNNNR)(C₈H₁₂PEt₃)(PEt₃)	4-NO₂C₆H₄	2.169/2.159	1.385/1.391	1.263	~0	~0	165
Pt(RNNNNR)(C≡CPh)₂(PEt₃)₂	4-NO₂C₆H₄	2.119/2.11	1.41/1.40	1.30	~0	~0	85

[a] In angstroms (Å).
[b] In degrees (°).
[c] Dihedral angles between MN₄ plane and the planes of the two aromatic groups R.

3. Nuclear Magnetic Resonance Spectra

Routine NMR data have been reported and analyzed for most of the complexes discussed in Section III,E. Multinuclear (^{1}H, ^{13}C, and ^{31}P) NMR studies have proved evidence of fluxional behavior and carbonyl ligand exchange in $Fe(MeNNNNMe)(CO)(PR_3)_2$ (R = Cy, Ph, or OMe) and $Fe(MeNNNNMe)(CO)_3$, respectively (114).

4. Nuclear Quadrupole Resonance Spectra

Cobalt-59 nuclear quadrupole resonance data have been recorded for $Co(C_6F_5NNNNC_6F_5)(C_5H_5)$ and some related cobalt complexes (156).

5. Electron Spin Resonance Spectra

ESR data have been used to help characterize the paramagnetic complexes $Ni(ArNNNNAr)(C_5H_5)$ (166, 168) and $[Co(ArNNNNAr)(C_5H_5)]^-$ (143).

6. Photoelectron Spectra

The rich structure observed in the low-energy photoelectron spectrum of $Fe(MeNNNNMe)(CO)_3$ is qualitatively in accord with the fact that there are eight orbitals—arising from the eight d electrons of the $Fe(CO)_3$ moiety, and the four π electrons plus two lone pairs of the MeNNNNMe ligand—with ionization potentials in the range 8.1–12.0 eV, and can be assigned on the basis of quantitative $X\alpha$ calculations (213). X-Ray photoelectron spectra have been used to help assign oxidation states in the complexes $Ni(ArNNNNAr)(C_5H_5)$ (166) and $Pt(ArNNNNAr)\{\overline{CHC(PEt_3)HCH_2CH_2CH}{=}CHCH_2CH_2\}(PEt_3)$ (165).

7. Mass Spectra

The high thermal stability of tetrazadiene complexes is reflected in their mass spectra, which generally show very intense parent ion peaks. Thus ionization of $Co(MeNNNNMe)(C_5H_5)$ affords a parent ion that fragments with loss of N_2 and Me groups to yield ions which include $[Co(N_2Me)(C_5H_5)]^+$ and $[CoH(C_5H_5)]^+$ (164). Field desorption mass spectra have been used to investigate ligand exchange and rearrangement reactions involving tetrazadiene complexes of cobalt, nickel, and platinum (169, 171).

D. CHEMICAL REACTIVITY

In marked contrast to the parent ligands, which are unknown in the free state, tetrazadiene complexes are extremely stable. In particular they are much less susceptible to oxidation, thermal decomposition, or hydrolysis than the analogous complexes of the isoelectronic 1,4-diazadienes. Substitution reactions usually involve the ancillary ligands or occur only when the incoming ligand can attack and destabilize the coordinated tetrazadiene ligand. The latter point is illustrated by the behavior of the complexes $Ni(ArNNNNAr)_2$, which react with Bu^tNC under drastic conditions to afford $Ni(ArNNNNAr)(Bu^tNC)_2$, but resist attacks by CO, PPh_3, or bipyridyl ligands (167, 168). There is good chemical and spectroscopic evidence that tetrazadiene ligands are superior to 1,4-diazadienes and even rival carbon monoxide as π-acceptor ligands (143, 213). The ease with which the complexes $Co(ArNNNNAr)(C_5H_5)$ undergo electrochemical reduction to the corresponding monoanions indicates that tetrazabutadiene ligands are very effective in stabilizing electron-rich organometallics (143). The ability of alcohols and, in particular, BF_3 to dramatically increase the rate of CO substitution in $Fe(MeNNNNMe)(CO)_3$ has been attributed to the formation of adducts 42 and 43, respectively, which make the iron center more

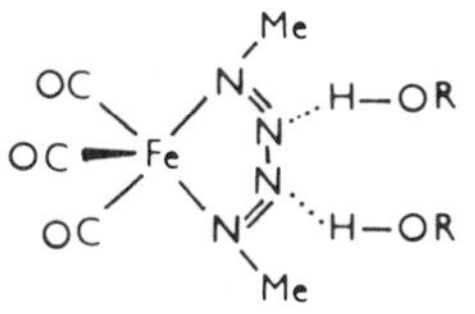

(42) 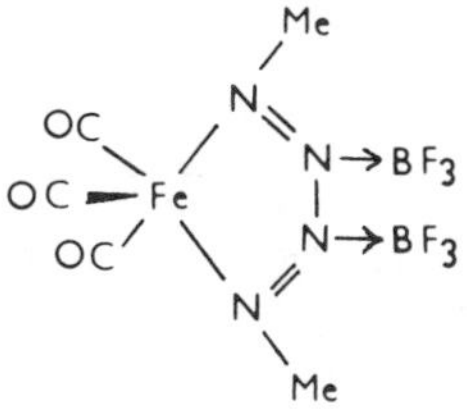   (43)

susceptible to nucleophilic attack (33). Cleavage of tetrazadiene ligands by HCl or HBF_4 to afford ArN_3 and $ArNH_2$ fragments has been reported for several cobalt (164), rhodium, iridium, and platinum derivatives (31, 134).

$$Pt(ArNNNNAr)(PPh_3)_2 + 2HCl \longrightarrow$$

$$PtCl_2(PPh_3)_2 + ArN_3 + ArNH_2 \quad (28)$$

Under the influence of UV radiation the cobalt complex $Co(PhNNNNPh)(C_5H_5)$ undergoes a remarkable transformation to the diimine derivative $Co(HNC_6H_4NPh)(C_5H_5)$ (93, 96). More details of this and some related reactions are given in Section III,E,3. Exchange reactions between tetrazadiene complexes and free aryl azides lead to partial or complete substitution (169).

$$Pt(ArNNNNAr)(C_8H_{12}) + Ar'N_3 \longrightarrow$$

$$Pt(Ar'NNNNAr)(C_8H_{12}) + Pt(Ar'NNNNAr')(C_8H_{12}) \quad (29)$$

Similar exchange reactions occur rather more slowly between pairs of complexes (*169*).

$$\text{Ni(Ar'NNNNAr')}_2 + \text{Ni(ArNNNNAr)}_2 \longrightarrow$$

$$\text{Ni(Ar'NNNNAr')(ArNNNNAr)} \xrightarrow{\text{very slow}} \text{Ni(Ar'NNNNAr)(Ar'NNNNAr)}, \text{etc.} \quad (30)$$

Finally, the ability of tetrazadiene ligands to transfer between metal centers via bridged intermediates has been utilized as a route to new tetrazadiene complexes (*170, 171*). Further discussion of these exchange and transfer reactions is given in Section III,E,4.

E. Group Survey

There appears to be no chemistry of tetrazenes or tetrazadienes with metals of the scandium, titanium, and vanadium groups.

1. Chromium, Molybdenum, and Tungsten

The tetrazene-bridged structure **44** has been tentatively proposed for the complex obtained from the reaction of phenylhydrazine with molybdate(VI)

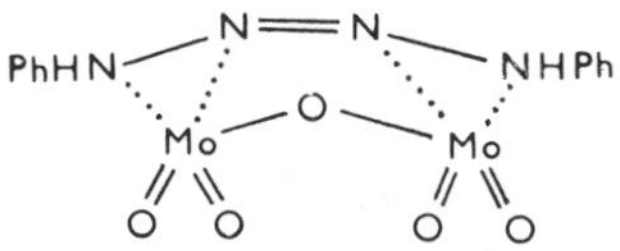

(44)

anions (*15*). More recently, a novel reaction between the tungsten nitride $W(N)Cl_3$ and the azide $W(N_3)Cl_5$ in the presence of $[Ph_4As]Cl$ has afforded $[Ph_4As]_2[Cl_5W(\mu\text{-}N_4)WCl_5]$, the first example of a complex containing the highly unusual Y-shaped μ-isotetrazenide(4$^-$) ligand, N_4^{4-} (*144*). The X-ray crystal structure (Fig. 10) reveals a bridging planar N_4 moiety linked to the tungsten atoms by very short W—N bonds [1.65(1) and 1.62(1) Å] (*144*).

No tetrazene or tetrazadiene complexes have been reported to date for members of the manganese, technetium, rhenium triad.

2. Iron, Ruthenium, and Osmium

The first tetrazadiene complex, $Fe(MeNNNNMe)(CO)_3$, which was isolated from the mixture of products obtained on treatment of $Fe_2(CO)_9$ with methyl azide, is a volatile orange solid of remarkable stability (*59*). On

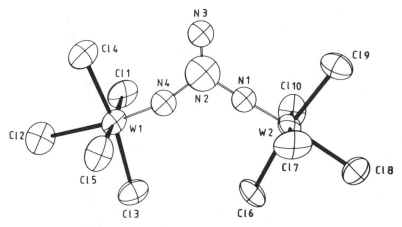

FIG. 10. An ORTEP diagram of the $[Cl_5W(\mu\text{-}N_4)WCl_5]^{2-}$ anion in the crystal of the tetraphenylarsonium salt.

the basis of infrared, NMR, and mass spectroscopic data the original authors proposed a "piano stool" structure analogous to that of butadiene iron tricarbonyl (59). However, an X-ray diffraction study quickly established that the MeNNNNMe ligand is bound in chelate fashion (Fig. 11) to generate an essentially planar FeN_4 metallocycle (65). Short Fe—N bonds (1.83 ± 0.03 Å) were taken to indicate considerable back donation from iron d orbitals to the π^* antibonding orbitals of the tetrazadiene ligand (65). CNDO calculations performed on the electronic structures of the two geometries considered above, in which the N_4 moiety is attached to the iron tricarbonyl fragment in σ or π fashion, confirmed that the chelate structure involving

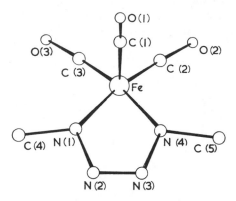

FIG. 11. The molecular configuration of $Fe(MeNNNNMe)(CO)_3$.

bonding between the iron atom and the terminal (N-1 and N-4) nitrogen atoms is the more stable one (2). More recently, Hückel molecular orbital and SCC-DV-Xα calculations have been performed on Fe(RNNNNR)(CO)$_3$ (R = Me or Ph) and a series of substituted products Fe(MeNNNNMe)(CO)$_{3-n}$L$_n$ [n = 1-3, L = PMe$_3$, PPh$_3$, or P(OMe)$_3$] (213). These are consistent with a structure in which the d^8 Fe(CO)$_3$ fragment possessing two $d\pi$ electrons interacts with the four $p\pi$ electrons of the N$_4$R$_2$ ligand to create a six π electron (Hückel aromatic) cyclic system. The SCC-DV-Xα calculations also provide a quantitative explanation of the electronic spectrum, which contains two intense transitions at 470–520 and 349–390 nm attributed to the presence of a low-lying unoccupied metallocycle π^* orbital. In sharp contrast with an earlier CNDO study, which showed a build up of electron density on iron and a decrease of π-electron density on the tetrazadiene ligand, the results of the SCC-DV-Xα calculations point to a flow of electron density from the metal to the tetrazadiene fragment (213). Support for this conclusion comes from infrared data (ν_{CO} frequencies) and there is evidence that RNNNNR ligands rival carbon monoxide as π acceptors (213). The vapor-phase He 1 photoelectron spectrum of Fe(MeNNNNMe)(CO)$_3$ displays ionizations in the 8- to 11-eV spectral region arising from orbitals containing dominant metal d character in addition to MeNNNNMe lone pair and $p\pi$ character (213).

Thermal carbon monoxide substitution in Fe(MeNNNNMe)(CO)$_3$ proceeds readily to form monosubstituted products Fe(MeNNNNMe)(CO)$_2$L (L = P or As donor, 4-CNpy, or Me$_3$CNC). With Me$_3$CNC bi- and trisubstituted species are also obtained (33). The substitutions proceed by a second-order mechanism with a rate law which is first order in complex and in entering ligand. Activation parameters, and rates that are strongly dependent upon the nature of the ligand L, particularly its size and basicity, provide further evidence that the substitution reactions proceed by an associative mechanism. A million-fold increase in the rate of substitution, brought about by the addition of excess BF$_3$, is attributed to the ability of the Lewis acid to withdraw electron density from the iron center and thus facilitate nucleophilic attack by the incoming ligand (33). The rate of reaction is also increased by use of polar solvents and, to a greater extent, by use of alcohols which are capable of H bonding to the nitrogens of the tetrazadiene ligands (33). In contrast to the corresponding thermal reactions, photo-substitution of CO in Fe(MeNNNNMe)(CO)$_2$L (L = CO or PPh$_3$) proceeds via a dissociative mechanism. Products obtained in this manner are of the form Fe(MeNNNNMe)(CO)$_{3-n}$L$_n$ [n = 1 or 2, L = PPh$_3$, PMe$_3$, or P(OMe)$_3$; n = 3, L = P(OMe)$_3$] (115). Quantum yields for CO substitution in the parent tricarbonyl increase exponentially as a function of excitation energy from 0.08 (at 578 nm) to 0.53 (at 313 nm) (115). Variable-temperature

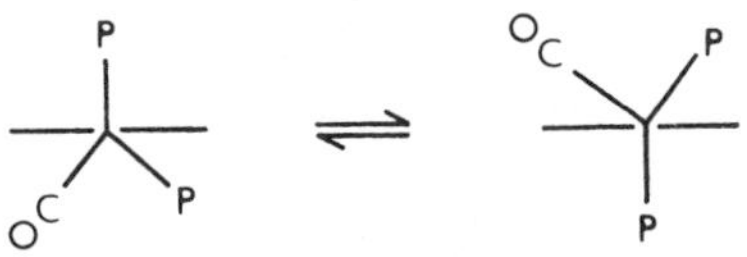

SCHEME 8. Fluxional process for complexes Fe(MeNNNNMe)(CO)(PR$_3$)$_2$. Reprinted with permission from *Inorg. Chem.* **21**, 427. Copyright (1982) American Chemical Society.

NMR studies have provided evidence of a fast rocking motion between two distorted square-pyramidal conformations (Scheme 8) for the complexes Fe(MeNNNNMe)(CO)(PR$_3$)$_2$, and of fast CO and P(OMe)$_3$ exchange in the complexes Fe(MeNNNNMe)(CO)$_3$ and Fe(MeNNNNMe){P(OMe)$_3$}$_3$, respectively (*114*).

3. Cobalt, Rhodium, and Iridium

Shortly after the synthesis of Fe(MeNNNNMe)(CO)$_3$, a similar reaction between organic azides, RN$_3$, and Co(C$_5$H$_5$)(CO)$_2$ was shown to afford air-stable deep green (R = Me) or brown (R = Ph) crystals of stoichiometry Co(RNNNNR)(C$_5$H$_5$) (*164*). Physical data and chemical reactivity patterns led to formulation of these complexes as tetrazenediyl (RN$^-$—N=N—N$^-$R) derivatives of cobalt(III). Mass spectra showed intense parent ion peaks (*164*). More recently the bis(perfluorophenyl)tetrazadiene derivative Co(C$_6$F$_5$NNNNC$_6$F$_5$)(C$_5$H$_5$) has been prepared and shown by X-ray diffraction methods to possess a metallocyclic structure (Fig. 12) (*94, 95*), analogous to that found for Fe(MeNNNNMe)(CO)$_3$. The extremely short Co—N distances [1.802(2) and 1.819(2) Å] are consistent with considerable multiple bond character (*94, 95*). An extended series of complexes Co(RNNNNR)(C$_5$H$_5$) (R = Me, Ph, C$_6$F$_5$, 2,4-F$_2$-C$_6$H$_3$, and 2,6-Me$_2$-C$_6$H$_3$) has been prepared and subjected to spectroscopic and theoretical study (*95*). Xα-Calculations predict that two $d\pi$ electrons on the Co(C$_5$H$_5$) fragment should interact strongly with an empty low-lying π^* orbital of the tetrazadiene ligand to yield metallocyclic π and π^* orbitals. The electronic spectra display three absorptions (600–670, 425–470, and 355–390 nm), which are attributed to one-electron transitions that terminate in the low-lying metallocycle π orbital (*95*). Although the complexes Co(ArNNNNAr)(C$_5$H$_5$) (Ar = Ph or C$_6$F$_5$) are thermally stable, they react on irradiation with visible or low-energy UV light to form diimine complexes Co(HNC$_6$X$_4$NC$_6$X$_5$)(C$_5$H$_5$) (X = H or F) (*96*). Evidence of the intramolecular nature of this rearrangement and nitrogen extrusion reaction was provided by cross-over experiments (*96*). The structure of the diimine product (X = F) has been confirmed by X-ray diffraction methods (*92*).

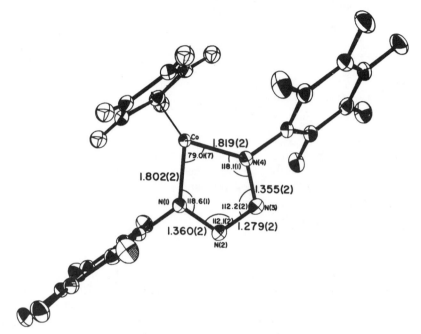

FIG. 12. Molecular structure of $Co(C_6F_5NNNNC_6F_5)(\eta^5\text{-}C_5H_5)$ showing selected bond lengths (in Å) and angles (degrees).

Similar reactions for the complexes $Co(ArNNNNAr)(C_5H_5)$ (Ar = $2,4\text{-}F_2\text{-}C_6H_3$ or $2,6\text{-}Me_2\text{-}C_6H_3$) yield the products shown in Scheme 9. Aromatic radical substitution processes are proposed to account for the unusual C—C and C—F bond breaking steps (93). Aryl azides ArN_3 react with $Co(ArNNNNAr)(C_5H_5)$ to form $Co(Ar'NNNNAr')(C_5H_5)$; a mechanism involving destabilization and substitution of a complete ArNNNNAr unit has been proposed (169). The ^{59}Co nuclear quadrupole resonance spectrum of $Co(PhNNNNPh)(C_5H_5)$ has been recorded (156). The paramagnetic d^9 complex anions $[Co(RNNNNR)(C_5H_5)]^-$ (R = Me, Ph, C_6F_5, $2,4\text{-}F_2\text{-}C_6H_3$, and $2,6\text{-}Me_2\text{-}C_6H_3$) have been obtained by Na/Hg reduction of the corresponding neutral complexes; one example has been isolated as the brick red dibenzo-18-crown-6-sodium salt $[C_{20}H_{24}O_6Na][Co(PhNNNNPh)(C_5H_5)]$. Reduction potentials (cyclic voltammetry in $CH_3CN/0.1\,M$ $Bu^n_4NBF_4$) are dependent upon the nature of R and range from -0.71 to -1.53 V. Each anion displays an isotropic ESR spectrum at ambient temperatures (g = 2.16–2.26, a_{iso} = 50–58 G), characteristic of cobalt–centered radicals. These results are consistent with the view that the tetrazadienes are best regarded as neutral π-acid ligands rather than dianions (143).

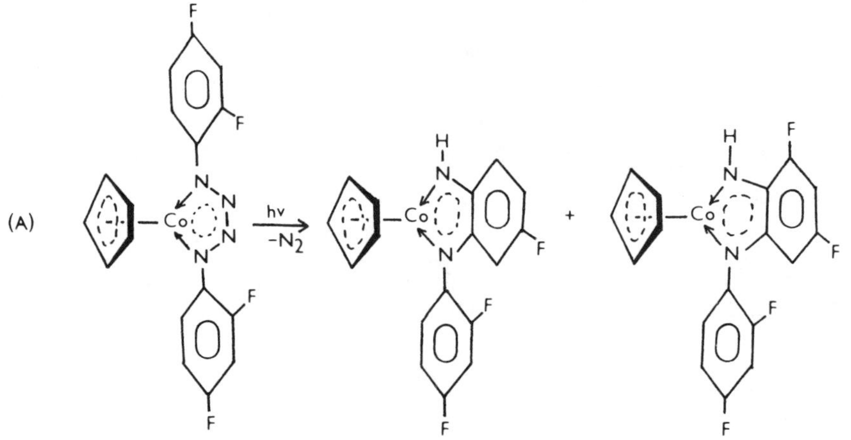

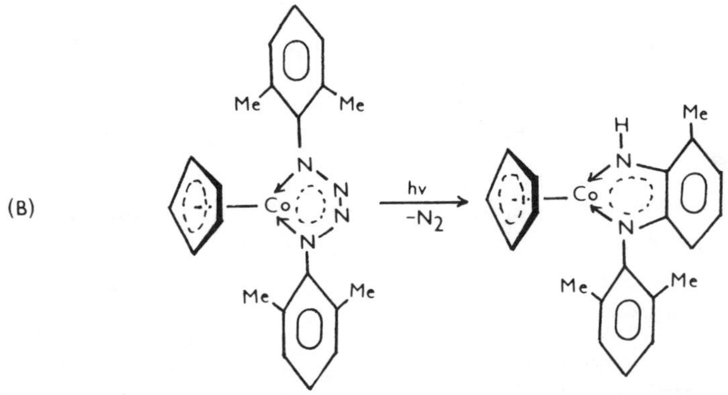

SCHEME 9. Photolytic rearrangements for complexes $Co(ArNNNNAr)(C_5H_5)$. Reprinted with permission from *Inorg. Chem.* **23**, 2968. Copyright (1984) American Chemical Society.

Rhodium and iridium complexes of the form $M(RNNNNR)(NO)(PPh_3)$ ($R = p\text{-Me-}C_6H_4\text{-SO}_2\text{-}$) have been obtained from $M(NO)(PPh_3)_3$ and p-toluenesulfonyl azide in benzene. Addition of ligands L (e.g., CO, PPh_3) affords pentacoordinated species $M(RNNNNR)(NO)(PPh_3)L$. Attempts to form analogous compounds from $Co(NO)(PPh_3)_3$ gave uncharacterizable products. NMR studies provided evidence that was interpreted in terms of an equilibrium (Scheme 10) involving opening of the metallocycle (*134*). However, a later report from the same laboratory concludes that the four-coordinate complexes possess a pseudo-tetrahedral structure and that the splitting of the methyl resonances observed to occur at low temperatures is attributable to a conformational effect (*31*). Cleavage with HCl affords RN_3

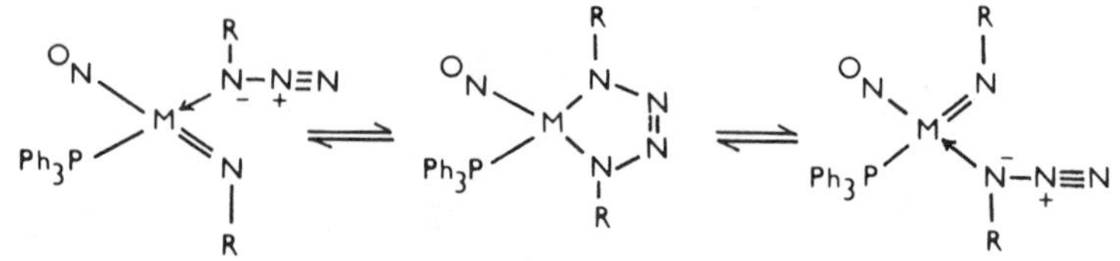

SCHEME 10. Proposed metallocycle cleaving equilibrium for complexes M(RNNNNR)-(NO)(PPh₃)(M = Rh or Ir). Adapted with permission from *J. Organomet. Chem.* **50**, 287 (1973).

and RNH_2 (*31, 134*). Products from the reactions of diazonium salts $[p\text{-}R\text{-}C_6H_4N_2]BF_4$ (R = H, F, Cl, Br, MeO, or CF_3) with Vaska's compound in benzene/ethanol solution include the bright red, air-stable salts $[Ir(p\text{-}R\text{-}C_6H_4NNNNC_6H_4\text{-}R\text{-}p)(CO)(PPh_3)_2][BF_4]$ (Scheme 11) (*76, 77, 80, 86*). The same salts have been obtained using $IrH(CO)(PPh_3)_3$ in place of $IrCl(CO)(PPh_3)_2$ (*80*). The X-ray crystal structure of one example (R = F) has been determined. The iridium coordination geometry approximates to square-pyramidal with an apical PPh_3 ligand. The tetrazadiene ligand is coordinated through N-1 and N-4 to generate an IrN_4 metallocycle. Bond distances [Ir—N-1, 1.941(13); Ir—N-4, 1.971(10), N-1—N-2, 1.400(16), N-3—N-4, 1.350(16), and N-2—N-3, 1.270(16) Å] are consistent with the tetrazenediyl $[RN^-\!-\!N\!=\!N\!-\!N^-R/Ir(III)]$ rather than the alternative $[RN\!=\!N\!-\!N\!=\!NR/Ir(I)]$ formulation (*76, 78*). This conclusion is supported by the results of a semiempirical (CNDO) calculation performed on the simplified model ion $[Ir(N_4H_2)(CO)(PH_3)_2]^+$ (*176*).

4. *Nickel, Palladium, and Platinum*

The first nickel tetrazadiene complex, $Ni(C_6F_5NNNNC_6F_5)(C_8H_{12})$, obtained from $Ni(C_8H_{12})_2$ and pentafluorophenyl azide, reacts with

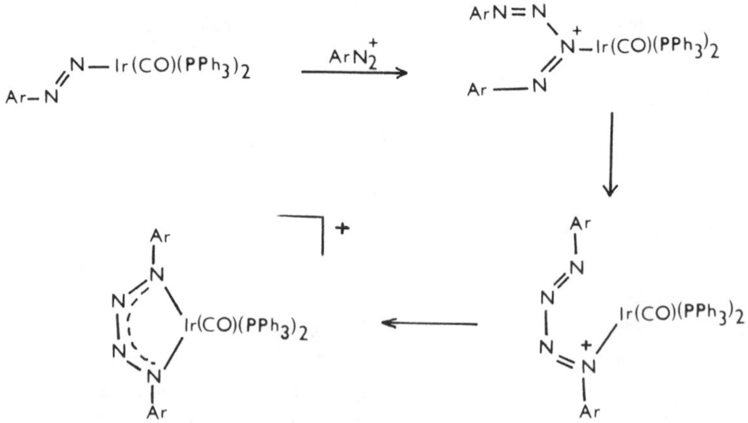

SCHEME 11. Tentative mechanism for the formation of iridium tetrazadiene complexes $[Ir(ArNNNNAr)(CO)(PPh_3)_2]^+$. Adapted with permission from *Can. J. Chem.* **52**, 3387 (1974).

bipyridyl (bipy) and various phosphorus donors [L = PPh$_3$, PPh$_2$Me, PPhMe$_2$, or P(OMe)$_3$] to form Ni(C$_6$F$_5$NNNNC$_6$F$_5$)(bipy) and Ni(C$_6$F$_5$NNNNC$_6$F$_5$)L$_2$, respectively (4). Proton NMR data for complexes containing PPhMe$_2$ and P(OMe)$_3$ have been interpreted in terms of d^8 nickel(II) (45) and d^{10} nickel(0) (46) structures, respectively. It has been

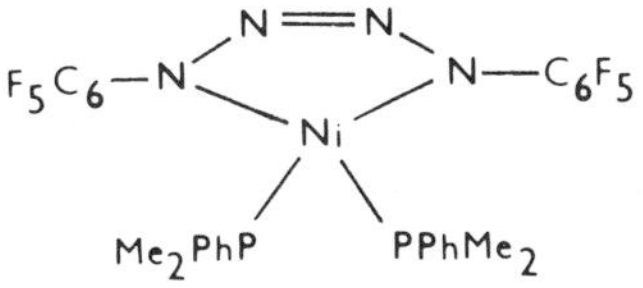

(45) 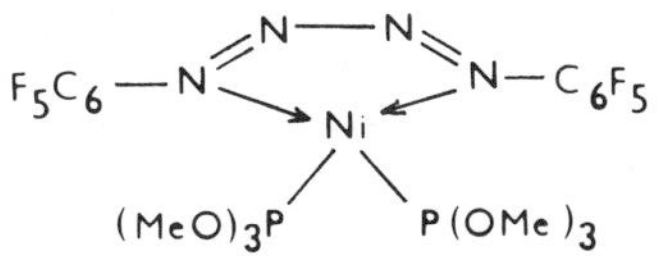   (46)

argued that the balance between the two forms is delicately poised and that the structures adopted reflect the good σ-donor and π-acceptor properties of PPhMe$_2$ and P(OMe)$_3$, respectively. Mechanisms for the formulation of these tetrazadiene complexes have been proposed (Scheme 12) (4). More recently Ni(C$_8$H$_{12}$)$_2$ has been shown to react exothermically with aryl azides ArN$_3$ (Ar = 4-Me-C$_6$H$_4$, 4-MeO-C$_6$H$_4$, 4-Cl-C$_6$H$_4$, or 3,5-Me$_2$-C$_6$H$_3$) to

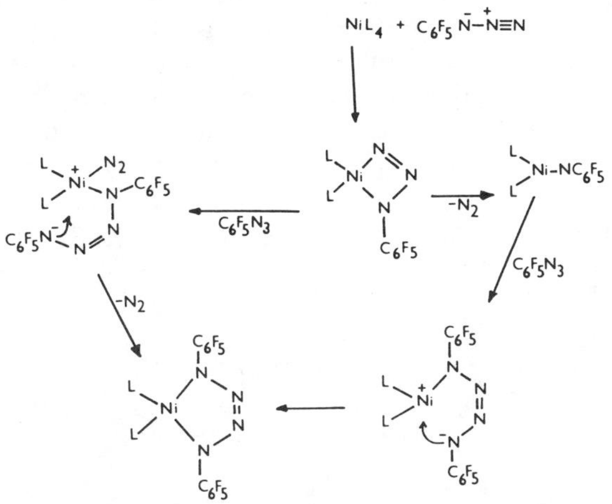

SCHEME 12. Mechanisms proposed for the formation of nickel tetrazadiene complexes Ni(RNNNNR)L$_2$. Adapted from *J. Chem. Soc., Dalton Trans.*, p. 1805 (1972).

form very stable dark purple diamagnetic bis(diaryltetrazadiene) complexes Ni(ArNNNNAr)$_2$ (167, 168). An X-ray diffraction study on one of these products (Ar = 3,5-Me$_2$-C$_6$H$_3$) reveals two mutually perpendicular NiN$_4$ metallocycles which impart pseudo-tetrahedral geometry to the nickel center (Fig. 13). On the basis of the nickel-nitrogen distances [1.851(2) Å] and the nearly equal nitrogen-nitrogen bond lengths (mean value 1.322 Å) the

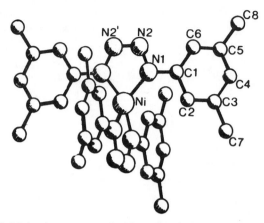

FIG. 13. Molecular structure of $Ni(3,5\text{-}Me_2C_6H_3NNNNC_6H_3Me_2\text{-}3,5)_2$.

complexes are formulated as d^{10} nickel(0) species (*167, 168*). The reaction of nickelocene with *p*-tolyl azide in boiling toluene affords a 10% yield of a black paramagnetic species $Ni(p\text{-tol-NNNN-tol-}p)(C_5H_5)$ together with a small amount of $Ni(p\text{-tol-NNNN-tol-}p)_2$ (*167, 168*). ESR spectra for the former product were originally interpreted in terms of a monomeric d^9 nickel(I) complex with an N-1,N-4-chelate tetrazadiene ligand (*168*). The proposed geometry of the complex has been confirmed by X-ray diffraction (*166*). However, the bond lengths for the tetrazadiene ligand [a short central N—N bond [1.278(2) Å] flanked by two longer N—N bonds [1.345(2) Å average] are more in keeping with a d^8 nickel(II)/ArNNNNAr$^-$ or d^7 nickel(III)/ArNNNNAr^{2-} formulation (Fig. 14). Further consideration of ESR and X-ray photoelectron spectroscopy (XPS) data led to the

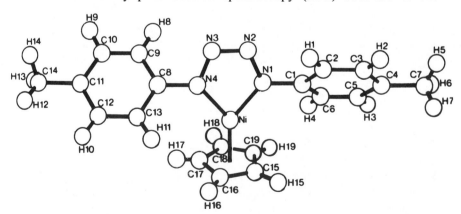

FIG. 14. PLUTO drawing of the molecular structure of $Ni(p\text{-tol-NNNN-tol-}p)(\eta^5\text{-}C_5H_5)$.

conclusion that the complex should be regarded as a d^7 nickel(III) species (*166*). The complex Ni(p-tol-NNNN-tol-p)(C_5H_5) reacts with p-tolyl azide and t-butylisocyanide to form Ni(p-tol-NNNN-tol-p)$_2$ and Ni(p-tol-NNNN-tol-p)(CNBu^t)$_2$, respectively (*166–168*). The latter product can also be obtained from a 1:1 mixture of Ni(p-tol-NNNN-tol-p)$_2$ and Ni(CNBu^t)$_4$ in the presence of excess CNBu^t or by treatment of Ni(p-tol-NNNN-tol-p)$_2$ with excess CNBu^t (*167*). It has been concluded from this work that 1,4-disubstituted tetrazadienes are better π-acceptor ligands than 1,4-disubstituted-1,4-diazadienes or 2,2'-dipyridyl (*166*). A novel route for the synthesis of metal tetrazadiene complexes has recently been reported; direct transfer of ArNNNNAr ligands from Ni(ArNNNNAr)$_2$ to Ni^0 and Pt^0 but not Pd^0 centers in the presence of CNBu^t affords the complexes M(ArNNNNAr)(CNBu^t)$_2$ (*170, 171*). These ligand-transfer reactions, which require CNBu^t as the co-ligand, are thought to involve binuclear tetraza-diene-bridged species Ni(ArNNNNAr)$_2$ML$_2$, for which tentative structures have been proposed (*171*). Using co-ligands other than CNBu^t it has proved possible to isolate bimetallic Ni/Pd and Ni/Pt complexes analogous to the proposed intermediates (see below). Reactions between pairs of complexes Ni(ArNNNNAr)$_2$ and Ni(Ar'NNNNAr')$_2$ initially afford mixed products Ni(ArNNNNAr)(Ar'NNNNAr'); after 3–6 days detectable amounts of species such as Ni(ArNNNNAr)(Ar'NNNNAr) are also formed (*169*). Simi-lar results are obtained when mixtures of Ni(ArNNNNAr)$_2$ and Ni(Ar'NNNNAr')(C_5H_5) are allowed to react. Reactions between nickel complexes Ni(ArNNNNAr)$_2$ or Ni(ArNNNNAr)(C_5H_5) and aryl azides Ar'N$_3$ are more rapid and fall into two categories: (1) those in which only complete tetrazadiene ligands are substituted, and (2) those involving forma-tion of mixed ligands Ar'NNNNNAr. Mechanisms proposed for these reactions are shown in Scheme 13 (*169*).

No simple palladium tetrazadiene complexes have been isolated to date; reactions between Pd(PPh$_3$)$_4$ and organic azides afford polymeric palladium phosphine complexes (*232*). The apparent instability of palladium tetraza-diene complexes has been tentatively attributed to the relatively low basicity of palladium, which does not permit sufficient π back donation to the tetrazadiene ligand (*32*). However, the bimetallic species Ni(p-tol-NNNN-tol-p)$_2$PdL$_2$ (L = Bu^{t}NC or PEt$_3$) have been obtained from reactions between Pd(norbornene)$_3$ and Ni(p-tol-NNNN-tol-p)$_2$ in the presence of Bu^{t}NC or PEt$_3$ at 0°C (*170, 171*).

The first platinum tetrazadiene complexes Pt(RNNNNR)(PPh$_3$)$_2$ (R = PhSO$_2$ or p-tolSO$_2$) were prepared by addition of the corresponding azides to Pt(PPh$_3$)$_n$ (n = 4 or, preferably, 3) or Pt(C_2H_4)(PPh$_3$)$_2$, but were origin-ally formulated as bis(aryldiazo) derivatives Pt(N$_2$R)$_2$(PPh$_3$)$_2$ (*7*). The tetrazadiene structure was later proposed by the same authors (*31*). The

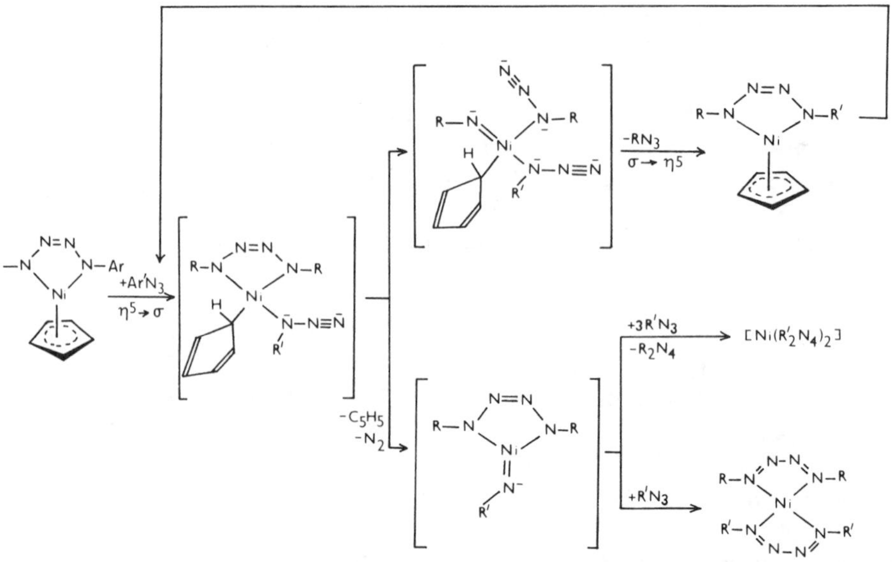

SCHEME 13. Proposed mechanisms for ligand scrambling processes in complexes $Ni(ArNNNNAr)_2$. Adapted with permission from *J. Chem. Soc., Dalton Trans.*, p. 1541 (1982).

complexes carbonylate to form $Pt(RNNNNR)(CO)(PPh_3)$ and are attacked by HBF_4 to liberate azide and amide species, but resist methyl iodide (*31*). Treatment of $Pt(CO)_2(PPh_3)_2$ or $Pt(CO)(PPh_3)_3$ with *p*-toluenesulfonyl azide in dry benzene affords a product $Pt(R_2N_4CO)(PPh_3)_2$ of unknown structure (*8*). Aryl azides ArN_3 ($Ar = p$-X-C_6H_4, $X = Cl$, NO_2, or Me) react with $Pt(C_8H_{12})_2$ to yield the products $Pt(ArNNNNAr)(C_8H_{12})$ (*168, 170*), which in turn react with Bu^tNC to form $Pt(ArNNNNAr)(Bu^tNC)_2$. The latter products can also be obtained from $Pt_3(CNBu^t)_6$ and the appropriate azide (*165, 170*) or by a ligand-transfer process involving $Ni(ArNNNNAr)_2$ and $Pt(C_8H_{12})_2$ in the presence of Bu^tNC at 60°C (*170*). Under milder conditions the ligand-transfer reactions afford bimetallic intermediates $Ni(ArNNNNAr)_2Pt(Bu^tNC)_2$ (*170*). Treatment of the complexes $Pt(ArNNNNAr)(C_8H_{12})$ with aryl azides $Ar'N_3$ leads to formation of products $Pt(ArNNNNAr')(C_8H_{12})$ and $Pt(Ar'NNNNAr')(C_8H_{12})$ containing partially and completely substituted tetrazadiene ligands, respectively (*169*). Infrared, NMR, and XPS spectroscopic data indicate that the complexes $Pt(ArNNNNAr)L_2$ ($L = Bu^tNC$, $L_2 = C_8H_{12}$) are best formulated as derivatives of platinum(II) (*165*). Triethylphosphine reacts with the complexes $Pt(ArNNNNAr)(C_8H_{12})$ to afford simple substitution products $Pt(ArNNNNAr)(PEt_3)_2$ ($Ar = p$-Me-C_6H_4, p-Cl-C_6H_4) or to attack the cycloocta-1,5-diene ring ($Ar = p$-O_2N-C_6H_4). The product

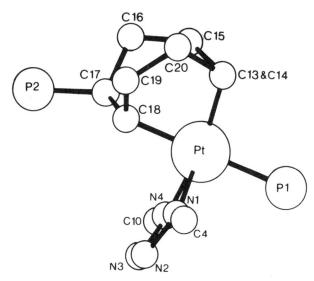

FIG. 15. View of the environment of the platinum atom in

$$\overline{\text{Pt}\{\text{CHC(PEt}_3\text{)H(CH}_2\text{)}_2\text{CH}=\text{CHCH}_2\text{CH}_2\}}(p\text{-}O_2\text{NC}_6\text{H}_4\text{NNNNC}_6\text{H}_4\text{NO}_2\text{-}p)(\text{PEt}_3).$$

of the latter reaction is an intensely blue complex Pt(ArNNNNAr)$\{\overline{\text{CHC(PEt}_3\text{)H(CH}_2\text{)}_2\text{CH}=\text{CHCH}_2\text{CH}_2}\}$(PEt$_3$), which has been found by X-ray diffraction methods to possess the structure shown in Fig. 15 (165). Reaction of trans-Pt(C≡CPh)$_2$(PEt$_3$)$_2$ with ArN$_3$ (Ar = p-O$_2$N-C$_6$H$_4$) generates the complex Pt(ArNNNNAr)(C≡CPh)$_2$(PEt$_3$)$_2$, which, on the basis of bond length data, has been formulated as an octahedral platinum(IV) tetrazene-1,4-diyl derivative (85).

To date no tetrazene or tetrazadiene complexes of the coinage metals—copper, silver, and gold—appear to have been reported.

5. Zinc, Cadmium, and Mercury

No examples of tetrazadiene complexes are known for members of this triad. However, all three elements form complexes with tetrazene ligands R$_2$N—N=N—NR$_2$.

Colorless tetramethyl-2-tetrazene complexes of zinc, ZnR$_2$(Me$_2$N—N=N—NMe$_2$), have been prepared from the corresponding organozinc compounds ZnR$_2$ (R = Bu^t, Ph, or C$_6$F$_5$) and the free ligand in hydrocarbon solvents. Whereas these products are reasonably stable, similar adducts with ZnBr$_2$ and, in particular, ZnCl$_2$ decompose explosively (160). Structures involving N-1,N-4-chelated tetrazene ligands were originally proposed (160), but an X-ray diffraction study (R = C$_6$F$_5$) later revealed that the coordinated tetrazene exists in the trans isomeric form and is bound to

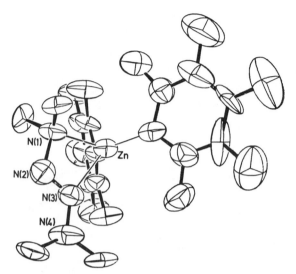

FIG. 16. ORTEP view of the $Zn(C_6F_5)_2(Me_2NN=NNMe_2)$ molecule.

the zinc through N-1 and N-3 (Fig. 16) (57). This result is in keeping with the report that the parent tetrazene $H_2N-N=N-NH_2$ also has a trans structure (224). Variable-temperature ^{1}H NMR spectra (178–298 K) show a single methyl resonance consistent with fluxional behavior in solution (57). Decomposition of $ZnCl_2(Me_2N-N=N-NMe_2)$ in THF affords Me_2N radicals, which can be trapped *in situ* by styrene to form $PhCH(NMe_2)CH_2(NMe_2)$ (153–155).

Cadmium complexes $CdX_2(Me_2N-N=N-NMe_2)$ (X = Cl, Br), obtained by addition of the free ligand to alcoholic solutions of cadmium halides, are considerably more stable than their zinc analogues (23). Although originally formulated as N-1,N-4 chelates (23), it now seems probable that they adopt the N-1,N-3- chelate structures analogous to that established for the related zinc complex (57).

The mercury(II) adduct $HgCl_2(Et_2N-N=N-NEt_2)$, first reported by Fischer over 100 years ago (82), and the more recently described examples $HgX_2(R_2N-N=N-NR_2)$ (X = Cl, Br; R = C_{1-4} alkyl) (23, 203), may well have similar N-1,N-3-chelate structures.

IV. Pentazadiene Complexes

The study of catenated nitrogen ligands has been extended to N_5 systems by the synthesis of transition metal complexes containing coordinated 1,5-diarylpentaza-1,4-diene-3-ide anions (6). Although substituted pentaza-1,4-

dienes have been known since 1894 (*228*) and are stable at room temperature, this exciting discovery affords the first examples of their participation in complex formation. Addition of metal amine complexes to solutions of 1,5-di-*p*-tolylpentaza-1,4-diene-3-ide anions in aqueous ammonia yields complexes of the transition metal ions Cu^I, Ag^I, Mn^{II}, Ni^{II}, Cu^{II}, Zn^{II}, Pd^{II}, Cd^{II}, and Co^{III}, as well as a thallium(I) derivative (*6*). The dimeric nickel product $\{Ni(p\text{-tol-}NNNNN\text{-tol-}p)_2\}_2$ was isolated as a paramagnetic ($\mu = 3.1$ BM per Ni atom) brown crystalline powder which decomposes at 120°C. An X-ray diffraction study on monoclinic crystals of a 1:1 THF adduct grown from THF/*n*-hexane solution reveals the structure shown in Fig. 17. Four N_5 zig-zag chains, whose longitudinal axes run parallel, each coordinate to both nickel atoms. Each octahedrally coordinated nickel atom is bound to the N-1 and N-3 atoms of two N_5 chains (Ni—N = 2.13 Å) and to the N-5 atoms of the other two (Ni—N = 2.07 Å).

The initial product of the copper reaction is a brown precipitate of stoichiometry $Cu(p\text{-tol-}NNNNN\text{-tol-}p)_2$, which, on heating, is reduced to deep red, air-stable $\{Cu(p\text{-tol-}NNNNN\text{-tol-}p)\}_3$. The latter product, which is weakly paramagnetic [μ ranges from 0.33 (113 K) to 1.52 BM (303 K)] and decomposes explosively at 160°C, has been found by X-ray diffraction methods to possess the trinuclear structure shown in Fig. 18. Three N_5 zig-zag chains coordinate three linearly arranged copper(I) ions through N-1, N-3, and N-5 atoms, such that each copper is in a trigonal-planar coordination environment. Mean copper–nitrogen distances are 2.036 Å for the outer copper atoms and 1.945 Å for the central copper atom. The copper–copper distances of 2.348 and 2.358 Å are the shortest yet recorded for copper(I) complexes (*6*).

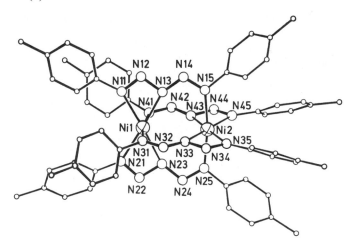

FIG. 17. Molecular structure of $Ni_2(p\text{-tol-}NNNNN\text{-tol-}p)_4$.

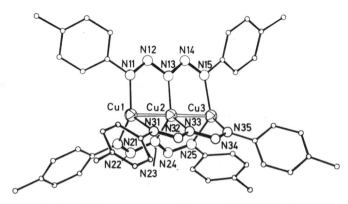

FIG. 18. Molecular structure of $Cu_3(p\text{-tol-NNNNN-tol-}p)_3$.

REFERENCES

1. Abel, E. W., and Towle, I. D. H., *J. Organomet. Chem.* **155**, 299 (1978).
2. Armstrong, D. R., Perkins, P. G., Scott, J. M., and Stewart, J. J. P., *Theor. Chim. Acta* **26**, 237 (1972).
3. Armstrong, J. E., and Walton, R. A., *Inorg. Chem.* **22**, 1545 (1983).
4. Ashley-Smith, J., Green, M., and Stone, F. G. A., *J. Chem. Soc., Dalton Trans.* p. 1805 (1972).
5. Bamberger, E., and Wulz, P., *Ber. Dtsch. Chem. Ges.* **24**, 2077 (1891).
6. Beck, J., and Strahle, J., *Angew. Chem. Int. Ed. Engl.* **24**, 409 (1985).
6a. Beck, J., and Strahle, J., *Angew. Chem. Int. Ed. Engl.* **25**, 95 (1986).
7. Beck, W., Bauder, M., La Monica, G., Cenini, S., and Ugo, R., *J. Chem. Soc. A* 113 (1971).
8. Beck, W., Rieber, W., Cenini, S., Porta, F., and La Monica, G., *J. Chem. Soc., Dalton Trans.*, p. 298 (1974).
9. Beck, W., Rieber, W., and Kirmaier, H., *Z. Naturforsch. B. Anorg. Chem. Org. Chem.* **32**, 528 (1977).
10. Benson, F. A., "The High Nitrogen Compounds," Wiley, New York (1984).
11. Bertho, A., *J. Prakt. Chem.* **116**, 101 (1927).
12. Bombieri, G., Immirzi, A., and Toniolo, L., *Acta Crystallogr. Sect. A* **31**, S141 (1975).
13. Bombieri, G., Immirzi, A., and Toniolo, L., *Inorg. Chem.* **15**, 2428 (1976).
14. Boschi, T., Belluco, U., Toniolo, L., Favez, R., and Roulet, R., *Inorg. Chim. Acta* **34**, 37 (1979).
15. Bozsai, I., *Talanta* **10**, 543 (1963).
16. Bradley, W., and Thompson, I. D., *Chimia* **15**, 147 (1961); *Chem. Abstr.* **60**, 7942 (1964).
17. Brinckman, F. E., Haiss, H. S., and Robb, R. A., *Inorg. Chem.* **4**, 936 (1965).
18. Brinckman, F. E., and Haiss, H. S., *Chem. Ind. (London)*, p. 1124 (1963).
19. Brown, I. D., and Dunitz, J. D., *Acta Crystallogr.* **14**, 480 (1961).
20. Brown, L. D., and Ibers, J. A., *Inorg. Chem.* **15**, 2788 (1976).
21. Brown, L. D., and Ibers, J. A., *Inorg. Chem.* **15**, 2794 (1976).
22. Brown, L. D., and Ibers, J. A., *J. Am. Chem. Soc.* **98**, 1597 (1976).
23. Bull, W. E., Seaton, J. A., and Audrieth, L. F., *J. Am. Chem. Soc.* **80**, 2516 (1958).
24. Burgess, K., Johnson, B. F. G., Lewis, J., and Raithby, P. R., *J. Chem. Soc., Dalton Trans.*, p. 2085 (1982).

25. Burgess, K., Johnson, B. F. G., Lewis, J., and Raithby, P. R., *J. Chem. Soc., Dalton Trans.*, p. 2119 (1982).
26. Burgess, K., Johnson, B. F. G., Lewis, J., and Raithby, P. R., *J. Organomet. Chem.* **224**, C40 (1982).
27. Calvin, M., and Bailes, R. H., *J. Am. Chem. Soc.* **68**. 949 (1946).
28. Campbell, T. W., and Day, B. F., *Chem. Rev.* **48**, 299 (1951).
29. Candeloro de Sanctis, S., Pavel, N. V., and Toniolo, L., *J. Organomet. Chem.* **108**, 409 (1976).
30. Candeloro de Sanctis, S., Toniolo, L., Boschi, T., and Deganello, G., *Inorg. Chim. Acta* **12**, 251 (1975).
31. Cenini, S., Fantucci, P., Pizzotti, M., and La Monica, G., *Inorg. Chim. Acta* **13**, 243 (1975).
32. Cenini, S., and La Monica, G., *Inorg. Chim. Acta* **18**, 279 (1976).
33. Chang, C.-Y., Johnson, C. E., Richmond, T. G., Chen, Y.-T., Trogler, W. C., and Basolo, F., *Inorg. Chem.* **20**, 3167 (1981).
34. Chetcuti, M. J., Chisholm, M. H., Folting, K., Haitko, D. A., and Huffman, J. C., *J. Am. Chem. Soc.* **104**, 2138 (1982).
35. Chetcuti, M. J., Chisholm, M. H., Folting, K., Huffman, J. C., and Janos, J., *J. Am. Chem. Soc.* **104**, 4684 (1982).
35a. Chisholm, M. H., Chiu, H. T., Huffman, J. C., and Wang, R. J., *Inorg. Chem.* **25**, 1092 (1986).
36. Chisholm, M. H., Folting, K., Huffman, J. C., and Rothwell, I. P., *Inorg. Chem.* **20**, 2215 (1981).
37. Chisholm, M. H., Haitko, D. A., Folting, K., and Huffman, J. C., *Inorg. Chem.* **20**, 171 (1981).
38. Chisholm, M. H., Haitko, D. A., Huffman, J. C., and Folting, K., *Inorg. Chem.* **20**, 2211 (1981).
39. Chisholm, M. H., Huffman, J. C., and Kelly, R. L., *Inorg. Chem.* **18**, 3554 (1979).
40. Chiu, K. W., Wilkinson, G., Thornton-Pett, M., and Hursthouse, M. B., *Polyhedron* **3**, 79 (1984).
41. Colson, S., and Robinson, S. D., unpublished results.
42. Connelly, N. G., Daykin, H., and Demidowicz, Z., *J. Chem. Soc., Dalton Trans.*, p. 1532 (1978).
43. Connelly, N. G., Finn, C. J., Freeman, M. J., Orpen, A. G., and Stirling, J., *J. Chem. Soc., Chem. Commun.*, p. 1025 (1984).
44. Connelly, N. G., and Demidowicz, Z., *J. Chem. Soc., Dalton Trans.*, p. 50 (1978).
45. Corbett, M., Hoskins, B. F., McLeod, N. J., and O'Day, B. P., *Acta Crystallogr. Sect. A* **28**, S76 (1972).
46. Corbett, M., Hoskins, B. F., McLeod, N. J., and O'Day, B. P., *Aust. J. Chem.* **28**, 2377 (1975).
47. Corbett, M., and Hoskins, B. F., *Aust. J. Chem.* **27**, 665 (1974).
48. Corbett, M., and Hoskins, B. F., *Inorg. Nucl. Chem. Lett.* **6**, 261 (1970).
49. Corbett, M., and Hoskins, B. F., *J. Am. Chem. Soc.* **89**, 1530 (1967).
50. Corbett, M., and Hoskins, B. F., *J. Chem. Soc., Chem. Commun.*, p. 1602 (1968).
51. Cotton, F. A., DeBoer, B. G., LaPrade, M. D., Pipal, J. R., and Ucko, D. A., *Acta Crystallogr. Sect. B* **27**, 1664 (1971).
52. Cotton, F. A., Ilsley, W. H., and Kaim, W., *Inorg. Chem.* **19**, 1450 (1980).
53. Cotton, F. A., Rice, G. W., and Sekutowski, J. C., *Inorg. Chem.* **18**, 1143 (1979).
54. Creswell, C. J., Queiros, M. A. M., and Robinson, S. D., *Inorg. Chim. Acta* **60**, 157 (1982).
55. Cuisa, R., and Pestalozza, U., *Atti. Accad. Naz. Lincei, Cl. Sci. Fis. Mat. Nat. Rend.* **18**, 92 (1911).
56. Cuisa, R., and Pestalozza, U., *Gazz. Chim. Ital* **41**, 391 (1911).

57. Day, V. W., Campbell, D. H., and Michejda, C. J., *J. Chem. Soc., Chem. Commun.*, p. 118 (1975).
58. De Roode, W. H., and Vrieze, K., *J. Organomet. Chem.* **153**, 345 (1978).
59. Dekker, M., and Knox, G. R., *J. Chem. Soc., Chem. Commun.*, p. 1243 (1967).
60. Dimroth, O., *Ber. Dtsch. Chem. Ges.* **36**, 909 (1903).
61. Dimroth, O., *Ber. Dtsch. Chem. Ges.* **38**, 673 (1905).
62. Dimroth, O., *Ber. Dtsch. Chem. Ges.* **38**, 681 (1905).
63. Dimroth, O., *Ber. Dtsch. Chem. Ges.* **39**, 3905 (1906).
64. Dimroth, O., *Ber. Dtsch. Chem. Ges.* **40**, 2376 (1907).
65. Doedens, R. J., *J. Chem. Soc., Chem. Commun.*, p. 1271 (1968).
66. Dolgoplosk, B. A., Erusalimsky, B. L., Krol, V. A., and Romanov, L. M., *Zh. Obshch. Khim.* **24**, 1775 (1954); *J. Gen. Chem. USSR. Engl. Transl.* **24**, 1745 (1954).
67. Dubicki, L., and Martin, R. L., personal communication (quoted in ref. *47*).
68. Dutta, R. L., and Sharma, R., *J. Sci. Ind. Res.* **40**, 715 (1981).
69. Dwyer, F. P., *Aust. Chem. Inst. J. Proc.* **6**, 348 (1939); *Chem. Abstr.* **34**, 733 (1940).
70. Dwyer, F. P., *Aust. Chem. Inst, J. Proc.* **6**, 362 (1939); *Chem. Abstr.* **34**, 734 (1940).
71. Dwyer, F. P., *J. Am. Chem. Soc.* **63**, 78 (1941).
72. Dwyer, F. P., *J. Soc. Chem. Ind. London Trans. Commun.* **58**, 110 (1939); *Chem. Abstr.* **33**, 5818 (1939).
73. Dwyer, F. P., and Earl, J. C., *Chem. Ind. (London)*, p. 136 (1940); *Chem. Abstr.* **34**, 3245 (1940).
74. Dwyer, F. P., and Mellor, D. P., *J. Am. Chem. Soc.* **63**, 81 (1941).
75. Edwards, P. G., Felix, F., Mertis, K., and Wilkinson, G., *J. Chem. Soc., Dalton Trans.*, p. 361 (1979).
76. Einstein, F. W. B., Gilchrist, A. B., Rayner-Canham, G. W., and Sutton, D., *J. Am. Chem. Soc.* **93**, 1826 (1971).
77. Einstein, F. W. B., Gilchrist, A. B., Rayner-Canham, G. W., and Sutton, D., *J. Am. Chem. Soc.* **94**, 645 (1972).
78. Einstein, F. W. B., and Sutton, D., *Inorg. Chem.* **11**, 2827 (1972).
79. Ershova, T. V., Rukhadze, E. G., and Terent'ev, A. P., *Zh. Obshch. Khim.* **39**, 59 (1969); *J. Gen. Chem. USSR Engl. Transl.* **39**, 51 (1969).
80. Farrell, N., and Sutton, D., *J. Chem. Soc., Dalton Trans.*, p. 2124 (1977).
81. Fischer, E., *Ann. Chim. Paris* **190**, 67 (1878).
82. Fischer, E., *Ann. Chim. Paris* **199**, 281 (1879).
83. Foner, S. N., and Hudson, R. L., *J. Chem. Phys.* **29**, 442 (1958); *Adv. Chem. Ser.* **36**, 34 (1962).
84. Furlani, C., and Di Tella, F., *Gazz. Chim. Ital.* **90**, 280 (1960).
85. Geisenberger, J., Nagel, U., Sebald, A., and Beck, W., *Chem. Ber.* **116**, 911 (1983).
86. Gilchrist, A. B., and Sutton, D., *Can. J. Chem.* **52**, 3387 (1974).
87. Graziani, R., Toniolo, L., Casellato, U., Rossi, R., and Magon, L., *Inorg. Chim. Acta* **52**, 119 (1981).
88. Griess, P., *Ann. Chim. Paris* **117**, 1 (1861).
89. Griess, P., *Ann. Chim. Paris* **121**, 257 (1862).
90. Griess, P., *Ann. Chim. Paris* **137**, 39 (1866).
91. Griess, P., *Proc. Roy. Soc. London* **9**, 594 (1859); *Ann. Chim. Paris* **117**, 1 (1861); *Ann. Chim. Paris* **121**, 257 (1862).
92. Gross, M. E., Ibers, J. A., and Trogler, W. C., *Organometallics* **1**, 530 (1982).
93. Gross, M. E., Johnson, C. E., Maroney, M. J., and Trogler, W. C., *Inorg. Chem.* **23**, 2968 (1984).
94. Gross, M. E., Trogler, W. C., and Ibers, J. A., *J. Am. Chem. Soc.* **103**, 192 (1981).
95. Gross, M. E., Trogler, W. C., and Ibers, J. A., *Organometallics* **1**, 732 (1982).
96. Gross, M. E., and Trogler, W. C., *J. Organomet. Chem.* **209**, 407 (1981).

97. Hantzsch, A., and Perkin, F. M., *Ber. Dtsch. Chem. Ges.* **30**, 1412 (1897).
98. Harris, C. M., Hoskins, B. F., and Martin, R. L., *J. Chem. Soc.*, p. 3728 (1959).
99. Harris, C. M., and Martin, R. L., *Proc. Chem. Soc. London*, p. 259 (1958).
100. Hayon, E., and Simic, M., *J. Am. Chem. Soc.* **94**, 42 (1972).
101. Henderson, R. A., Leigh, G. J., and Pickett, C. J., *Adv. Inorg. Chem. Radiochem.* **27**, 197 (1983).
102. Hendriks, P., Kuyper, J., and Vrieze, K., *J. Organomet. Chem.* **120**, 285 (1976).
103. Hendriks, P., Olie, K., and Vrieze, K., *Cryst. Struct. Commun.* **4**, 611 (1975).
104. Higginson, W. C. E., and Wright, P., *J. Chem. Soc.*, p. 1551 (1955).
105. Hillhouse, G. L., Goeden, G. V., and Haymore, B. L., *Inorg. Chem.* **21**, 2064 (1982).
106. Hillhouse, G. L., and Bercaw, J. E., *Organometallics* **1**, 1025 (1982).
107. Hillhouse, G. L., and Haymore, B. L., *J. Organomet. Chem.* **162**, C23 (1978).
108. Hofmann, K. A., and Buchner, K., *Ber. Dtsch. Chem. Ges.* **41**, 3084 (1908).
109. Immirzi, A., Bombieri, G., and Toniolo, L., *J. Organomet. Chem.* **118**, 355 (1976).
110. Immirzi, A., Porzio, W., Bombieri, G., and Toniolo, L., *J. Chem. Soc., Dalton Trans.*, p. 1098 (1980).
111. Jack, T., and Powell, J., *J. Organomet. Chem.* **27**, 133 (1971); *Inorg. Chem.* **11**, 1039 (1972).
112. Jaitner, P. E., Peringer, P., Huttner, G., and Zsolnai, I., *Transition Met. Chem.* (*Weinheim*) **6**, 86 (1981).
113. Johnson, B. F. G., Lewis, J., Raithby, P. R., and Sankey, S. W., *J. Organomet. Chem.* **228**, 135 (1982).
114. Johnson, C. E., and Trogler, W. C., *Inorg. Chem.* **21**, 427 (1982).
115. Johnson, C. E., and Trogler, W. C., *J. Am. Chem. Soc.* **103**, 6352 (1981).
116. Julliard, M., Vernin, G., Metzger, J., and Lopez, T. G., *Synthesis*, p. 49 (1982).
117. King, R. B., and Chen, K. N., *Inorg. Chem.* **16**, 2648 (1977).
118. King, R. B., and Nainan, K. C., *Inorg. Chem.* **14**, 271 (1975).
119. Kleinfeller, H., *J. Prakt. Chem.* **119**, 61 (1928).
120. Knoth, W. H., *Inorg. Chem.* **12**, 38 (1973).
121. Knowles, C. M., and Watt, G. W., *J. Am. Chem. Soc.* **64**, 935 (1942).
122. Kondrashev, Y. D., *Kristallografiya* **13**, 622 (1968); *Chem. Abstr.* **69**, 111050 (1968).
123. Kondrashev, Y. D., and Gladkova, V. F., *Kristallografiya* **17**, 33 (1972); *Chem. Abstr.* **76**, 132697m (1972).
124. Kops, R. T., Overbeek, A. R., and Schenk, H., *Cryst. Struct. Commun.* **5**, 125 (1976).
125. Kops, R. T., and Schenk, H., *Cryst. Struct. Commun.* **5**, 193 (1976).
126. Kravtsov, D. N., Nesmeyanov, A. N., Fedorov, L. A., Fedin, E. I., Peregudov, A. S., Borisov, E. V., Okulelevich, P. O., and Postovoy, S. A., *Dokl. Akad. Nauk. S.S.S.R.* **242**, 326 (1978); *Dokl. Acad. Sci. U.S.S.R.* (*Engl. Transl.*) **242**, 431 (1978).
127. Krigbaum, W. R., and Rubin, B., *Acta Crystallogr. Sect. B* **29**, 749 (1973).
128. Kubler, R., Luttke, W., and Weckherlin, S., *Z. Electrochem.* **64**, 650 (1960).
129. Kuyper, J., Van Vliet, P. I., and Vrieze, K., *J. Organomet. Chem.* **105**, 379 (1976).
130. Kuyper, J., Van Vliet, P. I., and Vrieze, K., *J. Organomet. Chem.* **96**, 289 (1975).
131. Kuyper, J., Vrieze, K., and Olie, K., *Cryst. Struct. Commun.* **5**, 179 (1976).
132. Kuyper, J., and Vrieze, K., *J. Organomet. Chem.* **107**, 129 (1976).
133. Kuz'mina, L. G., Struchkov, Y. T., and Kravtsov, D. N., *Zhur. Strukt. Khim.* **20**, 552 (1979); *J. Struct. Chem.* (*Engl. Transl.*) **20**, 470 (1979).
134. La Monica, G., Sandrini, P., Zingales, F., and Cenini, S., *J. Organomet. Chem.* **50**, 287 (1973).
135. Laing, K. R., Robinson, S. D., and Uttley, M. F., *J. Chem. Soc., Dalton Trans.*, p. 1205 (1974).
136. Maire, J. C., Baldy, A., Boyer, D., Llopiz, P., Vernin, G., and Bachlas, B. P., *Helv. Chim. Acta* **62**, 1566 (1979).

137. Majumdar, A. K., and Bhattacharyya, R. G., *J. Indian Chem. Soc.* **50**, 701 (1973).

138. Majumdar, A. K., and Saha, S. C., *J. Indian Chem. Soc.* **50**, 697 (1973).

139. Mandal, K. L., *Sci. Cult.* **6**, 59 (1940).

140. Mangini, A., *Gazzetta 65*, 298 (1935).

141. Mangini, A., *Gazzetta 67*, 384 (1937).

142. Mangini, A., and Dejudicibus, I., *Gazz. Chim. Ital.* **63**, 601 (1933).

142a. Marchi, A., Rossi, R., Duatti, A., Magon, L., Bertolasi, V., Ferretti, V., and Gilli, G., *Inorg. Chem.* **24**, 4744 (1985).

143. Maroney, M. J., and Trogler, W. C., *J. Am. Chem. Soc.* **106**, 4144 (1984).

144. Massa, W., Kujanek, R., Baum, G., and Dehnicke, K., *Angew. Chem. (Int. Ed. Engl.)* **23**, 149 (1984).

145. Medzhidov, A. A., Salimov, A. M., and Timakov, I. A., *Russ. J. Phys. Chem. (Engl. Transl.)* **51**, 1325 (1977).

146. Meldola, R., and Streatfeild, F. W., *J. Chem. Soc.* **51**, 434 (1887).

147. Meldola, R., and Streatfeild, F. W., *J. Chem. Soc.* **53**, 664 (1888).

148. Meldola, R., and Streatfeild, F. W., *J. Chem. Soc.*, p. 785 (1890).

149. Mellor, D. P., and Craig, D. P., *J. Proc. R. Soc. NSW* **74**, 495 (1940).

150. Meunier, L., *C.R. Hebd. Seances Acad. Sci.* **131**, 50 (1900).

151. Meunier, L., *C.R. Hebd. Seances Acad. Sci.* **137**, 1264 (1903).

152. Meunier, L., and Rigot, A., *Bull. Soc. Chim. Fr.* **23**, 103 (1900).

153. Michejda, C. J., and Campbell, D. H., *J. Am. Chem. Soc.* **101**, 7687 (1979).

154. Michejda, C. J., and Campbell, D. H., *J. Am. Chem. Soc.* **96**, 929 (1974).

155. Michejda, C. J., and Campbell, D. H., *J. Am. Chem. Soc.* **98**, 6728 (1976).

156. Miller, E. J., and Brill, T. B., *Inorg. Chem.* **22**, 2392 (1983).

157. Nesmeyanov, A. N., Borisov, E. V., Peregudov, A. S., Kravtsov, D. N., Fedorov, L. A., Fedin, E. I., and Postovoy, S. A., *Dokl. Akad. Nauk. S.S.S.R.* **247**, 1154 (1979); *Dokl. Acad. Sci. U.S.S.R. (Engl. Transl.)* **247**, 380 (1979).

158. Nesmeyanov, A. N., Fedin, E. I., Peregudov, A. S., Fedorov, L. A., Kravtsov, D. N., Borisov, E. V., and Kiryazev, F. Y., *J. Organomet. Chem.* **169**, 1 (1979).

159. Niementowski, S., and Roszkowski, J., *Z. Phys. Chem. (Leipzig)* **22**, 145 (1897).

160. Noltes, J. G., and Van Den Hurk, J. W. G., *J. Organomet. Chem.* **1**, 337 (1964).

161. O'Connor, J. E., Janusonis, G. A., and Corey, E. R., *J. Chem. Soc., Chem. Commun.*, p. 445 (1968).

162. Oddo, G., and Algerino, A., *Ber. Dtsch. Chem. Ges.* **69**, 279 (1936).

163. Omel'chenko, Y. A., and Kondrashev, Y. D., *Kristallografiya* **17**, 947 (1972); *Chem. Abstr.* **78**, 8896c (1973).

164. Otsuka, S., and Nakamura, A., *Inorg. Chem.* **7**, 2542 (1968).

165. Overbosch, P., Van Koten, G., Grove, D. M., Spek, A. L., and Duisenberg, A. J. M., *Inorg. Chem.* **21**, 3253 (1982).

166. Overbosch, P., Van Koten, G., Spek, A. L., Roelofsen, G., and Duisenberg, A. J. M., *Inorg. Chem.* **21**, 3908 (1982).

167. Overbosch, P., Van Koten, G., and Overbeek, O., *Inorg. Chem.* **21**, 2373 (1982).

168. Overbosch, P., Van Koten, G., and Overbeek, O., *J. Am. Chem. Soc.* **102**, 2091 (1980).

169. Overbosch, P., Van Koten, G., and Vrieze, K., *J. Chem. Soc., Dalton Trans.*, p. 1541 (1982).

170. Overbosch, P., Van Koten, G., and Vrieze, K., *J. Organomet. Chem.* **208**, C21 (1981).

171. Overbosch, P., and Van Koten, G., *J. Organomet. Chem.* **229**, 193 (1982).

172. Peringer, P., *Inorg. Nucl. Chem. Lett.* **16**, 461 (1980).

173. Peringer, P., *Monatsh. Chem.* **110**, 1123 (1979).

174. Peringer, P., *Z. Naturforsch. B: Anorg. Chem. Org. Chem.* **33**, 1091 (1978).

175. Peringer, P., *Inorg. Chim. Acta* **42**, 129 (1980).

176. Perkins, P. G., personal communication, cited in ref. *86.*

177. Pfeiffer, E., Kokkes, M. W., and Vrieze, K., *Transition Met. Chem. (Weinheim)* **4**, 389 (1979).
178. Pfeiffer, E., Kokkes, M. W., and Vrieze, K., *Transition Met. Chem. (Weinheim)* **4**, 393 (1979).
179. Pfeiffer, E., Kuyper, J., and Vrieze, K., *J. Organomet. Chem.* **105**, 371 (1976).
180. Pfeiffer, E., Oskram, A., and Vrieze, K., *Transition Met. Chem. (Weinheim)* **2**, 240 (1977).
181. Pfeiffer, E., Vrieze, K., and McCleverty, J. A., *J. Organomet. Chem.* **174**, 183 (1979).
182. Pfeiffer, E., and Olie, K., *Cryst. Struct. Commun.* **4**, 605 (1975).
183. Pfeiffer, E., and Vrieze, K., *Transition Met. Chem. (Weinheim)* **4**, 385 (1979).
184. Pochinok, V. Y., *Ukr. Khim. Zh. (Russ. Ed.)* **17**, 517 (1951); *Chem. Abstr.* **48**, 10640 (1954).
185. Pochinok, V. Y., Zaitseva, S. D., and El'gort, R. G., *Ukr. Khim. Zh. (Russ. Ed.)* **17**, 509 (1951); *Chem. Abstr.* **48**, 11392h (1954).
186. Pochinok, V. Y., *Zhur. Obshch. Khim.* **16**, 1303 (1946); *Chem. Abstr.* **41**, 3066h (1947).
187. Pochinok, V. Y., *Zhur. Obshch. Khim.* **16**, 1306 (1946); *Chem. Abstr.* **41**, 3066f (1947).
188. Pochinok, V. Y., and El'gort, R. G., *Ukr. Khim. Zh. (Russ. Ed.)* **15**, 311 (1949); *Chem. Abstr.* **48**, 3320 (1954).
189. Rigby, W., Lee, H.-B., Bailey, P. M., McCleverty, J. A., and Maitlis, P. M., *J. Chem. Soc., Dalton Trans.*, p. 387 (1979).
190. Robinson, S. D., and Uttley, M. F., *J. Chem. Soc., Chem. Commun.*, p. 1315 (1971).
191. Robinson, S. D., and Uttley, M. F., *J. Chem. Soc., Chem. Commun.*, p. 184 (1972).
192. Rossi, R., Duatti, A., Magon, L., Casellato, U., Graziani, R., and Toniolo, L., *J. Chem. Soc., Dalton Trans.*, p. 1949, (1982).
193. Rossi, R., Duatti, A., Magon, L., and Toniolo, L., *Inorg. Chim. Acta* **48**, 243 (1981).
193a. Rossi, R., Marchi, A., Duatti, A., Magon, L., and DiBernardo, P., *Transition Met. Chem. (Weinheim)* **10**, 151 (1985).
194. Rukhadze, E. G., Ershova, T. V., Fedorova, S. A., and Terent'ev, A. P., *Zhur. Obshch. Khim.* **39**, 303 (1969); *J. Gen. Chem. U.S.S.R. (Engl. Transl.)* **39**, 283 (1969).
195. Seidel, W., and Mark, H.-J., *Z. Anorg. Allg. Chem.* **416**, 83 (1975).
196. Shustorovich, E. M., Kagan, G. I., and Kagan, G. M., *Zh. Strukt. Khim.* **11**, 108 (1970); *J. Struct. Chem. (Engl. Transl.)* **11**, 95 (1970).
197. Shustorovich, E. M., *Zh. Strukt. Khim.* **10**, 159 (1969); *J. Struct. Chem. (Engl. Transl.)* **10**, 154 (1969).
198. Shustorovich, E. M., *Zh. Strukt. Khim.* **10**, 947 (1969); *J. Struct. Chem. (Engl. Transl.)* **10**, 839 (1969).
199. Sidgwick, N. V., "The Electronic Theory of Valency," p. 253. Oxford University Press, London and New York, 1927.
200. Skripnik, L. I., and Pochinok, V. Y., *Khim. Geterotsikl. Soedin.* **3**, 292 (1967); *Chem. Heterocycl. Compd. (Engl. Transl.)* **3**, 221 (1967).
201. Smart, L. E., and Woodward, P., personal communication, cited in ref. *44*.
202. Smith, P. A. S., "Open Chain Nitrogen Compounds," Vol. II, pp. 333–343. Benjamin, New York, 1966.
203. Sugiyama, K., *Kinki Daigaku Kogakubu Kenkyu Hokoku* **13**, 15 (1979); *Chem. Abstr.* **94**, 83364m (1981).
204. Süling, C., Aromatische Triazene und Hohere Azahomologe. In "Methoden der Organischen Chemie" (E. Muller, ed.), Vol. X/3, pp. 695–743. Georg Thieme Verlag, Stuttgart, 1965.
205. Terent'ev, A. P., Ershova, T. V., and Rukhadze, E. G., *Zhur. Obshch. Khim.* **36**, 1046 (1966); *J. Gen. Chem. U.S.S.R. (Engl. Transl.)* **36**, 1059 (1966).
206. Terent'ev, A. P., Rukhadze, E. G., and Ershova, T. V., *Dokl. Akad. Nauk. S.S.S.R.* **169**, 606 (1966); *Dokl. Acad. Sci. U.S.S.R. (Engl. Transl.)* **169**, 734 (1966).

207. Toniolo, L., Biscontin, G., Nicolini, M., and Cipollini, R., *J. Organomet. Chem.* **139**, 349 (1977).
208. Toniolo, L., DeLuca, G., and Panattoni, C., *Synth. Inorg. Met. Org. Chem.* **3**, 221 (1973).
209. Toniolo, L., Immirzi, A., Croatto, U., and Bombieri, G., *Inorg. Chim. Acta* **19**, 209 (1976).
210. Toniolo, L., *Inorg. Chim. Acta* **35**, L367 (1979).
211. Toniolo, L., and Cavinato, G., *Inorg. Chim. Acta* **26**, L5 (1978).
212. Toniolo, L., and Cavinato, G., *Inorg. Chim. Acta* **35**, L301 (1979).
213. Trogler, W. C., Johnson, C. E., and Ellis, D. E., *Inorg. Chem.* **20**, 980 (1981).
214. Van Der Linden, J. G. M., Dix, A. H., and Pfeiffer, E., *Inorg. Chim. Acta* **39**, 271 (1980).
215. Van Der Ploeg, A. F. M. J., Van Koten, G., and Vrieze, K., *Inorg. Chem.* **21**, 2026 (1982).
216. Van Der Ploeg, A. F. M. J., Van Koten, G., and Vrieze, K., *J. Organomet. Chem.* **226**, 93 (1982).
217. Van Vliet, P. I., Kokkes, M., Van Koten, G., and Vrieze, K., *J. Organomet. Chem.* **187**, 413 (1980).
218. Van Vliet, P. I., Kuyper, J., and Vrieze, K., *J. Organomet. Chem.* **122**, 99 (1976).
219. Van Vliet, P. I., Van Koten, G., and Vrieze, K., *J. Organomet. Chem.* **182**, 105 (1979).
220. Van Vliet, P. I., and Vrieze, K., *J. Organomet. Chem.* **139**, 337 (1977).
221. Vaughan, K., and Stevens, M. F. G., *Chem. Soc. Rev.* **7**, 377 (1978).
222. Vecchiotti, L., *Gazz. Chim. Ital.* **52**, 137 (1922).
223. Vecchiotti, L., and Capodacqua, A., *Gazz. Chim. Ital.* **55**, 369 (1925).
224. Veith, M., and Schlemmer, G., *Z. Anorg. Allg. Chem.* **494**, 7 (1982).
225. Vernin, G., Coen, S., and Poite, J. C., *C. R. Hebd. Seances Acad. Sci. Ser. C* **284**, 277 (1977).
226. Vernin, G., Siv, C., and Metzger, J., *Synthesis*, p. 691 (1977).
227. Von Pechmann, H., *Ber. Dtsch. Chem. Ges.* **26**, 1045 (1893).
228. Von Pechmann, H., and Frobenius, L., *Ber. Dtsch. Chem. Ges.* **27**, 899 (1894).
229. Walther, R. V., and Grieshammer, W., *J. Prakt. Chem.* **92**, 209 (1915).
230. Walz, D., and Fallab, S., *Helv. Chim. Acta* **43**, 540 (1960); **44**, 13 (1961).
231. Watt, G. W., and Fernelius, W. C., *Z. Anorg. Allg. Chem.* **221**, 187 (1934).
232. Werner, K. V., personal communication, cited in ref. 32.
233. Yamada, H., Ohta, H., Yamaguchi, M., and Tsumaki, T., *Bull. Chem. Soc. Jpn.* **34**, 241 (1961).
234. Zaitsev, B. E., Zaitseva, V. A., Batista, A., Ivanov-Emin, B. N., and Lisitsina, E. S., *Zh. Neorg. Khim.* **18**, 2063 (1973); *Russ. J. Inorg. Chem. (Engl. Transl.)* **18**, 1093 (1973).
235. Zaitsev, B. E., Zaitseva, V. A., Batista, A., and Ezhov, A. I., *Zhur. Neorg. Khim.* **20**, 709 (1975); *Russ. J. Inorg. Chem. (Engl. Transl.)* **20**, 397 (1975).
236. Zaitsev, B. E., Zaitseva, V. A., Ivanov-Emin, B. N., Lisitsyna, E. S., and Ezhov, A. I., *Russ. J. Inorg. Chem. (Engl. Transl.)* **18**, 30 (1973).
237. Zaitsev, B. E., Zaitseva, V. A., Molodkin, A. K., and Lisitsyna, E. S., *Zh. Neorg. Khim.* **22**, 909 (1977); *Russ. J. Inorg. Chem. (Engl. Transl.)* **22**, 504 (1977).

ADVANCES IN INORGANIC CHEMISTRY AND RADIOCHEMISTRY, VOL. 30

THE COORDINATION CHEMISTRY OF 2,2':6',2''-TERPYRIDINE AND HIGHER OLIGOPYRIDINES

E. C. CONSTABLE

University Chemical Laboratory, University of Cambridge, Cambridge CB2 1EW, England

I. Introduction

A. HISTORICAL INTRODUCTION

The potentially terdentate ligand 2,2':6',2''-terpyridine (Fig. 1) was first isolated by Morgan and Burstall as one of the numerous products from the reaction of pyridine with iron(III) chloride (323, 324, 326, 327). The higher oligopyridines, 2,2':6',2'':6'',2'''-quaterpyridine (Fig. 2) and 2,2':6',2'':6'',2''':6''',2''''-quinquepyridine (Fig. 3) were obtained in low yields from Ullmann reactions of bromopyridines (93). Improved methods for the synthesis of these interesting compounds are now available, and have been reviewed elsewhere (285). It was recognized from the first that these compounds should possess an interesting coordination chemistry. This review describes the coordination chemistry of these, and related, ligands.

Although a number of other reviews discussing general or specific aspects of the chemistry of these ligands have appeared, there has been none solely concerned with this class of compounds (69, 130, 131, 178, 210, 211, 285, 286, 291, 314, 401, 406, 415). There has been a considerable increase in the use of these ligands in recent years, prompted in part by the attractive photochemical and photophysical properties exhibited by complexes of the related ligands 2,2'-bipyridine (Fig. 4) and 1,10-phenanthroline (Fig. 5) (491). The availability of the ligands from facile, high-yielding syntheses suggests that their use will continue to increase (129, 132–144, 285).

It is also apparent that complexes of 2,2':6',2''-terpyridine differ in significant, and interesting, ways from those of 2,2'-bipyridine. Accordingly, a treatment of the coordination chemistry of 2,2':6',2''-terpyridine and the higher oligopyridines is felt to be justified and timely.

I have attempted to treat the subject as comprehensively as possible; errors and omissions will undoubtedly have occurred, and for these I proffer my

69

Copyright © 1986 by Academic Press, Inc.
All rights of reproduction in any form reserved.

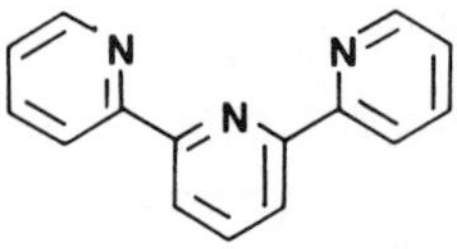

E. C. CONSTABLE

FIG. 1. 2,2′:6′,2″-Terpyridine.

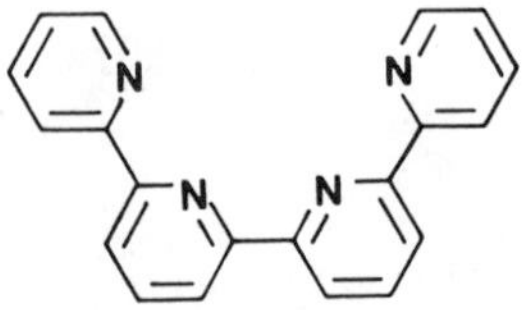

FIG. 2. 2,2′:6′,2″:6″,2‴-Quaterpyridine.

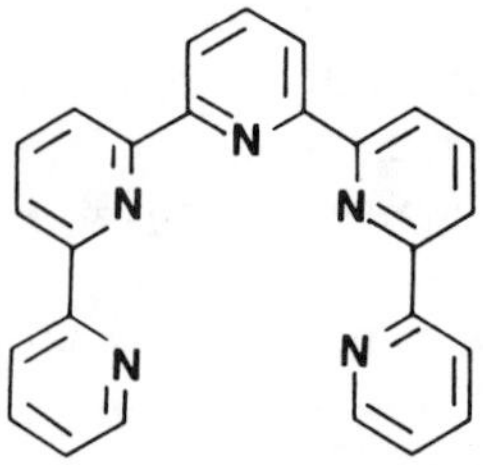

FIG. 3. 2,2′:6′,2″:6″,2‴:6‴,2⁗-Quinquepyridine.

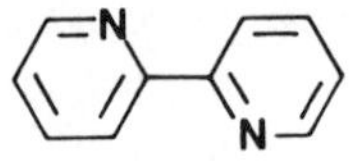

FIG. 4. 2,2′-Bipyridine.

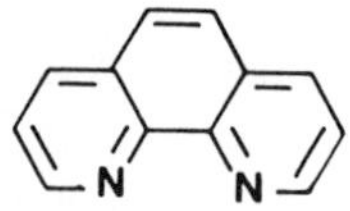

FIG. 5. 1,10-Phenanthroline.

apologies. Coverage of the primary journals is complete through November 1985, with *Chemical Abstracts* covered through Volume 103.

B. NOMENCLATURE

Nomenclature is always a vexing subject: the ligand (Fig. 1) has been known variously as 2,2′,2″-tripyridyl, 2,2′,2″-tripyridine, 2,6′-bis(2-pyridyl)-pyridine, and 2,2′:6′,2″-terpyridine. The latter is the systematic name (*262*) and will be used throughout this article. All other compounds will be named systematically, according to IUPAC recommendation A-54. In this article, the abbreviations terpy, quaterpy, quinquepy, bipy, and phen will be used for the ligands in Figs. 1–5, respectively. Substituted derivatives will not be similarly described; thus 4,4″-diethyl-4′-phenyl-2,2′:6′,2″-terpyridine (Fig. 6) will be denoted 4,4″-Et$_2$-4′-Phterpy.

C. STABILIZATION OF UNUSUAL COMPLEXES

The very concept of an "unusual" complex is suspect; what is uncommon or unusual to one person, or in one era, may be commonplace and unremarkable to another person or at another time. However, it is still a fact that the majority of transition metal complexes incorporate metal ions in the +2 or +3 oxidation state, in basically octahedral, tetrahedral, or square-planar geometries. The stabilization of other, higher or lower, oxidation states, and of other coordination numbers and geometries, is thus worthy of note. This class of ligands excels in all of these categories.

Low oxidation states are characterized by an excess of electron density at the metal atom; stabilization may be achieved by the use of ligands capable of reducing that electron density. One of the simplest ways to reduce the electron density is to design ligands with low-lying vacant orbitals of suitable symmetry for overlap with filled metal orbitals. This results in the transfer of electron density from the metal to the ligand (back-donation). In general, ligand nonbonding or π antibonding orbitals are of the correct symmetry for

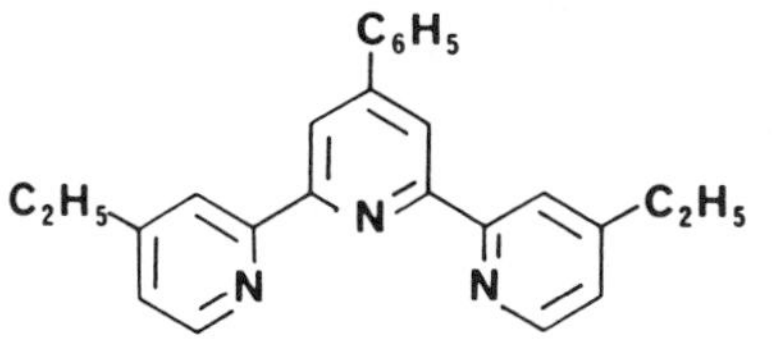

FIG. 6. 4,4″-Diethyl-4′-phenyl-2,2′:6′,2″-terpyridine.

such overlap. Similarly, metal ions in high oxidation states may be stabilized by powerful σ- or π-donor ligands.

The oligopyridines are ideally suited to such roles; they possess a filled highest occupied molecular orbital (HOMO) and a vacant lowest unoccupied molecular orbital (LUMO) of suitable energies for interaction with metal d orbitals (283). They are thus capable of stabilizing both high and low oxidation states of metal ions.

The geometries of complexes incorporating monodentate ligands are dominated by electronic, steric, and CFSE terms. In contrast, the geometries of complexes incorporating relatively inflexible polydentate ligands are predominantly determined by the configuration of the ligand; a planar pentadentate ligand *cannot* form octahedral complexes. This apparently facile observation has a number of far-reaching consequences. A compromise between the optimum geometry for the metal ion and that for the ligand must be reached. This may be expressed as a distorted geometry about the metal or as a distortion of the ligand. An extreme case of such a mismatch in the coordination requirements of the metal and the ligand will result in a polydentate ligand binding through only some of its potential sites; i.e., a bidentate terpy.

II. The Complexes—Types of Bonding

A. MONODENTATE, BIDENTATE, AND BRIDGING

In principle, terpy could act as a monodentate, bidentate, terdentate, or bridging ligand for a metal ion. Although monodentate and bidentate species have been widely assumed to be intermediates in the formation and dissociation of terpy complexes, there has been little evidence, until recently, that such species are isolable. On the basis of their vibrational spectra, the complexes $[M(CO)_3(terpy)X]$ (M = Mn, X = Br; M = Re, X = NO_3) were thought to contain bidentate terpy ligands (4, 205). It was also suggested, on the basis of ESR studies, that the radical anion terpy$^{\overline{\cdot}}$ coordinates to group IA and group IIA ions in a bidentate manner (79, 349). Solution NMR studies have indicated on–off terdentate–bidentate equilibria in $[Eu(terpy)_3]^{3+}$ (110). Conclusive proof of the bidentate bonding mode was provided by the X-ray structural analysis of the two (red and yellow) forms of $[Ru(terpy)(CO)_2Br_2]$, both of which contain a bidentate terpy ligand. The color differences arise from the orientation of the uncoordinated pyridine ring. In one case, a stacking interaction with the free pyridine ring of an adjacent molecule is observed in the solid state (162). No other examples of monodentate or bridging terpy ligands have yet been structurally characterized.

The higher oligopyridines are more likely to form complexes in which "dangling" pyridine rings are present (vide infra). There have been few structural studies of such complexes of the higher oligopyridines, but there is considerable evidence to suggest that they may act as bridging ligands. Lehn and co-workers demonstrated that the sterically demanding ligand 5,5',3",5"'-Me$_4$quaterpy forms a binuclear copper(I) complex, [Cu$_2$L$_2$][ClO$_4$]$_2$, in which the ligand acts as a bridging bis(bidentate) (290). A similar binuclear complex of quinquepy with copper has also been isolated (136).

B. TERDENTATE AND HIGHER POLYDENTATE

2,2':6',2"-Terpyridine commonly behaves as a chelating terdentate ligand, with the majority of complexes exhibiting 1:1 or 1:2 metal:ligand ratios. The 1:2 complexes are invariably based upon an octahedral geometry, and frequently exhibit D_{2d} local symmetry. These complexes are discussed at the appropriate points in the text.

Upon coordination to a metal center, the terpy ligand undergoes a number of significant changes. The most obvious results from the adoption of the cis,cis-configuration, in contrast to the trans,trans-equilibrium solution structure. It is also apparent that for terpy to act as an efficient terdentate, it is necessary to distort the ligand and reduce the interannular angle between the central and terminal pyridine rings (Fig. 7). It is regrettable that there appears to be no published crystallographic study of 2,2':6',2"-terpyridine itself. However, we have determined the crystal and molecular structure of 4'-Phterpy and the complex [Ni(4'-Phterpy)$_2$][PF$_6$]$_2$·10H$_2$O. The structure of the free ligand (Fig. 8a) displays slight differences in bond lengths from the coordinated ligand (Fig. 8b), but the major differences arise from distortion of the terminal rings. The dihedral angle between the terminal and central rings increases from 5.7 to 7.1° in the complex, while the C—C—N angles reduce from 116 in the free ligand to 114° in the complex. This is fully in accord with the prediction of increased strain in the complex (129).

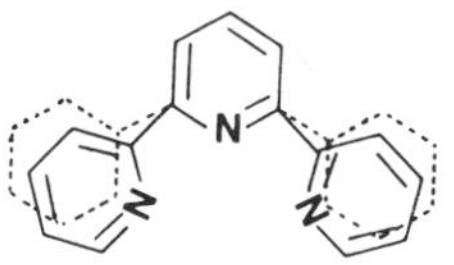

FIG. 7. The distortion experienced by a 2,2':6',2"-terpyridine ligand upon coordination to a metal ion in a planar terdentate mode.

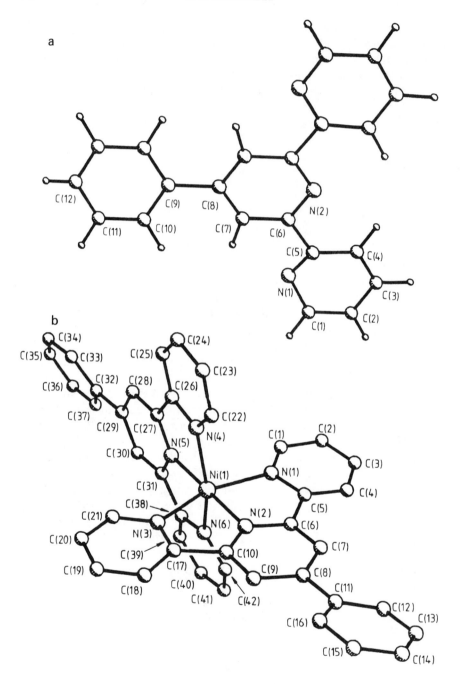

FIG. 8. The crystal and molecular structure of 4'-phenyl-2,2':6',2"-terpyridine (a) and its bis complex with nickel(II) (b) (*129*).

Numerous crystal structural analyses of complexes with the 1:1 metal: ligand stoichiometry have been reported. These are discussed in the appropriate sections. A wide variety of geometries are adopted, but it is clear that the essentially planar terdentate ligand imposes a steric requirement such that "uncommon" geometries are favored. In particular, the distorted trigonal-bipyramidal (Fig. 9a), square-pyramidal (Fig. 9b), and pentagonal-bipyramidal geometries (Fig. 9c) are commonly encountered.

The strain introduced into the ligand will be even higher in complexes of the higher oligopyridines, and it was at one time suggested that they could not exhibit their maximum denticity in monodentate complexes. It is now clear that quaterpyridine may act as a quaterdentate ligand, albeit with very unsymmetrical M—N distances.

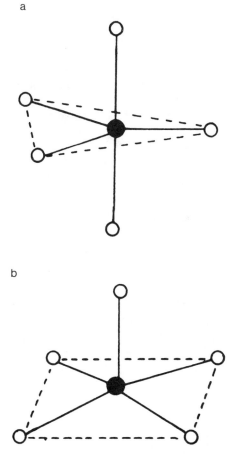

FIG. 9. The common five- (a,b) and seven-coordinate (c) geometries adopted in complexes of oligopyridines.

c

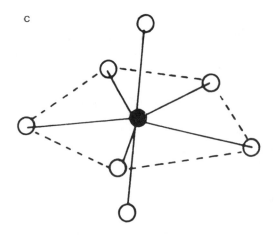

FIG. 9c. See legend on p. 75.

C. Cyclometallated

To date, the cyclometallated bonding mode has not been directly observed in complexes of terpy or higher oligopyridines. However, the unequivocal demonstration of a cyclometallated 2,2'-bipyridine moiety in iridium complexes suggests that such a possibility exists (71, 130, 131, 145, 233, 419, 420, 478). There is ^{1}H NMR and kinetic evidence to suggest that cyclometallated intermediates are not involved in the deuteration of [Ru(terpy)$_2$]$^{2+}$ at the 3, 3', 5', and 5" positions (138).

III. Coordination Compounds of 2,2':6',2"-Terpyridine

A. Group IA

It is not unreasonable to expect alkali metal ions to form complexes with the moderately hard nitrogen donor atoms of pyridine ligands, and it is now becoming apparent that a rich chemistry may exist in this area.

Vogtle et al. have demonstrated the formation of solid 1:1 adducts between terpy and MSCN (M = Li or Na); in contrast, [NH$_4$][SCN] forms a 2:1 adduct with terpy. Although stability constants were not reported, it was shown by ^{23}Na NMR studies that terpy is superior to bipy, but inferior to phen, as a ligand for the sodium cation (455). Presumably, the metal–nitrogen interaction is predominantly ionic in character.

Reaction of alkali metals (Li, Na, or Rb) with terpy in inert solvents results in the formation of alkali metal complexes of the terpy radical anion (349).

B. Group IIA

The complexes $Mg(NO_3)_2 \cdot terpy \cdot 2H_2O$, $Ca(SCN)_2 \cdot 2terpy \cdot H_2O$, and $Ba(SCN)_2 \cdot terpy \cdot H_2O$ have been reported, but are of unknown structure (455). ESR studies of the terpy radical anion obtained from the reaction of terpy with alkaline earth metals (Mg, Ca, Sr, or Ba) have indicated that a significant interaction occurs between the anion and the cation (79). Spectrophotometric measurements have indicated that magnesium sulfate forms a 1:1 complex with terpy in water, with $K = 5.858 \pm 0.023$ mol^{-1} (185, 186).

C. Group IIIB

Reaction of aluminum halides with acetonitrile solutions of terpy leads to the formation of $[Al(terpy)Cl_3]$ or $[Al(terpy)_2]X_3$ (X = Cl, Br, or I) (48). Spectroscopic and conductivity measurements suggest that these complexes are octahedral. Moore and co-workers have investigated the reaction of terpy with $[Al(DMSO)_6]^{3+}$ (DMSO, dimethylsulfoxide) by NMR methods; stepwise release of DMSO indicated sequential coordination of the pyridine rings, and it is noteworthy that the closure of the final ring is significantly slower than the other two, as predicted by our concepts of increasing strain in the chelated complex (78).

The complexes $[M(terpy)X_3]$ (M = Tl, X = Cl or Br; M = Ga or In, X = Cl, Br, or I) are all known, and may be prepared by the direct reaction of MX_3 with terpy (52, 53, 458–462, 467). The 2:1 complexes $[Tl(terpy)_2][TlX_4]$ (X = Cl, Br, or I) are also known. The $[M(terpy)Cl_3]$ compounds are isostructural, and a crystal structural analysis of $[Ga(terpy)Cl_3]$ revealed the expected distorted octahedral structure (52, 53).

The octahedral complexes $[In(terpy)_2]X_3$ (X = Cl, Br, I, ClO$_4$, or CNS) are readily prepared by the reaction of $InCl_3$ or indium perchlorate with terpy (99, 431, 432, 466, 474).

A number of organoindium complexes with terpy have been described, and include $[RIn(terpy)Cl_2]$ (R = Me or C$_6$F$_5$); it was proposed that these species incorporate a bidentate terpy ligand (120, 161).

D. Silicon, Germanium, Tin, and Lead

Silicon tetrachloride is reported not to react with carbon disulfide solutions of terpy; this is rather surprising in view of the facile reactions observed with the lower members of the group (37, 470). A number of

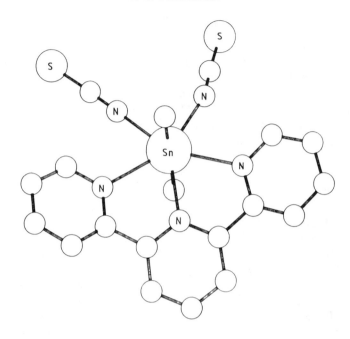

FIG. 10. The crystal and molecular structure of [Me$_2$Sn(terpy)(NCS)$_2$] (*346*).

workers have investigated the interaction of terpy with [R$_2$SnX$_2$] species, and have demonstrated the formation of 1:1 adducts (*468, 469*). Crystal structural analyses of a number of these complexes have been reported. The 1:2 adduct with [Me$_2$SnCl$_2$] should be formulated [Me$_2$Sn(terpy)Cl][Me$_2$SnCl$_3$], with a highly distorted octahedral cation (*149, 158, 163, 181–183, 193, 486*). Similar structures are expected for the 1:2 adducts with [Ph$_2$SnCl$_2$] (*163, 421*) and [(η^1-C$_5$H$_5$)$_2$SnCl$_2$] (*448*). A structural analysis has revealed [Me$_2$Sn(terpy)(NCS)$_2$] to possess a pentagonal-bipyramidal geometry, in which the methyl groups adopt the axial sites (Fig. 10) (*346*). This conclusion is also supported by ^{119}Sn NMR and Mössbauer spectroscopic studies (*310, 357*). Mössbauer studies have also been reported for some vinyl, butyl, and phenyl derivatives (*310, 347*), and for [Sn(terpy)Cl$_2$], which is thought to be five-coordinate (*170*). The complex SnCl$_4$·terpy·5H$_2$O is of unknown structure (*468*). The related lead complexes, [Ph$_2$Pb(terpy)X$_2$] (X = Cl, Br, or I) and [Ge(terpy)I$_3$]I, have also been described (*193*).

Tetraphenyldihalostannoles (Fig. 11) react with terpy to form [Ph$_4$C$_4$Sn(terpy)X][Ph$_4$C$_4$SnX$_3$] (*217*).

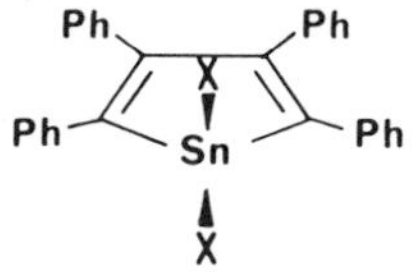

Ph, Ph
X
Ph Sn Ph
X

FIG. 11. Tetraphenyldihalostannole.

E. PHOSPHORUS, ARSENIC, ANTIMONY, AND BISMUTH

A number of complexes of the type $(EX_3)_n(terpy)_m$ (E = As, X = Cl, $n = 2$, $m = 1$; X = Br, $n = 1$, $m = 1$; E = Sb, X = Cl, Br, or I, $n = 3$, $m = 2$; E = Bi, $n = m = 1$) have been described, and are formulated as containing five-coordinate $[EX_2(terpy)]^+$ cations (74). The only other complex to have been reported is $[Bi(terpy)(S_2CNEt_2)I_2]$, which has been structurally characterized, and shown to possess a pentagonal-bipyramidal structure with axial iodine atoms (Fig. 12) (381).

F. SULFUR, SELENIUM, AND TELLURIUM

The 1:1 adducts of terpy with $[ECl_4]$ (E = Se or Te) behave as 1:1 electrolytes, and are formulated $[M(terpy)Cl_3]Cl$; the 1:2 adduct with $[TeCl_4]$ probably possesses the structure $[Te(terpy)Cl_2][TeCl_6]$ (151).

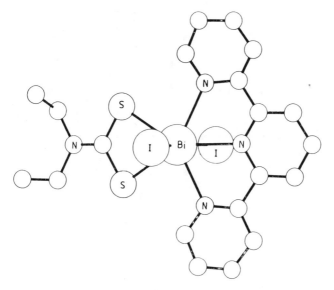

FIG. 12. The crystal and molecular structure of $[Bi(terpy)(S_2CNEt_2)I_2]$ (381).

G. Transition Metals

1. Early Transition Metals

The complex [Sc(NO$_3$)$_3$(terpy)] has been structurally characterized; the metal is in a nine-coordinate environment (Fig. 13), in which each nitrate acts as a chelating bidentate ligand (16).

Although the tetrahalides of titanium, zirconium, and hafnium might be expected to react with terpy to give adducts, these reactions do not appear to have been investigated. It is claimed that titanium(III) chloride reacts with acetonitrile solutions of terpy to give a paramagnetic 1:1 adduct, although the adduct has not been fully characterized (121). The reaction of [(cp)$_2$Ti(CO)$_2$] (cp, cyclopentadienyl) with excess terpy gives the formally zero-valent compound [Ti(terpy)$_2$] (43).

The orange or brown compounds [M(terpy)X$_5$] (M = Nb, X = Cl or Br; M = Ta, X = Br) are readily prepared by the reaction of terpy with the

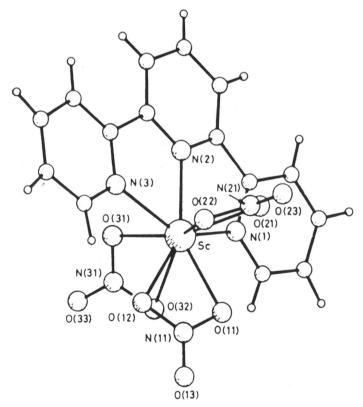

Fig. 13. The crystal and molecular structure of [Sc(NO$_3$)$_3$(terpy)] (16).

appropriate pentahalide in benzene. The only product isolated from the reaction with $TaCl_5$ was Ta_2Cl_{10}(terpy), which, like the previous compounds, is of unknown structure (38). The red vanadium(V) hydroxylamido complexes [V(terpy)(H$_2$NO)(NO)X]   (X = CN   or   N$_3$)   and [V(terpy)(H$_2$NO)(NO)X]Y (X = H$_2$O or NH$_3$; Y = I or Br) have also been prepared (481).

The adducts MCl_4(terpy) and M_2Cl_8(terpy) (M = Nb or Ta) have been described. The structures are not known with any certainty, although the mononuclear complexes are nonelectrolytes, whereas the bromo compounds are 1:1 electrolytes with extensive magnetic interactions between the paramagnetic centers (38).

The compound [VCl$_3$(terpy)] is a paramagnetic, octahedral complex formed in the reaction of VCl_3 (100) or [VCl$_3$('BuNC)$_3$] (408) with terpy. A bis(terpy)vanadium(III) complex is presumably the product of oxidation of [V(terpy)$_2$]$^{2+}$ (49). A red polynuclear complex is formed from the reaction of aqueous vanadium(III) solutions with one equivalent of terpy (49).

The complex [V(terpy)$_2$]I$_2$ may be isolated as a green air-sensitive solid from aqueous ethanolic solution (238). Taube and co-workers have investigated the electron-transfer reactions of [V(terpy)$_2$]$^{2+}$, prepared by the direct interaction of terpy with aqueous vanadium(II) solutions (49).

The reduction of [V(terpy)$_2$]I$_2$ by magnesium or Li[AlH$_4$] leads to the formation of [V(terpy)$_2$] (239), which may also be prepared by the reaction of terpy with [V(CO)$_4$(dppe)], [V(CO)$_2$(dppe)$_2$], [V(cp)$_2$], or [V(cp)(CO)$_4$] (40). The solid is black (μ_{eff} = 1.73 BM) and is thought to possess considerable ligand radical character, similar to the analogous bipy and phen complexes.

2. Chromium, Molybdenum, and Tungsten

The majority of studies of group VIA polypyridyl complexes has been concerned with the photophysical and electrochemical properties of the [Cr(terpy)$_2$]$^{3+}$ cation. However, a number of other complexes have been described.

2,2':6',2"-Terpyridine has been investigated as a reagent for the colorimetric detection of molybdenum(VI) (232). The molybdenum(IV) complex cation [Mo(terpy)(H$_2$NO)(NO)(H$_2$O)]$^{2+}$ is prepared by the reaction of terpy with [MoO$_4$]$^{2-}$ and [H$_3$NOH]Cl in water (484). The compound   is   diamagnetic,   and   reacts   with   cyanide   to   give [Mo(terpy)(NO)(CN)(NHO)] (483). This cyano complex has been structurally characterized, and shown to adopt a distorted pentagonal-bipyramidal geometry, with axial nitrosyl and cyanide ligands, and a bidentate N,O-bonded hydroxylamido ligand (482). The unusual complex cation [Mo(terpy)(H$_2$NO)(H$_2$NOH)(NO)]$^{2+}$ is formed from the reaction of

$[Mo(terpy)(H_2NO)(NO)(H_2O)]^{2+}$ with excess hydroxyammonium chloride (*483*).

The complex $[Cr(terpy)_2]^{3+}$ was first prepared by reaction of $CrCl_3$ with excess terpy in the presence of a reducing agent (*62, 328*), but is more readily obtained from the reaction of a chromium(II) source with terpy, followed by oxidation (*80, 189, 252–258, 406, 479, 480*). The photophysical properties of this yellow cation have excited considerable interest, and play an integral role in a number of potential solar energy conversion schemes. There is still some controversy over the precise photoexcited states involved in the photoaquation and other reactions of $[Cr(terpy)_2]^{3+}$ and related polypyridylchromium(III) cations, but the major species of importance appears to be the low-lying ^2E_g state, reached by intersystem crossing from the $^4T_{1g}$ or $^4T_{2g}$ states (ground state = $^4A_{2g}$) (*10, 80, 405*). The ^2E_g state is remarkably short-lived, and has $\tau = 0.05$ μsec [cf. $[Cr(bipy)_3]^{2+}$, $\tau = 63$ μsec], an observation which has been ascribed to an intimate association with solvent molecules approaching close to the metal in "interligand pockets." This is of obvious relevance to the photoaquation of the complexes, which has also been ascribed to direct nucleophilic attack of water upon the coordinated ligands. To some extent these differences have been settled by an X-ray crystal structure of the salt $[Cr(terpy)_2](ClO_4)_3 \cdot H_2O$ (Fig. 14) (*479, 480*). Although there was no evidence for the *water* molecule occupying a site close to the metal, perchlorate ions were found to have oxygen atoms within the proposed pockets (Cr—O, $\sim$4.5 Å), reminiscent of the outer–sphere complexes proposed for a number of transition metal α,α'-diimine complexes with more or less innocent anions. These observations provide strong support for the photoaquation reactions proceeding via attack at the metal to form a seven-coordinate intermediate, rather than at the ligand. The electron-

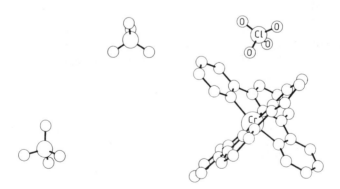

FIG. 14. The crystal and molecular structure of $[Cr(terpy)_2][ClO_4]_3 \cdot H_2O$ (*479, 480*).

transfer reaction between $[Cr(bipy)_3]^{2+}$ and $[Cr(terpy)_2]^{3+}$ was found to be too fast to follow ($k \geq 3 \times 10^6 \ M^{-1} \ sec^{-1}$) (189).

Related to the above studies are a number of investigations into the electrochemical properties of $[Cr(terpy)_2]^{3+}$ and some substituted derivatives (61, 252–258). The parent complex exhibits at least six reductions, the first four of which are fully reversible. It is suggested that the first three reductions are metal-centered, whereas the last is ligand-centered. In the absence of ESR evidence, these conclusions must be regarded as tentative. The complexes of 4'-Phterpy and 4,4"-Ph₂terpy have also been investigated. The complex $[Cr(terpy)Cl_3]$ has been described (77).

The chromium(II) cation $[Cr(terpy)_2]^{2+}$ is conveniently prepared by the interaction of aqueous chromium(II) solutions with excess terpy, and may be isolated as its perchlorate or iodide salt (239, 255, 299). The magnetic properties of $[Cr(terpy)_2][ClO_4]_2$ have been recorded over the temperature range 20–300 K; the complex is low spin its μ_{eff} exhibits a nearly linear temperature dependence over this range (299). The complex may be reduced electrochemically (255) or chemically (239). Herzog and Aul have isolated the various $[Cr(terpy)_2]^{n+}$ complexes that may be obtained by chemical reduction; $[Cr(terpy)_2]I_2$ (red-brown, $\mu_{eff} = 2.80$ BM), $[Cr(terpy)_2]I$ (wine-red, $\mu_{eff} = 1.85$ BM), and $[Cr(terpy)_2]$ (green, $\mu_{eff} = 0.63$ BM) (239).

The molybdenum(II) chloroclusters $[Mo_6Cl_8]X_4$ (X = Cl or I) react with terpy in tetrahydrofuran (THF) to give $[Mo_6Cl_8(terpy)X_3]X$, which are of unknown structure (191, 219). The compounds behave as 1:1 electrolytes (191); the X-ray photoelectron spectrum (PES) of $[Mo_6Cl_8(terpy)Cl_3]Cl$ has been reported (219).

The complex $[(C_3F_7)CrCl_2(py)_3]$ reacts with terpy to form $[(C_3F_7)Cr(terpy)Cl_2]$ (305).

The very air-sensitive compounds $[Mo(terpy)_2]$ (purple) and $[W(terpy)_2]$ (green) have been described by a number of workers (44, 174). The chromium compound may be prepared by the reaction of $[Cr(CO)_6]$ (44), $[Cr(CN)_6]^{6-}$ (42, 47), $[Cr(CO)_3(C_6H_6)]$ (47), or $[Cr(bipy)_3]$ (44) with two equivalents of terpy. The molybdenum (44, 174) and tungsten (44) compounds have been prepared from $[M(CO)_6]$ in a similar manner. Electrochemical studies on $[Mo(terpy)_2]$ indicate three one-electron oxidations (8). Photosubstitution reactions of $[M(CO)_6]$ in the presence of limited amounts of terpy lead to the formation of $[M(CO)_4(terpy)]$ (M = Cr, Mo, or W) (205). It is unlikely that these compounds are seven-coordinate, and they may well provide examples of a bidentate terpy. $[Mo(CO)_3(terpy)]$ may be prepared by the reaction of terpy with $[Mo(CO)_3(mesitylene)]$, but attempts to prepare the other group VI complexes of this stoichiometry lead to the formation of $[M(CO)_4(terpy)]$ (205).

3. Manganese, Technetium, and Rhenium

Although no terpy complexes of manganese(VII) have been described, the perrhenate, [terpyH][ReO$_4$], has been reported by Morgan and Davies (329). These authors also described the rhenium(IV) salt [terpyH$_2$][ReCl$_6$]. The complex [Mn(terpy)Cl$_3$] may be isolated; magnetic measurements indicate that the compound possesses a simple octahedral structure (214). In contrast to the mononuclear manganese(III) complex, a number of trirhenium(III) clusters have been described. The purple complexes [Re$_3$Cl$_9$(terpy)$_n$] (n = 1, 2, or 1.33) result from the reaction of terpy with [Re$_3$Cl$_9$]. Conductivity measurements indicate that the first two compounds should be formulated [Re$_3$Cl$_8$(terpy)]Cl and [Re$_3$Cl$_6$(terpy)]Cl$_3$, respectively; the terpy is thought to bind to one rhenium atom as a planar terdentate, perpendicular to the plane of the trirhenium triangle (192).

Manganese halides react with terpy to form well-characterized 1:1 complexes of the type [Mn(terpy)X$_2$] (X = Cl, Br, or I), which have been shown to be isomorphous with the corresponding zinc complexes. They are thus thought to possess a five-coordinate trigonal-bypyramidal environment (225, 271). The magnetic moments of these complexes are close to the expected spin-only value (271). There have been a number of investigations of the formation of the 1:1 complexes in water (245), methanol, or DMSO solution (50, 51, 83, 84). In water, the Eigen–Wilkins mechanism, in which the rate of formation of the complex is controlled by the rate of exchange of coordinated water with the bulk solvent, appears to operate (243, 245). In methanolic solution, the system is not so simple, and there is a suggestion that the formation of the second metal-to-nitrogen bond (i.e., formation of a chelate ring) may be the rate-determining step (50, 51). The reaction in DMSO is complex; [Mn(DMSO)$_6$]$^{2+}$ reacts with terpy in a fast step to form [(DMSO)$_4$Mn(terpy)]$^{2+}$ (bidentate terpy), which reacts rapidly with excess [Mn(DMSO)$_6$]$^{2+}$ to give a binuclear species [(DMSO)$_4$Mn(terpy)Mn(DMSO)$_5$]$^{4+}$ (bridging-terdentate terpy). The rate-determining step in the reaction is formation of the final chelate ring to give [Mn(DMSO)$_3$(terpy)]$^{2+}$ (83, 84). The complexes [Mn(terpy)X$_2$] dissociate rapidly in aqueous solution to give the [Mn(terpy)$_2$]$^{2+}$ cation (243). Studies show that the activation parameters for the reaction of manganese(II) with terpy in aqueous solution indicate an associative mechanism, in common with a number of other reactions of early transition metal compounds (322).

A number of workers have investigated the electrochemical behavior of the [Mn(terpy)$_2$]$^{2+}$ cation (6, 61, 256, 330, 377). In acetonitrile solution, there is a well-defined oxidation to the manganese(III) state at +1.28 V relative to the standard calomel electrode; this species is stable, in contrast to the tris(bipy) and tris(phen) analogues, which are converted to binuclear dimanganese(III,IV) species (256, 330, 377). The complex also shows at least five reduction waves (6, 61, 256, 377). The final reductions are probably due to free terpy, liberated in dissociative processes.

The complex [Mn(CO)$_3$(terpy)Br] has been isolated, and is thought to contain a bidentate terpy ligand, on the basis of the similarity of its infrared spectrum to that of [Mn(CO)$_3$(bipy)Br] (205). The reaction of [Re(CO)$_5$(NO$_3$)] with terpy leads to the formation of [Re(CO)$_3$(terpy)(NO$_3$)]; this complex is a nonelectrolyte in nitromethane, and is also thought to contain a bidentate terpy ligand (4).

4. Iron, Ruthenium, and Osmium

2,2':6',2"-Terpyridine and its derivatives have been widely proposed as analytical reagents for the detection of iron in biological and other materials (146, 164, 199, 218, 293, 331, 333, 358, 400–404, 416, 422, 450, 451, 485, 496–498). These applications have been reviewed elsewhere, and will not be considered further (69, 285, 291, 314, 401).

With the exception of salts of the [Fe(terpy)$_2$]$^{3+}$ cation (vide infra), iron(III) complexes of terpy are remarkably sparse. The complexes [Fe(terpy)Cl$_3$] and [(terpy)FeOFe(terpy)][NO$_3$]$_4$·H$_2$O are prepared by addition of chloride or nitrate ions to solutions containing [Fe(terpy)$_2$]$^{3+}$ (383, 384). Mössbauer, magnetic, and other spectroscopic properties of the oxy-bridged complexes are consistent with a sextet ($S = \frac{5}{2}$) ground state.

The complexes of stoichiometry [Fe(terpy)X$_2$] (X = Br, I, or NCS) are probably five-coordinate trigonal-bipyramidal species (385), whereas "[Fe(terpy)Cl$_2$]," which was originally thought to possess a similar structure (75, 76, 243, 271, 388, 389), has been shown to be [Fe(terpy)$_2$][FeCl$_4$] (385).

The purple complex cation [Fe(terpy)$_2$]$^{2+}$ and its blue oxidation product [Fe(terpy)$_2$]$^{3+}$ have been known since the first description of terpy by Morgan and Burstall (93, 176, 282, 286, 323, 326, 418). The crystal structural analysis of [Fe(terpy)$_2$][ClO$_4$]$_2$·H$_2$O reveals the cation to possess the expected distorted octahedral geometry (22). The lattice water and the perchlorate anions are oriented toward the interligand pockets, reminiscent of the structure of [Cr(terpy)$_2$][ClO$_4$]$_3$·H$_2$O. The stability constants for the iron(II) complexes are high (1gβ_2 = 18–21) (70, 85, 306). The stability constants and rates of formation and dissociation of the complexes are both pH and water dependent, an observation which has been interpreted in terms of covalent hydration of the ligand or of more conventional effects involving protonation of dissociated pyridines (85–87, 89, 190, 200, 243, 245, 457). There have been a number of studies involving the use of [Fe(terpy)$_2$]$^{3+}$ as an oxidizing agent. There is evidence that the oxidation of [Fe(H$_2$O)$_6$]$^{2+}$ and organotin compounds by [Fe(terpy)$_2$]$^{3+}$ follows a non-Marcus pathway in addition to the expected outer-sphere electron transfer (152, 360). Peroxodisulfate, peroxodiphosphate, or peroxide oxidation of [Fe(terpy)$_2$]$^{2+}$ may proceed by parallel Marcus and non-Marcus pathways; the latter *may* involve the oxidant approaching an interligand pocket (*35, 88, 106*). As with the ruthenium and osmium analogues, the luminescent properties of the bis complexes have suggested applications as potential photocatalysts (*115, 116, 153, 198, 352, 363*). A novel application of these complexes is seen in the extraction of long-chain alkyl surfactants by $[Fe(terpy)_2]^{2+}$ and derivatives (*440, 441*).

The $[Fe(terpy)_2]^{2+}$ cation is low-spin, as demonstrated by Mössbauer, electronic, ^1H NMR, and resonance Raman spectroscopy and magnetic measurements (*20, 184, 187, 228, 266*). Similarly, spectroscopic studies of the iron(III) cation have indicated a low-spin (^2B) ground term (*382*). There have been numerous electrochemical studies of the bis complexes (*177, 200, 256, 298, 332, 344, 373, 378, 379, 397, 398*). Ligand-centered reductions to formal oxidation states of iron(I), iron(0), and iron(−1) and oxidations to iron(III) are observed. The complex $[Fe(terpy)L][ClO_4]_2$ [L = tris(2'-pyridyl)1,3,5-triazine (Fig. 15)] has been prepared (*399, 442*).

The iron(0) complex $[Fe(terpy)(CO)_2]$ is readily prepared by the reaction of terpy with $[Fe(cot)(CO)_3]$ (cot, cyclooctatetraene) (*41, 46*); in contrast, $[Fe(terpy)_2]$ may be obtained as an air-sensitive, paramagnetic solid from the reduction of $[Fe(terpy)_2]I_2$ with lithium benzophenone ketyl (*240*).

Until recently there has been surprisingly little interest in high oxidation state complexes of terpy. Meyer and co-workers have demonstrated that the ruthenium(IV) complex $[Ru(terpy)(bipy)O]^{2+}$ is an effective active catalyst for the electrocatalytic oxidation of alcohols, aromatic hydrocarbons, or olefins (*335, 443, 445, 446*). The redox chemistry of the $[M(terpy)(bipy)O]^{2+}$ (M = Ru or Os) systems has been studied in some detail, and related to the electrocatalytic activity (*437, 445, 446*). The complexes are prepared by oxidation of $[M(terpy)(bipy)(OH_2)]^{2+}$. The related osmium(VI) complex $[Os(terpy)(O)_2(OH)]^+$ exhibits a three-electron reduction to $[Os(terpy)(OH_2)_3]^{3+}$ (*365, 366*). The complex $[Ru(terpy)(bipy)(H_2NCHMe_2)]^{2+}$ undergoes two sequential two-electron

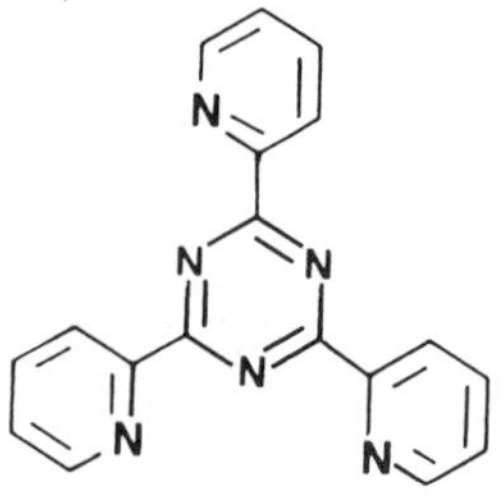

FIG. 15. Tris(2'-pyridyl)-1,3,5-triazine.

oxidations to produce $[Ru(terpy)(bipy)(HN=CMe_2)]^{2+}$ and $[Ru(terpy)-(bipy)(NCMe_2)]^{3+}$; this corresponds to the net oxidation of coordinated isopropylamine (3). In contrast, $[Ru(terpy)(bipy)(NH_3)]^{2+}$ undergoes a clean *chemical* oxidation to the corresponding Ru(III) compound (64, 345, 387, 495), but *electrochemical* oxidation results in the oxidation of coordinated ammonia to nitrate via coordinated imine, hydroxylamine, nitrosyl, and nitrite (444). The reverse reaction, the net six-electron reduction of $[Ru(terpy)(bipy)(NO)]^{3+}$ to $[Ru(terpy)(bipy)(NH_3)]^{2+}$ has also been effected electrochemically (339). The nitrite complex $[Os(terpy)-(bipy)(NO_2)]^{2+}$ disproportionates to yield $[Os(terpy)(bipy)(NO_2)]^+$, $[Os(terpy)(bipy)(NO)]^{3+}$, and $[Os(terpy)(bipy)(ONO_2)]^{2+}$ (365). Electrochemical studies of the behavior of $[Ru(terpy)(bipy)(OH_2)]^{3+}$ at Nafion-coated electrodes have been reported (312). Reductive cleavage of diphenylacetylene occurs upon heating aqueous solutions with $[Ru(terpy)(PPh_3)-(H_2O)_2]^{2+}$ to yield $[Ru(terpy)(PPh_3)(CO)(CH_2Ph)]^+$ and toluene (428).

The complex $[Ru(terpy)Cl_3]$ is obtained as a red solid from the reaction of terpy with ethanolic "$RuCl_3 \cdot 3H_2O$" (429); in dimethylformamide (DMF), carbonyl abstraction occurs, and $[Ru(terpy)Cl_2(CO)]$ is obtained (117). It is, perhaps, surprising that $[Os(terpy)Cl_3]$ is reported as the product of the reaction of terpy with $K_2[OsCl_6]$ in DMF (81). The complexes $[Ru(terpy)(NO)Cl_2]_2[RuCl_5(NO)]$ and $[Ru(terpy)(NO)Cl_2]Cl \cdot 3.5H_2O$ result from the reaction of terpy with $K_2[RuCl_5(NO)]$ (325, 327–329).

Intense room temperature charge-transfer emission was detected in terpy complexes for the first time in the complexes $[Os(terpy)LL']^{n+}$ (L = $Ph_2PCH=CHPPh_2$, L′ = Cl, n = 1; L′ = py, MeCN, or CO, n = 2; L = $Ph_2PCH_2PPh_2$, L′ = Cl, n = 1) (7). Phosphines react with $[Ru(terpy)Cl_3]$ in $CHCl_3$ in the presence of Et_3N to give *trans*-$[Ru(terpy)Cl_2L]$, which undergoes a thermal rearrangement to the *cis* complex (also available by the reaction of terpy with $[RuL_3Cl_2]$) (429). The reaction of $[Ru(bipy)_2(CO)_2]Cl_2$ with terpy in the presence of amine oxide results in the formation of the $[Ru(terpy)(bipy)Cl]^+$ cation (56). Substitution reactions of $[Ru(terpy)(bipy)(H_2O)]^{2+}$ with a range of nucleophiles have been reported (159).

Both *cis*- and *trans*-$[Ru(terpy)(CO)Cl_2]$ have been structurally characterized (162); the *cis* complex is obtained by reaction of $[Ru(terpy)(CO)_2Cl_2]$ with Me_3NO, and the *trans* by reaction of terpy with "$RuCl_3 \cdot 3H_2O$" in DMF (117, 162, 429). The two (red and yellow) forms of $[Ru(CO)_2Br_2(terpy)]$ have also been structurally characterized (Fig. 16); each contains a *bidentate* terpy ligand. The color differences arise from differences in intermolecular ligand contacts (162).

The complexes $[M(terpy)(bipy)L]^{n+}$ (M = Ru or Os; L = NH_3, NO, NO_2, H_2O, H_2NCHMe_2, phenothiazine, Cl, Br, I, *N*-methylphenothiazine,

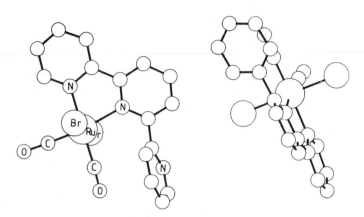

FIG. 16. The crystal and molecular structures of the two crystal forms of [Ru(terpy)(CO)$_2$Br$_2$] (*162*).

thianthrene, OH, py, thiourea, N$_3$, SCN, 4-vinylpyridine, or substituted pyridines) are readily prepared by the reaction of [M(terpy)(bipy)Cl] with L in a variety of solvents and conditions (*3, 64, 81, 82, 95, 160, 179, 216, 261, 312, 335, 345, 365, 366, 387, 395, 428, 437, 444–446, 495*). Numerous studies of the photophysical, photochemical, electrochemical, and kinetic properties of these complexes have been reported. A number of complexes of the type [Ru(terpy)(bipy)L]$^{2+}$ (L = sulfur, selenium, or tellurium donor) have been prepared; the peroxide oxidation of [Ru(terpy)(bipy)(SMe$_2$)]$^{2+}$ gives [Ru(terpy)(bipy)(S(O)Me$_2$)]$^{2+}$ by nucleophilic attack of the coordinated sulfur upon the oxygen—oxygen bond (*394*).

Red [Ru(terpy)$_2$]Cl$_2$ and green [Os(terpy)$_2$]Cl$_2$ were first prepared by the reaction of terpy with mixtures of metal chloride and metal at elevated temperatures (*328*). More convenient procedures involve the reaction of terpy with K$_2$[RuCl$_5$(H$_2$O)] in the presence of H$_2$PO$_2$ (*118, 292*), the reaction of "RuCl$_3$·3H$_2$O" with excess terpy in DMF (*63*) or ethanol (*418*), or the reaction of terpy with "ruthenium blue solutions" (*424*). Numerous electrochemical studies of these systems have been reported (*24, 63, 107, 118, 177, 319, 320, 332, 398, 447*). The ruthenium(II) complex exhibits a reversible oxidation and four reduction waves; ESR studies indicate that the first three reductions are ligand centered, whereas the last is associated with an ECE process (*332*). The photophysical properties have also attracted some interest; the ruthenium(II) complex is very weakly luminescent in liquid aqueous or ethanolic solution, although flash photolysis in the presence of a quencher indicates electron transfer from an excited state followed by a back reaction. The lifetime of the excited state is short (1.2 nsec < τ_0 < 5.0 nsec) (*63, 153, 197, 292, 318, 320, 494*). A negative enthalpy of activation for the quenching of $[Ru(terpy)_2]^{2+*}$ by $[Fe(H_2O)_6]^{3+}$ indicates a non-Marcus electron-transfer process. At lower temperatures in frozen glasses, efficient photoluminescence occurs, with charge-transfer emission from at least three strongly spin-orbit coupled excited states (5, 165, 166, 197, 280, 424). The absorption spectra of $[Ru(terpy)_2]^{n+}$ are pH dependent in the range pH 0–6, and the ruthenium(II) complex exhibits an apparent pK_a of 2.9; these observations are clearly related to the anomalous pH dependence of ligand dissociation from $[Fe(terpy)_2]^{n+}$ (118, 275). Ruthenium complexes of substituted terpy ligands and $[Os(terpy)_2]^{2+}$ exhibit similar photophysical properties (172, 173, 231, 359). The FD mass spectrum of $[Os(terpy)_2][PF_6]_2$ exhibits weak peaks due to the monocation (105). The ^1H NMR spectra of $[M(terpy)_2]^{2+}$ (M = Ru or Os) salts have been reported; DMSO-d_6 solutions of the ruthenium complex undergo a specific base-catalyzed deuterium-exchange reaction at the 3, 3′, 5′, and 3″ positions (138, 300). Salts of $[Ru(terpy)_2]^{2+}$ exhibit potent bacteriostatic properties, and inhibit the action of acetylcholinesterase; effects similar to those exhibited by d-tubocurarine are observed (180, 281, 282). The bifunctional osmium complexes $[(bipy)ClOsLOs(bipy)Cl]^{2+}$ and $[(bipy)(py)OsLOs(bipy)(py)]^{4+}$ (L = Fig. 17) also exhibit potent curariform activity (439).

The Strasbourg group extended their studies to highly hindered derivatives of terpy. The ligand 6,6″-Ph_2terpy is very hindered, but forms a bis complex with ruthenium(II); the NMR and photophysical properties of this complex have been reported (168, 279). The steric requirements of the ligand are expressed in dissociative photoanation reactions.

5. Cobalt, Rhodium, and Iridium

The majority of interest in cobalt(III) complexes of terpy has centered on the $[Co(terpy)_2]^{3+}$ cation, although a few 1:1 complexes have been reported. A crystal structural analysis of the complex $[Co(terpy)(CO_3)(OH)]\cdot4H_2O$, obtained by the reaction of $[Co(terpy)_2]^{2+}$ with aqueous carbonate, has revealed the expected distorted octahedral geometry about the metal ion, with a bidentate chelating carbonate and *meridional* terpy ligands (287). The

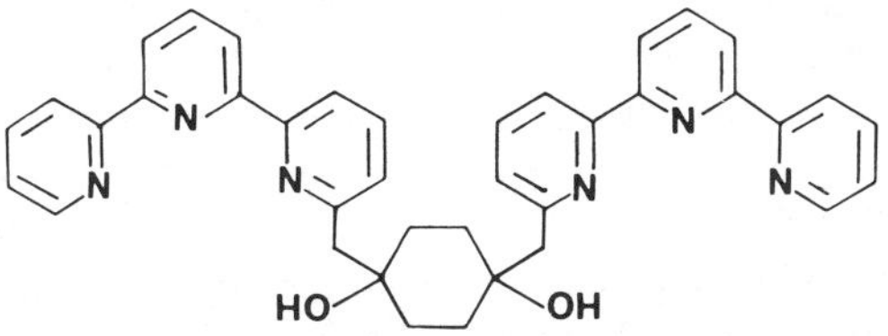

FIG. 17. A bridging bis-2,2′:6′,2″-terpyridine ligand (439).

complexes $[Co(terpy)L(H_2O)]^{2+}$ [L = phen, bipy, or bis(2-pyridyl)ketone]
react   readily   and   reversibly   with   dioxygen   to   form   adducts
$[(terpy)LCo(O_2)CoL(terpy)]^{4+}$, which may be oxidized to the superoxo
species $[(terpy)LCo(O_2)CoL(terpy)]^{5+}$ (223, 250, 251, 321, 356). If the
mononuclear complex is stabilized in an inert zeolite Y matrix, the superoxide adduct $[Co(bipy)(terpy)O_2]^{2+}$ may be characterized (321).

Complexes of the lower members of the group with terpy have attracted
little attention. A number of salts of the $[Rh(terpy)_2]^{3+}$ cation have been
characterized, as have the complexes $[Rh(terpy)X_3]$ (X = Cl, Br, or I)
(55, 224, 328). In each case, the compounds are thought to possess a distorted
octahedral geometry. The luminescent 1:1 chelate formed between irid-
ium(III) and terpy has been studied, and may be of some application in the
determination of iridium (197, 198, 328).

The complex cation $[Co(terpy)_2]^{3+}$ is readily prepared by the oxidation of
the cobalt(II) complex (304, 328), and has been widely used as an oxidizing
agent. Reduction by superoxide (65, 313), dithionite (65, 315), titanium(III)
(21), ascorbate (487), metallophthallocyanines (353, 355), galactose oxidase
(267), methylene blue (355), or deoxyhemerythrin (220) has been investigated,
and the results are generally in accord with a Marcus outer-sphere mechanism. Pulse radiolysis of $[Co(terpy)_2]^{3+}$ produces a transient ligand-centered
radical, which collapses to "normal" $[Co(terpy)_2]^{2+}$ (36). The structure of
$[Co(terpy)_2]Cl_3 \cdot 11H_2O$ has only recently been determined; it possesses the
expected distorted octahedral structure (194–196).

Salts of $[Co(terpy)_2]^{2+}$ are obtained by the reaction of cobalt(II) salts with
excess terpy. Structural analyses of $[Co(terpy)_2]X_2 \cdot nH_2O$ (X = ClO_4, n =
1.3; X = I, n = 2; X = ClO_4, n = 0.5; X = Br, n = 3; X = SCN, n = 2) have
been reported; this apparent overkill is explained by the spin-crossover
properties of the salts (195, 196, 235, 307, 380). It is clear that the spin-state
crossover between high- and low-spin forms of these cobalt(II) complexes is
dictated by lattice forces. The perchlorate is close to the high-spin limit, but
the temperature dependence of the moment is critically dependent upon the
degree of hydration (196, 235). The complex $[Co(terpy)_2][SCN]_2 \cdot 2H_2O$
undergoes an X-ray-induced phase transition, apparently linked with a
crossover phenomenon (380). The spin-crossover properties of these salts
have been investigated by magnetic, ESR, and Mössbauer methods (169,
221, 269, 284, 302, 303, 402, 425). The cobalt(II) complexes are moderately stable, but are acid-labile (190, 243, 245, 278, 372). Oxidation to the
cobalt(III)   complexes   is   facile   and   has   been   studied   intensively
(12, 155, 189, 208, 276, 311, 374, 423, 493). The redox properties of the
$[Co(terpy)_2]^{2+/3+}$ system have been scrutinized by the solar energy fraterni-
ty (116). Electrochemical studies indicate that reduction to formal oxidation
states of cobalt(I) and cobalt(0) is possible (6, 256, 301, 373, 377, 378, 398). A

colorimetric method utilizing terpy has been proposed for the determination of cobalt (334).

Numerous 1:1 complexes [Co(terpy)X$_2$] (X = SCN, SeCN, Cl, Br, I, CN, NO$_2$, or F) have been described (157, 212, 225, 268, 270, 272, 277). It has long been suspected that these complexes are five-coordinate, and structural analyses of [Co(terpy)(NCO)$_2$] (277) and [Co(terpy)Cl$_2$] (21) have confirmed this formulation. The isocyanato complex is of interest in exhibiting a weak hydrogen-bonding interaction between an isocyanate oxygen atom and H-3,3' of the terpy (277). The displacement of terpy from the complexes [Co(terpy)L$_3$]$^{2+}$ (L = H$_2$O or hmpa) has been studied; in the acid-catalyzed dissociation of [Co(terpy)(H$_2$O)$_3$]$^{2+}$ there is some evidence for a monodentate intermediate [Co(terpyH$_2$)(H$_2$O)$_n$]$^{4+}$ (1, 2, 249).

The cobalt(I) complex [Co(terpy)(BH$_4$)] has been shown to possess a five-coordinate structure in which the borohydride acts as a chelating bidentate ligand (Fig. 18) (148). The bridging hydride atoms were located in a neutron-diffraction study of the compound. Although the solid is weakly paramagnetic, solutions are essentially diamagnetic, and ^{1}H NMR studies indicate a high barrier to the interconversion of the bridging and terminal hydridic sites (488).

The complex [Co(terpy)$_2$][Co(CO)$_4$] has also been reported (45).

6. Nickel, Palladium, and Platinum

Although there have been numerous reports of nickel(II), palladium(II), and platinum(II) complexes of 2,2':6',2"-terpyridine, complexes in higher oxidation states are all but unknown. Morgan and Burstall described the

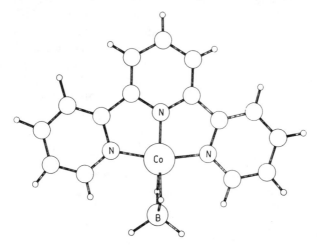

FIG. 18. The crystal and molecular structure of [Co(terpy)(BH$_4$)] (148).

platinum(IV) complex [Pt(terpy)Cl$_3$]Cl·3H$_2$O as the product from the chlorine oxidation of [Pt(terpy)Cl]Cl (*324*); although the compound has not been investigated since, there seems to be little doubt that it is an octahedral platinum(IV) species with a terdentate 2,2′:6′,2″-terpyridine ligand. The dimethylplatinum(IV) complex [PtMe$_2$(terpy)(PMe$_2$Ph)$_2$][PF$_6$]$_2$ has also been described; the structure is not known, but a bidentate 2,2′:6′,2″-terpyridine seems to be more likely than a seven-coordinate platinum(IV) center (*119*). Electrochemical oxidation of [Ni(terpy)$_2$]$^{2+}$ leads to the formation of [Ni(terpy)$_2$]$^{3+}$, which may be isolated as its tris(perchlorate) salt (*373*).

There are a number of distinct types of complexes with the stoichiometry M(terpy)X$_2$. Nickel(II) complexes with a 1:1 stoichiometry may be prepared by addition of terpy to solutions of nickel(II) halides; the hydrated forms appear to be typical octahedral nickel(II) compounds (*294*). Magnetic measurements and spectroscopic studies have indicated that these 1:1 complexes obtained from solution, usually hydrated, are best formulated as the cation–anion species [Ni(terpy)$_2$][NiX$_4$] (*222*). These complexes may be prepared by addition of terpy to acetone solutions containing the appropriate tetrahalonickel(II) (*222*). The complexes Ni(terpy)X$_2$ (X = CN, N$_3$, or CNS) have also been reported. The cyano and azido compounds are best described as [Ni(terpy)$_2$][NiX$_4$], on the basis of their spectroscopic and magnetic properties (*222*). In contrast, the complex Ni(terpy)(NCS)$_2$ is polymeric, with bridging thiocyanates and a ferromagnetic interaction between the nickel atoms (*222*). The anhydrous complexes Ni(terpy)X$_2$ (X = Cl, Br, or I) are prepared by pyrolysis of the [Ni(terpy)$_2$]X$_2$ species, and are of rather more interest (*243, 268–271, 294*). The spectral and magnetic properties of these complexes are not compatible with octahedral or square-planar structures, and give a best fit with five-coordinate trigonal-bipyramidal or square-based pyramidal schemes (*294*). This formulation is supported from a study of the X-ray powder photographs obtained from the complexes: [Ni(terpy)Cl$_2$] is isomorphous with [Zn(terpy)Cl$_2$] (form II); [Ni(terpy)Br$_2$] and [Ni(terpy)I$_2$] are isomorphous with the appropriate zinc halide (forms II and I, respectively). The solution chemistry of these nickel(II) complexes is deferred until later in this section.

The reactions of [MCl$_4$]$^{2-}$ (M = Pd or Pt) with terpy are complex, the major products being [M(terpy)Cl]$_2$[MCl$_4$] and [M(terpy)Cl]Cl (*324, 328*). These complexes are diamagnetic, square-planar compounds, in contrast to the five-coordinate anhydrous nickel(II) complexes. The crystal structures of [Pd(terpy)Cl]$_2$[PdCl$_4$] (*259*) and [Pd(terpy)Cl]Cl·2H$_2$O (*260*) have been reported. The terpy ligand does not have an entirely suitable bite for square-planar coordination, and some distortion of the interring bonds and of the square plane results. In each case the bond to the central ring is shorter than

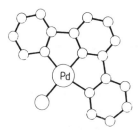

FIG. 19. The crystal and molecular structure of [Pd(terpy)Cl]Cl (*260*).

those to the terminal pyridines (Fig. 19), as observed in the majority of terpy complexes. It is of interest that, in each case, there is a stacking interaction, in which the planar cations are at an average separation of 3.4 Å. This observation is of relevance to the properties of the related [Pt(terpy)X]⁺ complexes, which have not been structurally characterized. Morgan and Burstall described a number of other related complexes (*324*).

Burmeister and colleagues have described the related pseudohalogen derivatives $M(terpy)X_2$ (X = SCN or SeCN) (*90–92*). The platinum compound exhibits the two thiocyanate stretching frequencies expected for a square-planar complex, and is formulated [Pt(terpy)(NCS)][NCS]. However, the palladium complexes are less easily formulated, exhibiting absorptions due to coordinated ECN (E = S or Se) only. These observations were interpreted in terms of a square-planar structure, with a bidentate terpy ligand; in view of the known ability for palladium and platinum diimine complexes to form five-coordinate species, this formulation must also be considered. In the absence of definitive structural evidence, the formulation as five-coordinate species must be regarded as speculative.

The substitution reactions of square-planar complexes of the type [M(terpy)X]⁺:

$$[M(terpy)X]^+ + Y \rightleftharpoons [M(terpy)Y]^+ + X$$

obey the expected rate law:

$$\text{rate} = k_{obs}[M(terpy)X^+]$$

where

$$k_{obs} = k_1 + k_2[Y]$$

However, the rates of substitution by pyridine are considerably faster than might be expected on comparison with other terdentate complexes. Specifically, rate enhancements of three to four orders of magnitude are observed upon comparison with complexes of 1,5-diamino-3-azapentane (diethylene-triamine, dien) (*34, 156, 337, 338*). There is some disagreement over the dependence of the k_1 term on the nature of the incoming nucleophile, although all workers agree on the rate enhancement. Explanations of this rate enhancement have invoked attack by the incoming nucleophile at the coordinated 2,2′:6′,2″-terpyridine ligand, or an extensive interaction between π-bonding nucleophiles and the planar 2,2′:6′,2″-terpyridine. A π-bonding or stacking interaction, as postulated in reactions of $[Ni(terpy)L_3]^{2+}$ complexes, would seem to be most likely; in particular, there is now *no* evidence for attack by nucleophiles at the ligand in platinum(II) complexes of 2,2′-bipyridine, 1,10-phenanthroline, or 2,2′:6′,2″-terpyridine.

Removal of chloride from $[Pt(terpy)Cl]^+$ may be achieved with silver(I) compounds, although the aqua or hydroxo compounds produced are relatively unstable (337, 338). Mureinik and Bidani reported the formation of $[Pt(bipy)(SCN)_2]$ in the reaction of $[Pt(terpy)Cl]Cl$ with silver(I) followed by excess thiocyanate (337); this remarkable observation might indicate attack by hydroxide (free or coordinated) on the coordinated terpy, although the reaction certainly deserves reinvestigation. Much of the recent interest in complexes of platinum(II) and palladium(II) with terpy has stemmed from the discovery that they show specific interactions with nucleic acids. Planar aromatic compounds may show "stacking" interactions, in which configurations with the rings parallel, and interactions between the π systems, represent minima on the potential energy surface. Square-planar complexes with aromatic ligands may also show such interactions, which are almost certainly of more general occurrence than hitherto recognized. Such stacking interactions are of importance in the solid-state structures adopted by complexes, and are invoked to explain the observed crystal structures for $[Pd(terpy)Cl]_2[PdCl_4]$ and $[Pd(terpy)Cl]Cl \cdot 2H_2O$. It is also apparent that stacking interactions are important solution phenomena, and have been invoked to explain the reaction rates of terpy and related complexes. Nucleic acids are characterized by sequences of planar heterocyclic bases arranged in a mutually parallel fashion, and a number of planar aromatic drugs (e.g., acridines, phenanthridines) have been shown to insert between them in a stacking fashion. This process is known as intercalation, and may provide a mechanism for the action of such compounds. Planar transition metal complexes are also expected to show intercalation behavior with nucleic acids.

Initial studies on $[Pt(terpy)Cl]^+$ revealed that the complex showed specific interactions with calf thymus DNA, as shown by inhibition of the fluorescence of the ethidium bromide–DNA intercalate (247). The complex contains a labile halide, however, and shows covalent interactions with the bases in addition to the expected intercalation. These covalent interactions are probably responsible for the observed chain scission of PM-2 DNA (247). The palladium complex shows the same behavior (247). These covalent interactions may be utilized in introducing heavy metal atoms onto nucleic acids as an aid to imaging in electron microscopy. Thus, $[Pt(terpy)X]^+$ (X = Cl or H_2O) binds covalently to the sulfur atoms of adenosine monophosphorothioate and uridine monophosphorothioate, and those of the polynucleotides $poly(_sA-U)$ (426). Binding of modified polynucleotides to sulfur atoms is also of importance with yeast tRNA^{Phe} modified at the C_s-C_s-A terminus, which binds 2 mol[Pt(terpy)Cl]Cl (433). The 2:2 intercalation product of [Pt(terpy)Cl]Cl with adenosine 5'-monophosphate has been structurally characterized (490). The intercalate exhibits an unusual head-to-head stacking interaction. It is of interest that [Pt(terpy)Cl]Cl stacks with itself in solution, with a dimerization constant of $4 \pm 2 \times 10^3$ (264). This may explain the observed concentration dependence of the ^1H NMR spectra of D_2O solutions (264).

In order to avoid the ambiguities resulting from competing intercalation and covalent interactions, a number of $[Pt(terpy)X]^{n+}$ complexes with nonlabile ligands have been investigated. The most widely studied complex ion is $[Pt(terpy)(SCH_2CH_2OH)]^+$, which has been shown to intercalate with calf thymus DNA (265). The platinum atoms are separated by 10.2 Å in the calf thymus DNA intercalate, which is thought to undergo unwinding of $22 \pm 6°$ per intercalator (57). In the absence of covalent interactions, $[Pt(terpy)(SCH_2CH_2OH]^+$ does not produce chain scission with PM-2 DNA (242, 247). Similar interactions occur with salmon sperm DNA (203), and are possibly responsible for the observed inhibition of genetic recombination in pneumococci (407). Intercalation also occurs with polynucleotides, and the duplex structure of polyA·polyU is stabilized by the platinum intercalate (32). Although useful ^{31}P NMR spectra are not obtained from normal nucleic acid samples, such spectra may be obtained from the short (~ 200 base pairs) double-helical sequences obtained from prolonged sonication (489). The ^{31}P NMR spectra of $[Pt(terpy)(SCH_2CH_2OH)]^+$ intercalates of such DNA samples are compatible with an unwinding angle of $18 \pm 2°$ (489).

Crystal structures of $[Pt(terpy)(SCH_2CH_2OH)][NO_3]$ (264) and its intercalates with deoxymethoxy-pTpA and deoxy-CpG (463, 464) have been described. The parent compound shows two types of stacking interactions in the solid, with the planes separated by 3.4 Å (264). The intercalate with deoxymethoxy-pTpA contains the novel trinuclear cation $[Pt(H_2NCH_2CH_2SPt(terpy))_2]^{4+}$, which has been prepared independently by Lippard and co-workers (167). A structural investigation of the tetra-fluoroborate salt (Fig. 20) revealed the two Pt(terpy) units to be parallel and stacked. Since the sulfur atoms are sp^3 hybridized, the molecules are chiral.

A number of other 1:1 complexes with 2,2':6',2''-terpyridine have also been reported. Complexes with monothiodiketonates, $[NiL_2(terpy)]$, have been

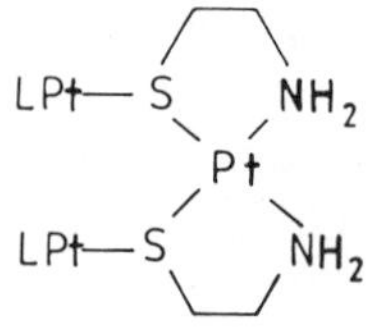

FIG. 20. The molecular structure of the cation $[Pt(H_2NCH_2CH_2SPt(terpy)_2]^{4+}$ (*167*).

described for a range of substituted ligands (*111, 241, 296*). The electronic spectra of these complexes indicate that they adopt an octahedral geometry, and the infrared spectra are compatible with a formulation containing a tridentate 2,2′:6′,2″-terpyridine ligand, and one monodentate S-bonded and one bidentate O,S-bonded thiodiketone. Related trinuclear species have also been described. Other ternary complexes incorporating 2,2′:6′,2″-terpyridine include $[NiL_2(terpy)]$ $[HL = (RO)_2P(S)SH]$, in which only one sulfur is coordinated to the (octahedral) metal (*297*), and $[NiL(terpy)]$ (H_2L = 2,6-pyridinedicarboxylic acid) (*246*).

The solution properties of the 1:1 complexes of nickel(II) with 2,2′:6′,2″-terpyridine are poorly characterized; this is particularly unfortunate, since the nickel(II)–2,2′:6′,2″-terpyridine system has been widely studied in investigations into the binding of 2,2′:6′,2″-terpyridine to metal ions. The rate of the reaction:

$$Ni(terpy)^{2+} + terpy \rightleftharpoons [Ni(terpy)_2]^{2+}$$

is apparently faster than a simple water loss (Eigen–Wilkins) mechanism would predict, and numerous studies in water and mixed solvent systems have been described (*101, 103, 112–114, 125–128, 236, 243, 245, 317, 375, 386, 434–436*). It is particularly relevant that the rate is affected by added halide ion. It is apparent that a number of 1:1 species may be formed in aqueous (and nonaqueous) solutions. Freshly prepared solutions of the five-coordinate complex $[Ni(terpy)Cl_2]$ in water possess absorption spectra different from those of aged samples and 1:1 mixtures containing 2,2′:6′,2″-terpyridine and nickel(II). Similar observations are obtained with DMSO solutions, in which equilibria of the type:

$$[Ni(terpy)Cl_2(DMSO)] \rightleftharpoons [Ni(terpy)Cl(DMSO)_2]^+$$

$$[Ni(terpy)Cl(DMSO)_2]^+ \rightleftharpoons [Ni(terpy)(DMSO)_3]^{2+}$$

are postulated.

The crystal structure of $[Ni(terpy)_2][PF_6]_2$ has been described (*17*). Several authors have reported the spectroscopic and magnetic properties of $[Ni(terpy)_2]^{2+}$ species (*236, 243, 269, 278, 328*). The electrochemical reduction of these complexes has also been investigated (*6, 373*). To date, only two

2,2':6',2"-terpyridine complexes of nickel(0) have been described: [Ni(terpy)(PPh$_3$)$_2$] (452) and [Ni(terpy)$_2$] (39). The latter complex is prepared by reaction of nickelocene with terpy, and may well be a ligand-radical complex of nickel(I) or nickel(II).

7. Copper, Silver, and Gold

Although Morgan and Burstall had difficulties in isolating 1:2 complexes of terpy with copper(II), salts of [Cu(terpy)$_2$]$^{2+}$ are readily prepared with excess terpy (8, 18, 236, 243, 309). Crystal structural analyses of [Cu(terpy)$_2$]X$_2$ (X = NO$_3$ or PF$_6$) have been reported (8, 18, 309); the cation is Jahn–Teller distorted. As a result of the initial D_{2d} local symmetry, the two terpy ligands become nonequivalent upon Jahn–Teller distortion. ESR studies indicate the presence of a static Jahn–Teller distortion (8, 62, 236).

The 1:1 complexes [Cu(terpy)X$_2$] (X = F, Cl, Br, I, or NCS) have all been reported (209, 225, 227, 243). These complexes possess a distorted trigonal-bipyramidal geometry, as indicated by ESR studies (23, 213, 234) and structural determinations of [Cu(terpy)X$_2$] (X = Cl, Br, or I) (147, 225, 234, 393, 396). Gagné et al. reported the PES spectra of [Cu(terpy)Cl$_2$] (204). One of the few coordination compounds of copper(II) fluoride is [Cu(terpy)F$_2$]·2H$_2$O, which is assumed to be five-coordinate (209). The Jahn–Teller distortion of these compounds has stimulated numerous structural studies, and has also resulted in a certain ambiguity in describing the structures as trigonal-bipyramidal, tetragonal-pyramidal, or square-pyramidal. Treatment of solutions of [Cu(terpy)Cl$_2$], which is five-coordinate in the solid state, with hexafluorophosphate results in the formation of [Cu(terpy)Cl][PF$_6$]. Anomalies in the ESR and magnetic properties led to the formulation of the cation as a dimeric chloro-bridged species [(terpy)CuCl$_2$Cu(terpy)]$^{2+}$ (Fig. 21) (392). The related complexes [Cu(terpy)CN][NO$_3$]·H$_2$O and [Cu(terpy)CN][ClO$_4$] also exhibited anomalous magnetic properties (289). A structural analysis of the nitrate revealed a distorted five-coordinate geometry about the metal, with the fifth contact provided by the nitrogen atom of the cyanide of an adjacent cation (11).

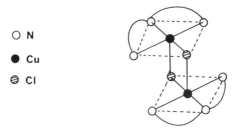

○ N

● Cu

◒ Cl

FIG. 21. The proposed structure of the cation [(terpy)CuCl$_2$Cu(terpy)]$^{2+}$ (392).

Similar magnetic anomalies are associated with another chain compound [Cu(terpy)L][ClO$_4$]$_2$ (L = pyrazine), incorporating bridging pyrazine groups; once again a five-coordinate structure is proposed (*150*). The five-coordinate square-pyramidal geometry is also observed in the structurally characterized complex [Cu(terpy)(ONO)(H$_2$O)][NO$_2$]·H$_2$O (*9*).

It seems likely that the complex cation [Cu(terpy)(bipy)]$^{2+}$ is also five-coordinate (*15, 227*). Structural studies of [Cu(terpy)-(O$_2$CCH$_2$ECH$_2$CO$_2$)]·nH$_2$O (E = NH, O, S, or Se) have been reported, and the two compounds exhibit interesting differences. The oxy diacetate complex consists of mononuclear octahedral metal centers, linked by water molecules (*73*); in contrast, the thio diacetate thioether sulfur is a weaker donor to copper(II) than an ether oxygen, and is noncoordinated. The metal is in a distorted trigonal-bipyramidal arrangement with bridging thio diacetate groups (Fig. 22) (*60*). Numerous mixed-ligand complexes with other polydentate species have been described (*13, 14, 58–60, 72, 73, 102, 215, 246, 288, 343, 492*).

The formation of copper(II) complexes with terpy has been investigated fairly intensively. The interaction is pH dependent, and numerous hydroxy, aqua, and polynuclear species are present in aqueous solution (*94, 245, 278*). In general, an Eigen–Wilkins mechanism appears to be operative, although the kinetics are complicated by ligand-protonation equilibria (*263, 390, 391*). In acidic solution, 1:1 complexes predominate (*361*). A number of substituted terpyridine ligands have been evaluated as potential colorimetric reagents for copper (*400*). The adsorption behavior of copper(II)-terpy complexes at silica surfaces has been studied (*499*). Such complexes are reasonably active as catalysts for the hydrolysis of fluorophosphate esters (*456*).

2,2′:6′,2″-Terpyridine is reported to form a 1:1 complex with copper(I) in aqueous solution (*362*), and [Cu(terpy)Cl] has been isolated (*336*). These complexes have been found to be surprisingly effective oxygen-transfer

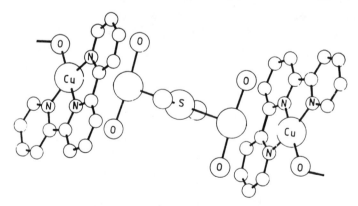

FIG. 22. The crystal and molecular structure of [Cu(terpy)(O$_2$CCH$_2$SCH$_2$CO$_2$)] (*60*).

catalysts for the oxidation of thiosulfate to sulfate (108, 109). The complex [Cu(terpy)Cl] only displays weak oxidase activity for the oxidation of ethanol to acetaldehyde (336). In aqueous acetonitrile, $[Cu(terpy)_2]^+$ undergoes a two-electron oxidation by dioxygen to yield $[Cu(terpy)_2]^{2+}$ and hydrogen peroxide (154). It was suggested that both terpy ligands are bidentate in $[Cu(terpy)_2]^+$.

Morgan and Burstall described the silver(I) complexes [Ag(terpy)X] (X = NO_3 or ClO_4), and the silver(II) species Ag(terpy)X_2 (X = NO_3, ClO_4, $\frac{1}{2}S_2O_6$, or $\frac{1}{2}S_2O_8$) of unknown structure (328). The complexes are unstable in aqueous solution and give the corresponding silver(I) complexes (226). More recently, $[Ag(terpy)_2][S_2O_8]$ has been prepared; PES spectra of this and other octahedral $[Ag(terpy)L]^{2+}$ species have been reported (340–342). Salts of $[Ag(terpy)_2]^{2+}$ are obtained by electrochemical oxidation of silver(I) in the presence of terpy (453, 501).

The only gold complexes of terpy to have been reported are $[Au(terpy)Cl]Cl_2 \cdot 3H_2O$ and the mixed-valence compound $[Au(terpy)Cl]_2[AuCl_2]_3[AuCl_4]$, both of which have been structurally characterized (244). The complex $[Au(terpy)Cl]Cl_2 \cdot 3H_2O$ possesses the expected distorted square-planar cation, whereas the mixed-valence compound exhibits a chainlike structure with three spiral $[AuCl_2]$ units bridging the two [Au(terpy)Cl] centers, with tetrachloroaurate(III) chlorine atoms occupying the remaining axial sites of the now octahedral gold(III) sites (Fig. 23) (244).

8. Zinc, Cadmium, and Mercury

Zinc or cadmium halides react readily with terpy to form a series of compounds $[M(terpy)X_2]$. The complex $[Zn(terpy)Cl_2]$ has been most intensively investigated, and exists in two crystal modifications (147). Similar modifications are observed for the majority of $[M(terpy)X_2]$ complexes. It was initially proposed that the compounds were four-coordinate and of the type [M(terpy)Cl]Cl (328), although it later became apparent that this formulation was not entirely satisfactory. The complexes $[M(terpy)X_2]$ (M = Zn or Cd; X = Cl, Br, or I) behave as nonelectrolytes in nitrobenzene(171). A crystal structural analysis revealed $[Zn(terpy)Cl_2]$ (form II) to be a distorted trigonal-bipyramidal five-coordinate species (147). A later refinement of this structure has been reported; there is considerable variation in the Zn—N distances, with appreciable shortening of the bond to the central pyridine ring (182). The Zn—Cl distances are similar, and there is some distortion of the ligand, resulting in a nonplanar configuration. The crystal structure of form I has also been described; this is also a distorted trigonal-bipyramidal species (454). Gerloch suggested that the structure is better described in terms of a distorted square-based pyramid, with a basal

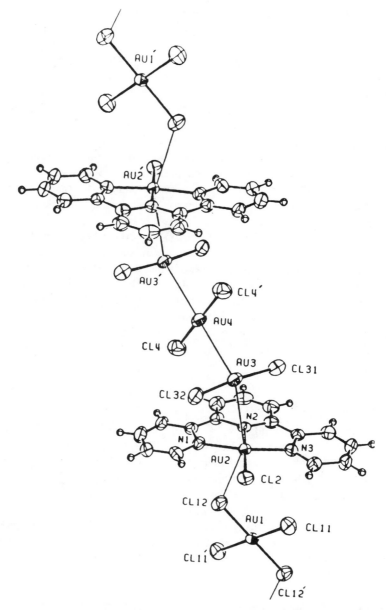

FIG. 23. The crystal and molecular structure of $[Au(terpy)Cl]_2[AuCl_2]_3[AuCl_4]$ (*244*).

N_3Cl plane (207). The other halids and the cadmium complexes are isomorphous with the appropriate forms, and undoubtedly possess related five-coordinate structures. A number of studies of the vibrational spectra of the $[M(terpy)X_2]$ species have been described, all of which support the formulation as isomorphous five-coordinate complexes (124, 171, 225, 371). The copper(II) complexes $[Cu(terpy)Cl_2]$ are isostructural, and a number of studies of the paramagnetic species doped into a host matrix of $[Zn(terpy)Cl_2]$ have been reported (23, 213). The zinc complex $[Zn(terpy)Cl_2]$ exhibits absorption maxima at 22,650 and 18,000 cm^{-1} (23, 348). The mercury(II) halide adducts are not so well characterized, but may be prepared by the direct reaction or HgX_2 with terpy (171) or by trans-metallation of $[Ph_2Sn(terpy)Cl_2]$ with $HgCl_2$ (471). They are thought to possess similar, five-coordinate structures. The structures of the 1:1 adducts of the nitrates $M(NO_3)_2$·terpy are not known with any certainty (171, 328). A ^{113}Cd NMR study of $Cd(NO_3)_2$·terpy has been reported (430).

A related trigonal-bipyramidal structure is observed in the heterotrinuclear complexes $[(terpy)Cd(Mn(CO)_5)_2]$, which are readily prepared by the reaction of $[Cd(Mn(CO)_5)_2]$ with 2,2':6',2"-terpyridine (122, 123, 248). The five-coordinate trigonal-bipyramidal structure is encountered in $[Hg(terpy)(CF_3)_2]$, which has been structurally characterized (274).

A number of alkylmercury(II) and arylmercury(II) complexes of terpy have been described, and appear to have a pronounced tendency for the ligand to adopt a bidentate, rather than terdentate, mode. The reaction of $[MeHgOH]$ with terpy has been studied by T-jump techniques; the complex shows a very high reaction rate with hydroxide, and it is proposed that a bidentate intermediate is involved (206). The complexes $[MeHg(terpy)][NO_3]$ and $[MeHg(4,4',4"-Et_3terpy)][NO_3]$ are also thought to possess bidentate terpy ligands in solution; specifically, the values of $^2J(Hg-H)$ are compatible only with a bidentate N_2 donor (97, 98). In the solid state, the 4,4',4"-Et$_3$terpy complex, and presumably that with terpy also, is terdentate; a crystal structural analysis has established unequivocally that the metal is in an approximately planar four-coordinate environment, with a weak interaction between the metal and the nitrate group (97, 98). The structure of $[(C_6F_5)_2Hg(terpy)]$ is not known, but it is proposed that the ligand acts in a bidentate or bridging bidentate mode (96). It is of interest that the complex $[ZnL_2(terpy)]$ [HL = 4,4',4"-trifluoro-3-oxo-1-(2-thienyl)butane-1-thione] is six-coordinate, with the 2,2':6',2"-terpyridine acting as a terdentate, and one of the other ligands adopting a monodentate S-bonded mode (241).

The complex $[(terpy)CdFe(CO)_4]$, prepared by the reaction of terpy with $[\{CdFe(CO)_4\}_4]$, is probably monomeric, although the structure is not known with any certainty (188).

Formation of the 1:1 complexes with zinc and cadmium has been studied in aqueous solution, and the complexes are more stable than the corresponding 1:1 adducts with bipy (245). Ternary mercury(II)–hydroxo–terpy complexes are active catalysts for hydrolysis of esters and amides (477).

The 1:2 complexes $[M(terpy)_2]^{2+}$ are known, and are presumably octahedral (243).

The reaction of terpy with zinc amalgam in 2-methyltetrahydrofuran results in the formation of the ligand radical complex $[Zn(terpy)_2]$, which has been characterized by ESR spectroscopy (79).

H. LANTHANIDES AND ACTINIDES

The lanthanides form a range of complexes with terpy, with stoichiometries of 1:1, 1:2, and 1:3, although the 1:1 compounds are the most widely investigated. A number of workers have reported difficulties in the preparation of the 1:2 complexes, particularly with the lighter lanthanides, and it seems to be essential that only weakly coordinating counterions are present.

The only complexes of lanthanum or cerium to be described are $[La(terpy)_3][ClO_4]_3$ (175) and $Ce(terpy)Cl_3 \cdot H_2O$ (411). The lanthanum compound is a 1:3 electrolyte in MeCN or MeNO$_2$, and is almost certainly a nine-coordinate mononuclear species; the structure of the cerium compound is not known with any certainty. A number of workers have reported hydrated 1:1 complexes of terpy with praseodymium chloride (376, 411, 438), and the complex $PrCl_3(terpy) \cdot 8H_2O$ has been structurally characterized (376). The metal is in nine-coordinate monocapped square-antiprismatic $[Pr(terpy)Cl(H_2O)_5]^{2+}$ cations (Fig. 24). Complexes with a 1:1 stoichiometry have also been described for neodymium (33, 409, 411, 413, 417), samarium (33, 411, 412), europium (33, 316, 411, 414, 417), gadolinium (33, 411), terbium (316, 410, 414), dysprosium (33, 410, 412), holmium (33, 410), erbium (33, 410, 417), thulium (410, 412), and ytterbium (410). The 1:2 stoichiometry has only been observed with the later lanthanides, europium (33, 411, 414), gadolinium, dysprosium, and erbium (33).

Hart and co-workers have demonstrated that the nine-coordinate cations $[M(terpy)_3]^{3+}$ may be prepared in the absence of coordinating counterions in the cases of europium, samarium, lanthanum, and lutetium (175, 201, 202). The most widely investigated compound in this series is $[Eu(terpy)_3][ClO_4]_3$, which has been structurally characterized. The metal is in a nine-coordinate tricapped trigonal-prismatic arrangement (Fig. 25) (201). The distortion from D_3 symmetry to C_2 is explained by the nonplanarity of the terpy ligands, and is predicted from spectroscopic observations. It is not clear how the above observations may be correlated with a report that

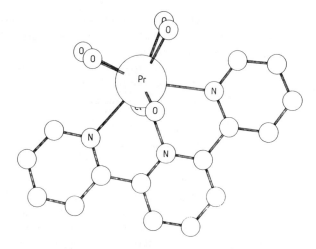

FIG. 24. The crystal and molecular structure of [Pr(terpy)Cl(H₂O)₅]Cl₂·3H₂O (376).

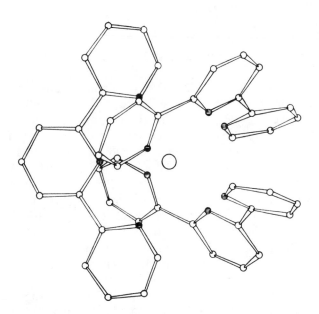

FIG. 25. The crystal and molecular structure of [Eu(terpy)₃][ClO₄]₃ (201).

acetonitrile solutions of $[Eu(terpy)_3]^{3+}$ exhibit on–off equilibria ·of the pyridine rings, giving a predominantly eight-coordinate solution structure (*110*). An interesting ligand-transfer reaction occurs when $[Eu(terpy)_3]^{3+}$ reacts with Tb^{3+}; the products are $Eu(terpy)_2^{3+}$ and $Tb(terpy)^{3+}$, formed after rate-determining loss of terpy from the tris complex (*202*).

All of the complexes are intensely fluorescent, and have been subjects of numerous spectroscopic investigations (*33, 175, 202, 273, 316, 409–414, 427, 438, 500*). 2,2′:6′,2″-Terpyridine has been investigated as a colorimetric reagent for the determination of cerium(IV), praseodymium(III), and gadolinium(III) (*19*).

A single example of a europium(II) complex has been described by Hart and Zhu, who obtained $[Eu(terpy)Cl_2]$ from the reaction of $EuCl_2$ with terpy in MeCN (*229, 230*). It seems unlikely that the compound is monomeric.

A number of adducts of terpy with thorium(IV) nitrate or chloride (*472, 476*) or uranium (*434, 465*) have been reported. No structural details are available for these poorly characterized species.

IV. Coordination Compounds of Higher Oligopyridines

A. 2,2′:6′,2″:6″,2‴-QUATERPYRIDINE

The majority of complexes of the higher oligopyridines have been reported for 2,2′:6′,2″:6″,2‴-quaterpyridine (Fig. 2). In their pioneering studies of the coordination chemistry of this group of ligands, Morgan and Burstall investigated the complexes of 2,2′:6′,2″:6″,2‴-quaterpyridine (*327*). In these studies, they reported the complexes with silver(I), iron(II), cobalt(II), cobalt(III), nickel(II), copper(II), zinc(II), cadmium(II), iridium(III), and platinum(II). Although these authors speculated upon the structure of these complexes, they presented no unambiguous evidence. In a later paper, they reported the compounds $[Ru(NO)Cl(quaterpy)]Cl_2 \cdot 5H_2O$ and $[Ru(NO)Cl(quaterpy)][RuCl_5(NO)]$ (*325*). Brandt (*69*) questioned the ability of quaterpy to act as a planar quaterdentate, on the basis of a study of molecular models. This is indeed true, and it is necessary to introduce considerable strain in the ligand *or* to have a very distorted coordination. After these initial reports, there was very little interest in the ligand until 1964, when the interaction of quaterpy with iron(II) and iron(III) was reinvestigated (*54*). It was shown that 1:1 and 1:2 complexes were formed with iron(II) and a 1:1 complex with iron(III). Presumably, the 1:2 complex contains a quaterpy acting in a terdentate or bidentate mode. Lip and Plowman (*295*) reinvestigated the properties of $[Pt(quaterpy)]X_2$ (X = I,

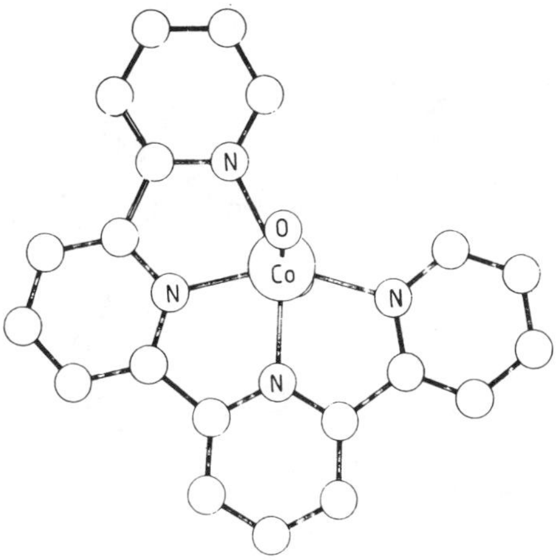

FIG. 26. The crystal and molecular structure of [Co(quaterpy)(H₂O)₂][NO₃]₂ (237).

ClO_4, BPh_4, or NO_3), and suggested, on the basis of conductivity measurements, that quaterpy was acting as a tetradentate ligand. The first unequivocal evidence that quaterpy could act as a tetradentate came from the structural analysis of [Co(quaterpy)(OH₂)(SO₃)][NO₃]·H₂O, which revealed the quaterpy to be acting as a planar tetradentate ligand (308). The bonds to the terminal pyridine rings were longer than those to the central ones. Crystal structural analyses have also been reported for the complexes [Cu(quaterpy)(H₂O)(NO₃)][NO₃]·H₂O and [Co(quaterpy)(H₂O)₂][NO₃]₂ (Fig. 26) (237). Once again, the quaterpy was shown to act as an asymmetric planar tetradentate.

There have been numerous reports concerning the properties of iron complexes of quaterpy by an Italian group. Treatment of iron(II) sulfate with quaterpy in air led to the formation of the oxy-bridged complex [(quaterpy)FeOFe(quaterpy)][SO₄]₂·7H₂O (66, 104). The complex ions [Fe(quaterpy)X₂]$^{n+}$ have been shown to exhibit specific interactions with a wide range of biological substrates. Poly(L-glutamate) forms a helical structure upon reaction with [Fe(quaterpy)(OH)₂]$^+$ (67, 68). Once the [Fe(quaterpy)] unit has been bound to the poly(L-glutamate) it is found to be an active catalyst for the hydrogen peroxide oxidation of L-adrenaline (367, 368, 370) or ascorbate (27–29, 367). The polymer-bound complex is also active as a catalyst for the decomposition of hydrogen peroxide (25, 26, 30, 31, 369).

The Strasbourg group has investigated the coordination chemistry of 5,5',3",5'''-Me₄-quaterpy, and reported a crystal structural analysis of the complex [Cu₂L₂][ClO₄]₂·H₂O (Fig. 27) (290). The methyl groups prevent the ligand from acting as a planar tetradentate, and a twist about the central

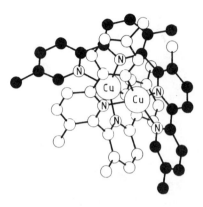

FIG. 27. The crystal and molecular structure of $[Cu_2(5,5',3'',5'''-Me_4quaterpy)_2]-$
$[ClO_4]_2 \cdot H_2O$ (290).

C—C bond enables it to act as a bis bidentate. Each copper(I) is bound to
two bidentate quaterpyridine moieties.

B. HIGHER OLIGOPYRIDINES

There have been very few reports concerning the coordination chemistry of
the higher oligopyridines. Oepen amd Vogtle reported the formation of a 1:1
adduct of quinquepy with $Li[ClO_4]$ (351). This observation indicates that
there may well be a hitherto to unsuspected coordination chemistry of alkali
metal cations with pyridine derivatives. It is worthy of note that cyclosexipyri-
dines and other planar sexidentate nitrogen-donor ligands form very stable
alkali metal complexes. Complexes of $4',4'''-Ph_2$-quinquepy with cobalt(II),
nickel(II), and cadmium(II) have been reported (135). The structures of these
compounds are not known in any detail, although it is tempting to propose a
planar quinquedentate donor set. A combination of electrochemical and ESR
studies has indicated that metal-centered nickel(I) or cobalt(I) compounds
are obtained upon one-electron reduction. A brown mixed oxidation state
complex of the type $[Cu_2(quinquepy)(O_2)][PF_6]_2$ has also been described
(132).

V. Coordination Compounds of Macrocyclic Derivatives

The archetypal macrocyclic oligopyridine is cyclosexipyridine (Fig. 28).
The synthesis of this molecule was achieved by Newkome and co-workers
(350). The elegant synthesis reported appears to be low-yield, and, as yet, no
coordination chemistry has been reported. However, $4,4'''$-diaryl derivatives

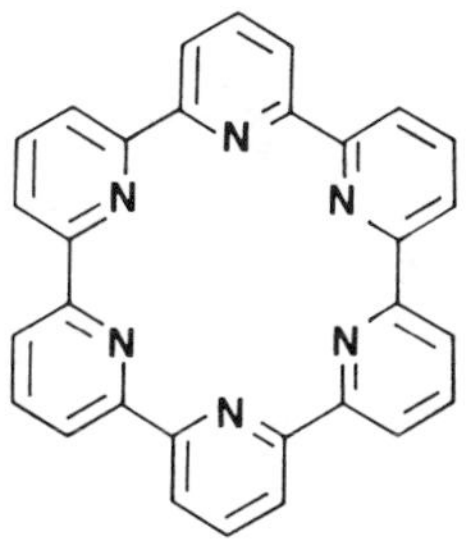

FIG. 28. Cyclosexipyridine.

have been prepared in a moderately high-yield (42 %) two-step reaction (*449*), and the development of the coordination chemistry of cyclosexipyridine is due. It is interesting that the 4,4'''-diaryl derivatives were isolated as sodium complexes; no sodium was present in the reaction mixtures, and one is led to speculate upon the possibility of the ligands leaching alkali metal ions from the glassware.

The Cambridge group has recently developed the chemistry of macrocyclic ligands based upon 2,2':6',2"-terpyridine (*133, 135, 136, 139–143*). The macrocycles (Figs. 29 and 30) are prepared by template condensations of

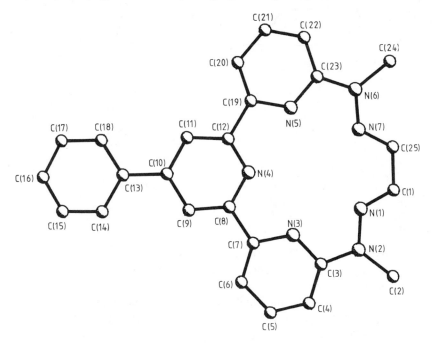

FIG. 29. A pentadentate macrocycle based upon 2,2':6',2"-terpyridine.

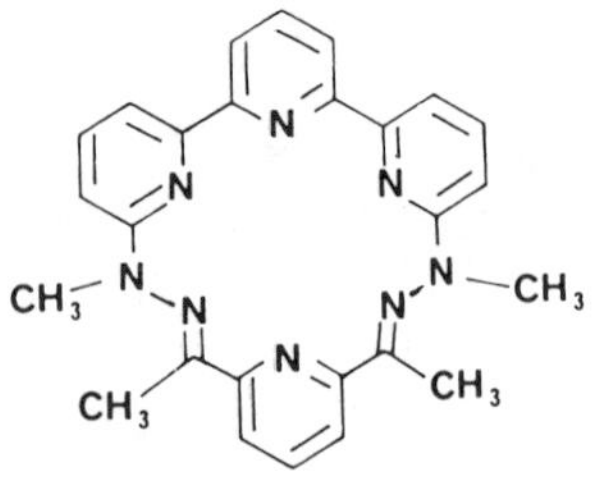

FIG. 30. A sexidentate macrocycle based upon 2,2′:6′,2″-terpyridine.

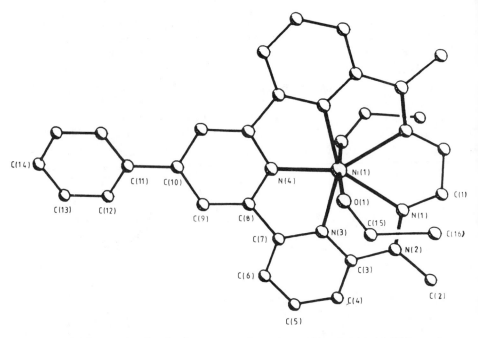

FIG. 31. The crystal and molecular structure of a pentagonal-bipyramidal nickel(II) complex of the ligand shown in Fig. 29.

6,6″-bis(hydrazino)-2,2′:6′,2″-terpyridines with appropriate dicarbonyls. In general, pentagonal-bipyramidal complexes are formed (Fig. 31), in which the axial sites are occupied by solvent molecules. The electrochemistry of these compounds is of interest, and metal-centered reductions are observed. Some novel template effects, by which free ligands are obtained, have been observed in these systems.

ACKNOWLEDGMENTS

I should like to thank Dr. Sharon Bellard of the Cambridge Crystallographic Data Base[1] and Dr. Paul Raithby for their invaluable assistance in obtaining structural data, and Dr. Ken Seddon for kindling and nurturing an interest in these ligands.

REFERENCES

1. Abe, Y., and Wada, G., *Bull. Chem. Soc. Jpn.* **55**, 3645 (1982).
2. Abe, Y., and Wada, G., *Bull. Chem. Soc. Jpn.* **54**, 3334 (1981).
3. Adcock, P. A., and Keene, R. F., *J. Am. Chem. Soc.* **103**, 6494 (1981).
4. Addison, C. C., Davis, R., and Logan, N., *J. Chem. Soc.*, p. 2071 (1974).
5. Agnew, S. F., Stone, M. I., and Crosby, G. A., *Chem. Phys. Lett.* **85**, 57 (1982).
6. Aihara, M., Kishita, H., and Misumi, S., *Bull. Chem. Soc. Jpn.* **48**, 680 (1975).
7. Allen, G. H., Sullivan, B. P., and Meyer, T. J., *J. Chem. Soc., Chem. Commun.*, p. 793 (1981).
8. Allmann, R., Henke, W., and Reinen, D., *Inorg. Chem.* **17**, 378 (1978).
9. Allmann, R., Kremer, S., and Kucharzcyk, D., *Inorg. Chim. Acta* **85**, L19 (1984).
10. Allsopp, S. R., Cox, A., Kemp, T. J., Reed, W. J., Sostero, S., and Traverso, O., *J. Chem. Soc., Faraday Trans. 1* **76**, 162 (1980).
11. Anderson, O. P., Packard, A. B., and Wicholas, M., *Inorg. Chem.* **15**, 1613 (1976).
12. Anson, F. C., Ohsaka, T., and Saveant, J.-M., *J. Am. Chem. Soc.* **105**, 4883 (1983).
13. Arena, G., Bonomo, R. P., Musumeci, S., Rizzarelli, E., and Sammartano, S., *Transition Met. Chem. (Weinheim. Ger.)* **7**, 29 (1982).
14. Arena, G., Bonomo, R. P., Rizzarelli, E., and Seminara, A., *Inorg. Chim. Acta* **30**, 13 (1978).
15. Arena, G., Bonomo, R. P., Musumeci, S., Purrello, R., Rizzarelli, E., and Sammartano, S., *J. Chem. Soc., Dalton Trans.*, p. 1279 (1983).
16. Arif, A. M., Hart, F. A., Hursthouse, M. B., Thornton-Pett, M., and Zhu, W., *J. Chem. Soc., Dalton Trans.*, p. 2449 (1984).
17. Arriortua, M. I., Rojo, T., Amigo, J. M., Germain, G., and Declercq, J. P., *Bull. Soc. Chim. Belg.* **91**, 337 (1982).
18. Arriortua, M. I., Rojo, T., Amigo, J. M., Germain, G., and Declercq, J. P., *Acta Crystallogr., Sect. B* **B38**, 1323 (1982).
19. Awadallah, R. M., Sherif, M. K., and El Desouky, T. I., *J. Indian Chem. Soc.* **61**, 403 (1984).
20. Baggio-Saitovitch, E., and De Paoli, M. A., *Inorg. Chim. Acta* **27**, 15 (1978).
21. Bakac, A., Marcec, R., and Orhanovic, M., *Inorg. Chem.* **15**, 3133 (1977).
22. Baker, A. T., and Goodwin, H. A., *Aust. J. Chem.* **38**, 207 (1985).
23. Banerjee, K., and Bose, S. K., *J. Indian Chem. Soc.* **54**, 751 (1977).
24. Barnes, G. T., Backhouse, J. R., Dwyer, F. P., and Gyarfas, E. C., *Proc. R. Soc. N.S.W.*, p. 151 (1956).
25. Barteri, M., Farinella, M., and Pispisa, B., *Biopolymers* **16**, 2569 (1977).
26. Barteri, M., Farinella, M., and Pispisa, B., *J. Inorg. Nucl. Chem.* **40**, 1277 (1978).
27. Barteri, M., and Pispisa, B., *J. Chem. Soc., Faraday Trans. 1* **78**, 2073 (1982).

[1] Allen, F. H., Kennard, O., and Taylor, R., *Acc. Chem. Res.* **16**, 146 (1983).
Allen, F. H., Bellard, S., Brice, M. D., Cartwright, B. A., Doubleday, A., Higgs, H., Hummelink, T., Hummelink-Peters, B. G., Kennard, O., Motherwell, W. D. S., Rodgers, J. R., and Watson, D. G., *Acta Crystallogr., Sect. B* **B35**, 2331 (1979).

28. Barteri, M., and Pispisa, B., *J. Chem. Soc., Faraday Trans. 1* **78**, 2085 (1982).
29. Barteri, M., Pispisa, B., and Primiceri, M. V., *Biopolymers* **18**, 3115 (1979).
30. Barteri, M., Farinella, M., Pispisa, B., and Splendorini, L., *J. Chem. Soc., Faraday Trans. 1* **73**, 288 (1977).
31. Barteri, M., Farinella, M., Pispisa, B., and Splendorini, L., *Inorg. Chem.* **17**, 3366 (1978).
32. Barton, J. K., and Lippard, S. J., *Biochemistry* **18**, 2661 (1979).
33. Basille, L. J., Gronert, D. L., and Ferraro, J. R., *Spectrochim. Acta Part A*, p. 24A (1968).
34. Basolo, F., Gray, H. B., and Pearson, R. G., *J. Am. Chem. Soc.* **82**, 4201 (1960).
35. Baxendale, J. H., *in* "The Kinetics and Mechanism of Inorganic Reactions in Solution" (K. W. Sykes, ed.) The Chemical Society, London, 1954.
36. Baxendale, J. H., and Fiti, M., *J. Chem. Soc., Dalton Trans.*, p. 1995 (1972).
37. Beattie, I. R., and Leigh, G. J., *J. Inorg. Nucl. Chem.* **23**, 55 (1961).
38. Begolli, B., Valjak, V., Allegretti, V., and Katovic, V., *J. Inorg. Nucl. Chem.* **43**, 2785 (1981).
39. Behrens, H., and Meyer, K., *Z. Naturforsch., Teil B*, **21B**, 489 (1966).
40. Behrens, H., Brandl, H., and Lutz, K., *Naturforsch. B. Anorg. Chem. Org. Chem.* **22B**, 99 (1967).
41. Behrens, H., Feilner, H.-D., and Lindner, E., *Z. Anorg. Allg. Chem.* **385**, 321 (1971).
42. Behrens, H., and Müller, A., *Z. Anorg. Allg. Chem.* **341**, 124 (1965).
43. Behrens, H., and Brandl, H., *Z. Naturforsch. B: Anorg. Chem. Org. Chem.* **22B**, 121 (1967).
44. Behrens, H., and Anders, U., *Z. Naturforsch. B: Anorg. Chem. Org. Chem.* **19B**, 767 (1964).
45. Behrens, H., and Aquila, W., *Z. Anorg. Allg. Chem.* **356**, 8 (1967).
46. Behrens, H., and Aquila, W., *Z. Naturforsch. B: Anorg. Chem. Org. Chem.* **22B**, 454 (1967).
47. Behrens, H., Meyer, K., and Müller, A., *Z. Naturforsch. B: Anorg. Chem. Org. Chem.* **20B**, 74 (1965).
48. Bellavance, P. L., Corey, E. R., Corey, J. Y., and Hey, G. W., *Inorg. Chem.* **16**, 462 (1977).
49. Bennett, L. E., and Taube, H., *Inorg. Chem.* **7**, 254 (1968).
50. Benton, D. J., and Moore, P., *J. Chem. Soc., Chem. Commun.*, p. 717 (1972).
51. Benton, D. J., and Moore, P., *J. Chem. Soc., Dalton Trans.*, p. 399 (1973).
52. Beran, G., Carty, A. J., Patel, H. A., and Palenik, G. J., *J. Chem. Soc., Chem. Commun.*, p. 222 (1970).
53. Beran, G., Dymock, K., Patel, H. A., Carty, A. J., and Boorman, P. M., *Inorg. Chem.* **11**, 896 (1972).
54. Bergh, A., Offenhartz, P. O'D., George, P., and Haight, G. P., Jr., *J. Chem. Soc.*, p. 1533 (1964).
55. Bhayat, I. J., and McWhinnie, W. R., *Spectrochim. Acta, Part A* **28A**, 743 (1982).
56. Black, D. St. C., Deacon, G. B., and Thomas, N. C., *Inorg. Chim. Acta* **65**, L75 (1982).
57. Bond, P. J., Langridge, R., Jennette, K. W., and Lippard, S. J., *Proc. Natl. Acad. Sci.* **72**, 4825 (1975).
58. Bonomo, R. P., Rizzarelli, E., Sammartano, S., and Riggi, F., *Inorg. Chim. Acta* **43**, 11 (1980).
59. Bonomo, R. P., Bruno, G., Rizzarelli, E., Seminara, A., and Siracusa, G., *Thermochim. Acta* **18**, 207 (1977).
60. Bonomo, R. P., Rizzarelli, E., and Bresciani-Pahor, N., *J. Chem. Soc., Dalton Trans.*, p. 681 (1982).
61. Bonomo, R., Musumeci, S., Rizzarelli, E., and Sammartano, S., *Gazz. Chim. Ital.* **104**, 1067 (1974).
62. Bonomo, R. P., and Riggi, F., *Transition Met. Chem.* (*Weinheim. Ger.*) **9**, 308 (1984).
63. Braddock, J. N., and Meyer, T. J., *J. Am. Chem. Soc.* **95**, 3158 (1973).
64. Braddock, J. N., Cramer, J. L., and Meyer, T. J., *J. Am. Chem. Soc.* **97**, 1972 (1975).
65. Bradic, Z., and Wilkins, R. G., *J. Am. Chem. Soc.* **106**, 2236 (1984).

66. Branca, M., Pispisa, B., and Aurisicchio, C., *J. Chem. Soc., Dalton Trans.*, p. 1543 (1976).
67. Branca, M., Marini, M. E., and Pispisa, B., *Biopolymers* **15**, 2219 (1976).
68. Branca, M., and Pispisa, B., *J. Chem. Soc., Faraday Trans. 1* **73**, 213 (1979).
69. Brandt, W. W., Dwyer, F. P., and Gyarfas, E. C., *Chem. Rev.* **54**, 959 (1954).
70. Brandt, W. W., and Wright, J. P., *J. Am. Chem. Soc.* **76**, 3082 (1954).
71. Braterman, P. S., Heath, G. A., MacKenzie, A. J., Noble, B. C., Peacock, R. D., and Yellowlees, L. J., *Inorg. Chem.* **23**, 3425 (1984).
72. Bresciani-Pahor, A., Nardin, G., Bonomo, R. P., and Rizzarelli, E., *J. Chem. Soc., Dalton Trans.*, p. 2625 (1984).
73. Bresciani-Pahor, N., Nardin, G., Bonomo, R. P., and Rizzarelli, E., *J. Chem. Soc., Dalton Trans.*, p. 1797 (1983).
74. Brodie, A. M., and Wilkins, C. J., *Inorg. Chim. Acta* **8**, 13 (1974).
75. Brodie, A. M., Personal communication.
76. Broomhead, J. A., and Dwyer, F. P., *Aust. J. Chem.* **14**, 250 (1961).
77. Broomhead, J. A., Evans, J., Grumley, W. D., and Sterns, M., *J. Chem. Soc., Dalton Trans.*, p. 173 (1977).
78. Brown, A. J., Howarth, O. W., Moore, P., and Parr, W. J. E., *J. Chem. Soc., Dalton Trans.*, p. 1776 (1978).
79. Brown, I. M., Weissman, S. I., and Snyder, L. C., *J. Chem. Phys.* **42**, 1105 (1965).
80. Brunschwig, B., and Sutin, N., *J. Am. Chem. Soc.* **100**, 7568 (1978).
81. Buckingham, D. A., Dwyer, F. P., and Sargeson, A. M., *Aust. J. Chem.* **17**, 622 (1964).
82. Buckingham, D. A., Dwyer, F. P., and Sargeson, A. M., *Inorg. Chem.* **5**, 1243 (1966).
83. Buck, D. M. W., and Moore, P., *J. Chem. Soc., Chem. Commun.*, p. 61 (1974).
84. Buck, D. M. W., and Moore, P., *J. Chem. Soc., Dalton Trans.*, p. 638 (1976).
85. Bullock, J. I., and Simpson, P. W. G., *J. Chem. Soc., Faraday Trans. 1* **77**, 1991 (1981).
86. Burgess, J., and Prince, R. H., *J. Chem. Soc.*, p. 6061 (1965).
87. Burgess, J., and Twigg, M. V., *J. Chem. Soc., Dalton Trans.*, p. 2032 (1974).
88. Burgess, J., and Prince, R. H., *J. Chem. Soc. (A)*, p. 1772 (1966).
89. Burgess, J., Chambers, J. G., and Haines, R. I., *Transition Met. Chem. (Weinheim. Ger.)* **6**, 145 (1981).
90. Burmeister, J. L., and Grayling, H. J., *Inorg. Chim. Acta* **1**, 100 (1967).
91. Burmeister, J. L., and Basolo, F., *Inorg. Chem.* **3**, 1587 (1964).
92. Burmeister, J. L., Hassel, R. I., Johnson, K. A., and Lim, J. C., *Inorg. Chim. Acta* **9**, 23 (1974).
93. Burstall, F. H., *J. Chem. Soc.*, p. 1662 (1937).
94. Cali, R., Rizzarelli, E., Sammartano, S., and Siracusa, G., *Transition Met. Chem. (Weinheim. Ger.)* **4**, 328 (1979).
95. Calvert, J. M., Schmehl, R. H., Sullivan, B. P., Facci, J. S., Meyer, T. J., and Murray, R. W., *Inorg. Chem.* **22**, 2151 (1983).
96. Carty, A. J., and Deacon, G. B., *Aust. J. Chem.* **24**, 489 (1971).
97. Carty, A. J., Chaichit, N., Gatehouse, B. M., George, E. E., and Hayhurst, G., *Inorg. Chem.* **20**, 2414 (1981).
98. Carty, A. J., Hayhurst, G., Chaichit, N., and Gatehouse, B. M., *J. Chem. Soc., Chem. Commun.*, p. 316 (1980).
99. Carty, A. J., and Tuck, D. G., *J. Chem. Soc.*, p. 6013 (1964).
100. Casey, A. T., and Clark, R. J. H., *Transition Met. Chem. (Weinheim. Ger.)* **2**, 76 (1977).
101. Cassatt, J. C., Johnson, W. A., Smith, L. M., and Wilkins, R. G., *J. Am. Chem. Soc.* **94**, 8399 (1972).
102. Cassol, A., Maggiore, R., Musumeci, S., Rizzarelli, E., and Sammartano, S., *Transition Met. Chem. (Weinheim. Ger.)* **1**, 252 (1976).
103. Cayley, G. R., and Margerum, D. W., *J. Chem. Soc., Chem. Commun.*, p. 1002 (1974).

104. Cerdonio, M., Mogno, F., Pispisa, B., Romani, G. L., and Vitale, S., *Inorg. Chem.* **16**, 400 (1977).

105. Cerny, R. L., Sullivan, B. P., Bursey, M. M., and Meyer, T. J., *Anal. Chem.* **55**, 1954 (1983).

106. Chaffee, E., Creaser, I. I., and Edwards, J. O., *J. Inorg. Nucl. Chem.* **7**, 1 (1971).

107. Chandra, B. P., and Majumdar, B., *Cryst. Res. Technol.* **18**, 237 (1983).

108. Chandra, M., O'Driscoll, K. F., and Rempel, G. L., *Prepr. Can. Symp. Catal.* **6**, 44 (1979).

109. Chandra, M., O'Driscoll, K. F., and Rempel, G. L., *J. Mol. Catal.* **8**, 339 (1980).

110. Chapman, R. D., Loda, R. T., Riehl, J. P., and Schwartz, R. W., *Inorg. Chem.* **23**, 1652 (1984).

111. Chaston, S. H. H., Livingstone, S. E., and Lockyer, T. N., *Aust. J. Chem.* **19**, 1401 (1966).

112. Chattopadhyay, P. K., and Coetzee, J. F., *Inorg. Chem.* **12**, 113 (1973).

113. Chattopadhyay, P. K., and Coetzee, J. F., *Inorg. Chem.* **15**, 400 (1976).

114. Chattopadhyay, P. K., and Kratochvil, B., *Can. J. Chem.* **55**, 3449 (1977).

115. Chen, Y.-W. D., Santhanam, K. S. V., and Bard, A. J., *J. Electrochem. Soc.* **128**, 1460 (1981).

116. Chen, Y.-W. D., Santhanam, K. S. V., and Bard, A. J., *J. Electrochem. Soc.* **129**, 61 (1982).

117. Choudhury, D., Jones, R. F., and Cole-Hamilton, D., *J. Chem. Soc., Dalton Trans.*, p. 1143 (1982).

118. Ciantelli, G., Legittimo, P., and Pantani, F., *Anal. Chim. Acta* **53**, 303 (1971).

119. Clark, H. C., and Manzer, L. E., *Inorg. Chem.* **11**, 2749 (1972).

120. Clark, H. C., and Pickard, A. L., *J. Organomet. Chem.* **13**, 61 (1968).

121. Clark, R. J. H., "The Chemistry of Titanium and Vanadium." Elsevier, Amsterdam, 1968.

122. Clegg, W., and Wheatley, P. J., *J. Chem. Soc., Chem. Commun.*, p. 760 (1972).

123. Clegg, W., and Wheatley, P. J., *J. Chem. Soc., Dalton Trans.*, p. 90 (1973).

124. Coates, G. E., and Ridley, D., *J. Chem. Soc.*, p. 166 (1964).

125. Cock, P. A., Cottrell, C. E., and Boyd, P. K., *Can. J. Chem.* **50**, 402 (1972).

126. Coetzee, J. F., and Gilles, D. M., *Inorg. Chem.* **15**, 405 (1976).

127. Coetzee, J. F., and Karakatsamis, C. G., *Inorg. Chem.* **15**, 3112 (1976).

128. Coetzee, J. F., and Umemoto, K., *Inorg. Chem.* **15**, 3109 (1976).

129. Constable, E. C., Liptrot, M. C., and Raithby, P. R., *J. Chem. Soc., Dalton Trans.*, in press.

130. Constable, E. C., *Polyhedron* **2**, 551 (1983).

131. Constable, E. C., *Polyhedron* **3**, 1037 (1984).

132. Constable, E. C., and Corr, S., *J. Chem. Soc., Dalton Trans.*, in press.

133. Constable, E. C., and Lewis, J., *Polyhedron* **1**, 303 (1982).

134. Constable, E. C., Lewis, J., and Schroder, M., *J. Chem. Soc., Dalton Trans.*, in press.

135. Constable, E. C., Lewis, J., and Schroder, M., *Polyhedron* **1**, 311 (1982).

136. Constable, E. C., and Corr, S., *Inorg. Chim. Acta*, in press.

137. Constable, E. C., Lewis, J., and Schroder, M., *Polyhedron* **1**, 311 (1982).

138. Constable, E. C., *J. Chem. Soc., Dalton Trans.*, p. 2687 (1985).

139. Constable, E. C., Holmes, J. M., and McQueen, R. C. S., *J. Chem. Soc., Dalton Trans.*, in press.

140. Constable, E. C., and Holmes, J. M., *J. Chem. Soc., Dalton Trans.*, in press.

141. Constable, E. C., Khan, F. K., Lewis, J., Liptrot, M. C., and Raithby, P. R., *J. Chem. Soc., Dalton Trans.*, p. 333 (1985).

142. Constable, E. C., Lewis, J., Liptrot, M. C., Raithby, P. R., and Schroder, M., *Polyhedron* **2**, 301 (1983).

143. Constable, E. C., Lewis, J., and Marquez, V. E., *J. Chem. Soc., Dalton Trans.*, in press.

144. Constable, E. C., Lewis, J., Liptrot, M. C., and Raithby, P. R., *J. Chem. Soc., Dalton Trans.*, p. 2177 (1984).

145. Coombe, V. T., Heath, G. A., MacKenzie, A. J., and Yellowlees, L. J., *Inorg. Chem.* **23**, 3423 (1984).

146. Cooper, L. H. N., *Proc. R. Soc. London, Ser. B* **118**, 419 (1935).

147. Corbridge, D. E. C., and Cox, E. G., *J. Chem. Soc.*, p. 594 (1956).
148. Corey, E. J., Cooper, N. J., Canning, W. M., Lipscomb, W. N., and Koetzle, T. F., *Inorg. Chem.* **21**, 192 (1982).
149. Cornwell, A. B., Harrison, P. G., and Richards, J. A., *J. Organomet. Chem.* **140**, 273 (1977).
150. Coronado, E., Drillon, M., and Beltran, D., *Inorg. Chim. Acta* **82**, 13 (1984).
151. Couch, D. A., Elmes, P. S., Fergusson, J. E., Greenfield, M. L., and Wilkins, C. J., *J. Chem. Soc. A*, p. 1813 (1967).
152. Cramer, J. L., and Meyer, T. J., *Inorg. Chem.* **13**, 1250 (1974).
153. Creutz, C., Chou, M., Netzel, T. L., Okumura, M., and Sutin, N., *J. Am. Chem. Soc.* **102**, 1309 (1980).
154. Crumbliss, A. L., and Poulos, A. T., *Inorg. Chem.* **14**, 1529 (1975).
155. Cummins, D., and Gray, H. B., *J. Am. Chem. Soc.* **99**, 5158 (1977).
156. Cusumano, M., Guglelmo, G., and Ricevuto, V., *Inorg. Chim. Acta* **27**, 197 (1978).
157. Das, B., and Roy, P. C., *J. Inst. Chem. (India)* **56**, 145 (1984).
158. Das, M. K., Buckle, J., and Harrison, P. G., *Inorg. Chim. Acta* **6**, 17 (1972).
159. Davies, N. R., and Mullins, T. L., *Aust. J. Chem.* **21**, 915 (1968).
160. Davies, N. R., and Mullins, T. L., *Aust. J. Chem.* **20**, 657 (1967).
161. Deacon, G. B., and Parrott, J. C., *Aust. J. Chem.* **27**, 2547 (1974).
162. Deacon, G. B., Patrick, J. M., Skelton, B. W., Thomas, N. C., and White, A. H., *Aust. J. Chem.* **37**, 929 (1984).
163. Debye, N. W. G., Rosenberg, E., and Zuckerman, J. J., *J. Am. Chem. Soc.* **90**, 3234 (1968).
164. Deggau, E., Krohnke, F., Schnalke, K. E., Staudinger, H. J., and Weis, W., *Z. Klin. Chem.* **3**, 102 (1965).
165. Demas, J. N., and Crosby, G. A., *J. Am. Chem. Soc.* **93**, 2841 (1971).
166. Demas, J. N., and Crosby, G. A., *J. Mol. Spectrosc.* **26**, 72 (1968).
167. Dewan, J. C., Lippard, S. J., and Bauer, W. R., *J. Am. Chem. Soc.* **102**, 858 (1980).
168. Dietrich-Buchecker, C. O., Marnot, P. A., Sauvage, J.-P., Kintzinger, J. P., and Maltese, P., *Nouv. J. Chim.* **8**, 573 (1984).
169. Dose, E. V., Hoselton, M. A., Sutin, N., Tweedle, M. F., and Wilson, L. J., *J. Am. Chem. Soc.* **100**, 1141 (1978).
170. Doskey, M. A., and Curran, C., *Inorg. Chim. Acta* **3**, 169 (1969).
171. Douglas, J. E., and Wilkins, C. J., *Inorg. Chim. Acta* **3**, 635 (1979).
172. Dressick, W. J., Hauenstein, B. L., Jr., Gilbert, T. B., Demas, J. N., and DeGraff, B. A., *J. Phys. Chem.* **88**, 3337 (1984).
173. Dressick, W. J., Raney, K. W., Demas, J. N., and DeGraff, B. A., *Inorg. Chem.* **23**, 875 (1984).
174. DuBois, D. W., Iwamoto, R. T., and Kleinberg, J., *Inorg. Chem.* **9**, 968 (1970).
175. Durham, D. A., Frost, G. H., and Hart, F. A., *J. Inorg. Nucl. Chem.* **31**, 833 (1969).
176. Dwyer, F. P., Gyarfas, E. C., and O'Dwyer, M. F., *J. Proc. R. Soc. N.S.W.* **89**, 146 (1956).
177. Dwyer, F. P., and Gyarfas, E. C., *J. Am. Chem. Soc.* **76**, 6320 (1954).
178. Dwyer, F. P., and Mellor, D. P., "Chelating Agents and Metal Chelates." Academic Press, Orlando, 1964.
179. Dwyer, F. P., Goodwin, H. A., and Gyarfas, E. C., *Aust. J. Chem.* **16**, 42 (1963).
180. Dwyer, F. P., Gyarfas, E. C., Rogers, W. P., and Koch, J. H., *Nature (London)* **170**, 190 (1952).
181. Einstein, F. W. B., and Penfold, B. R., *J. Chem. Soc. A*, p. 3019 (1968).
182. Einstein, F. W. B., and Penfold, B. R., *Acta Crystallogr.* **20**, 924 (1966).
183. Einstein, F. W. B., and Penfold, B. R., *J. Chem. Soc., Chem. Commun.*, p. 780 (1966).
184. Elsbernd, H., and Beattie, J. K., *J. Inorg. Nucl. Chem.* **34**, 771 (1972).
185. Emara, M. M., and Lin, C.-T., *J. Indian Chem. Soc.* **57**, 876 (1980).
186. Emara, M. M., Lin, C.-T., and Atkinson, G., *Bull. Soc. Chim. Fr., Pt. 1*, p. 173 (1980).

187. Epstein, L. M., *J. Chem. Phys.* **40**, 435 (1964).
188. Ernst, R. D., and Marks, T. J., *Inorg. Chem.* **17**, 1477 (1978).
189. Farina, R., and Wilkins, R. G., *Inorg. Chem.* **7**, 514 (1968).
190. Farina, R., Hogg, R., and Wilkins, R. G., *Inorg. Chem.* **7**, 170 (1968).
191. Fergusson, J. E., Robinson, B. H., and Wilkins, C. J., *J. Chem. Soc. A*, p. 486 (1967).
192. Fergusson, J. E., and Hickford, J. H., *Inorg. Chim. Acta* **2**, 475 (1968).
193. Fergusson, J. E., Roper, W. R., and Wilkins, C. J., *J. Chem. Soc.*, p. 3716 (1965).
194. Figgis, B. N., Kucharski, E. S., and White, A. H., *Aust. J. Chem.* **36**, 1527 (1983).
195. Figgis, B. N., Kucharski, E. S., and White, A. H., *Aust. J. Chem.* **36**, 1537 (1983).
196. Figgis, B. N., Kucharski, E. S., and White, A. H., *Aust. J. Chem.* **36**, 1563 (1983).
197. Fink, D. W., and Ohnesorge, W. E., *Anal. Chem.* **41**, 39 (1969).
198. Fink, D. W., and Ohnesorge, W. E., *J. Am. Chem. Soc.* **91**, 4995 (1969).
199. Fink, D. W., Pivnichny, J. V., and Ohnesorge, W. E., *Anal. Chem.* **41**, 833 (1969).
200. Ford-Smith, M. H., and Sutin, N., *J. Am. Chem. Soc.* **83**, 1830 (1961).
201. Frost, G. H., Hart, F. A., Heath, C. A., and Hursthouse, M. B., *J. Chem. Soc., Chem. Commun.*, p. 1421 (1969).
202. Frost, G. H., and Hart, F. A., *J. Chem. Soc., Chem. Commun.*, p. 836 (1970).
203. Gabbay, E. J., Adkins, M. A., and Yen., S., *Nucleic Acids Res.* **7**, 1081 (1979).
204. Gagné, R. R., Allison, J. L., Koval, C. A., Mialki, W. S., Smith, T. J., and Welton, R. A., *J. Am. Chem. Soc.* **102**, 1905 (1980).
205. Ganorkar, M. C., and Stiddard, M. H. B., *J. Chem. Soc.*, p. 5346 (1965).
206. Geier, G., Erni, I., and Steiner, R., *Helv. Chim. Acta* **60**, 9 (1977).
207. Gerloch, M., *J. Chem. Soc. A*, p. 1317 (1966).
208. German, E. D., *Izv. Akad. Nauk SSSR, Ser. Khim.*, p. 1967 (1983).
209. Gibbs, P. R., Graves, P. R., Gulliver, D. J., and Levason, W., *Inorg. Chim. Acta* **45**, L207 (1980).
210. Gillard, R. D., *Coord. Chem. Rev.* **16**, 67 (1975).
211. Gillard, R. D., *Coord. Chem. Rev.* **50**, 303 (1983).
212. Goldschmeid, E., and Stephenson, N. C., *Acta Crystallogr., Sect. B* **B26**, 1867 (1970).
213. Goodgame, D. M. L., and Brun, G., *Bull. Soc. Chim. Fr.*, p. 2236 (1973).
214. Goodwin, H. A., and Sylva, R. N., *Aust. J. Chem.* **20**, 629 (1967).
215. Graddon, D. P., and Ong, W. K., *Aust. J. Chem.* **24**, 741 (1974).
216. Guarr, T., McGuire, M., Strauch, S., and McLendon, G., *J. Am. Chem. Soc.* **105**, 616 (1983).
217. Gustavson, W. A., Principe, L. M., Rhee, W.-Z. M., and Zuckerman, J. J., *Inorg. Chem.* **20**, 3460 (1981).
218. Hadsovic, S., Nikolin, B., and Stern, P., *Eur. J. Pharmacol.* **1**, 15 (1967).
219. Hamer, A. D., and Walton, R. A., *Inorg. Chem.* **13**, 1446 (1974).
220. Harringtom, P. C., and Wilkins, R. G., *J. Inorg. Biochem.* **19**, 339 (1983).
221. Harris, C. M., Lockyer, T. N., Martin, R. L., Patil, H. R. H., Sinn, E., and Stewart, I. M., *Aust. J. Chem.* **22**, 2105 (1969).
222. Harris, C. M., and Lockyer, T. N., *Aust. J. Chem.* **23**, 1703 (1970).
223. Harris, W. R., McLendon, G. L., Martell, A. E., Bess, R. C., and Mason, M., *Inorg. Chem.* **19**, 21 (1980).
224. Harris, C. M., and McKenzie, E. D., *J. Inorg. Nucl. Chem.* **25**, 171 (1963).
225. Harris, C. M., Lockyer, T. N., and Stephenson, N. C., *Aust. J. Chem.* **19**, 1741 (1966).
226. Harris, C. M., and Lockyer, T. N., *Aust. J. Chem.* **23**, 1125 (1970).
227. Harris, C. M., and Lockyer, T. N., *Aust. J. Chem.* **23**, 673 (1970).
228. Harris, C. M., Patil, H. R. H., and Sinn, E., *Inorg. Chem.* **8**, 101 (1969).
229. Hart, F. A., and Zhu, W., *Inorg. Chim. Acta* **54**, L275 (1981).
230. Hart, F. A., and Zhu, W., *Rare Earths Mod. Sci. Technol.* **3**, 95 (1982).

231. Hauenstein, B. L., Jr., Dressick, W. J., Buell, S. L., Demas, J. N., and DeGraff, B. A., *J. Am. Chem. Soc.* **105**, 4251 (1983).
232. Havermans, E., Verbeek, F., and Hoste, J., *Anal. Chim. Acta* **26**, 326 (1962).
233. Hazell, A. C., and Hazell, R. G., *Acta Crystallogr., Sect. C.* **C40**, 806 (1984).
234. Henke, W., Kremer, S., and Reinen, D., *Inorg. Chem.* **22**, 2858 (1983).
235. Henke, W., and Kremer, S., *Inorg. Chim. Acta* **65**, L115 (1982).
236. Henke, W., and Reinen, D., *Z. Anorg. Allg. Chem.* **436**, 187 (1977).
237. Henke, W., Kremer, S., and Reinen, D., *Z. Anorg. Allg. Chem.* **491**, 124 (1982).
238. Herzog, S., and Aul, H., *Z. Chem.* **6**, 343 (1966).
239. Herzog, S., and Aul, H., *Z. Chem.* **6**, 382 (1966).
240. Herzog, S., and Weber, A., *Z. Chem.* **8**, 115 (1968).
241. Ho, R. K. Y., Livingstone, S. E., and Lockyer, T. N., *Aust. J. Chem.* **21**, 103 (1968).
242. Howe-Grant, M., and Lippard, S. J., *Biochemistry* **18**, 5763 (1979).
243. Hogg, R., and Wilkins, R. G., *J. Chem. Soc.*, p. 341 (1962).
244. Hollis, L. S., and Lippard, S. J., *J. Am. Chem. Soc.* **105**, 4293 (1983).
245. Holyer, R. H., Hubbard, C. D., Kettle, S. F. A., and Wilkins, R. G. A., *Inorg. Chem.* **5**, 622 (1966).
246. Hoof, D. L., and Walton, R. A., *Inorg. Chim. Acta* **12**, 71 (1975).
247. Howe-Grant, M., Wu, K. C., Bauer, W. R., and Lippard, S. J., *Biochemistry* **15**, 4339 (1976).
248. Hsieh, A. T. T., and Mays, M. J., *J. Chem. Soc. A*, p. 729 (1971).
249. Huchital, D. H., and Kiel, L. F., *J. Coord. Chem.* **11**, 45 (1981).
250. Huchital, D. H., and Martell, A. E., *Inorg. Chem.* **13**, 2966 (1974).
251. Huchital, D. H., and Martell, A. E., *J. Chem. Soc., Chem. Commun.*, p. 869 (1973).
252. Hughes, M. C., and Macero, D. J., *Inorg. Chem.* **13**, 2739 (1974).
253. Hughes, M. C., and Macero, D. J., *Inorg. Chem.* **15**, 2040 (1976).
254. Hughes, M. C., and Macero, D. J., *Inorg. Chem.* **9**, 2041 (1976).
255. Hughes, M. C., and Macero, D. J., *Inorg. Chim. Acta* **4**, 3277 (1970).
256. Hughes, M. C., Macero, D. J., and Rao, J. M., *Inorg. Chim. Acta* **49**, 241 (1981).
257. Hughes, M. C., Macero, D. J., and Rao, J. M., *Inorg. Chim. Acta* **49**, 241 (1981).
258. Hughes, M. C., Rao, J. M., and Macero, D. J., *Inorg. Chim. Acta* **18**, 127 (1976).
259. Intille, G. M., Pfluger, C. E., and Baker, W. A., Jr., *Cryst. Struct. Commun.* **2**, 217 (1973).
260. Intille, G. M., Pfluger, C. E., and Baker, W. A., Jr., *J. Cryst. Mol. Struct.* **3**, 47 (1973).
261. Isied, S. S., and Taube, H., *Inorg. Chem.* **15**, 3070 (1976).
262. International Union of Pure and Applied Chemistry, Nomenclature of Organic Chemistry, 1979 Ed. Pergamon Press, Oxford, England
263. James, B. R., and Williams, R. J. P., *J. Chem. Soc.*, p. 2007 (1961).
264. Jennette, K. W., Gill, J. T., Sadownick, J. A., and Lippard, S. J., *J. Am. Chem. Soc.* **98**, 6159 (1976).
265. Jennette, K. W., Lippard, S. J., Vassiliades, G. A., and Bauer, W. R., *Proc. Natl. Acad. Sci. U.S.A.* **71**, 3839 (1974).
266. Jensen, P. W., and Jorgensen, L. B., *J. Mol. Struct.* **79**, 87 (1982).
267. Johnson, J. M., Halsall, H. B., and Heineman, W. R., *Biochemistry* **24**, 1579 (1985).
268. Judge, J. S., and Baker, W. A., Jr., *Inorg. Chim. Acta* **1**, 245 (1967).
269. Judge, J. S., and Baker, W. A., Jr., *Inorg. Chim. Acta* **1**, 68 (1967).
270. Judge, J. S., and Baker, W. A., Jr., *Inorg. Chim. Acta* **1**, 239 (1967).
271. Judge, J. S., Reiff, W. M., Intille, G. M., Ballway, P., and Baker jun., W. A., *J. Inorg. Nucl. Chem.* **29**, 1711 (1967).
272. Kaden, T. A., Holmquist, B., and Vallee, B. L., *Inorg. Chem.* **13**, 2585 (1974).
273. Kallistratos, G., Kallistratos, U., and Muendner, H., *Chem. Chron.* **11**, 249 (1982).
274. Kamenar, B., Korpar-Colig, B., Herzold-Brundic, A., and Popovic, Z., *Acta Crystallogr., Sect. B* **B38**, 1593 (1982).

275. Kamra, L. C., and Ayres, G. H., *Anal. Chim. Acta* **81**, 117 (1976).
276. Karlin, K. D., and Yandell, J. K., *Inorg. Chem.* **23**, 1184 (1984).
277. Kepert, D. L., Kucharski, E. S., and White, A. H., *J. Chem. Soc., Dalton Trans.*, p. 1932 (1980).
278. Kim, K.-Y., and Nancollas, G. H., *J. Phys. Chem.* **81**, 948 (1977).
279. Kirchhoff, J. R., McMillin, D. R., Marnot, P. A., and Sauvage, J.-P., *J. Am. Chem. Soc.* **107**, 1138 (1985).
280. Klassen, D. M., and Crosby, G. A., *J. Chem. Phys.* **48**, 1851 (1968).
281. Koch, J. H., and Gallagher, C. H., *Biochem. Pharmacol.* **3**, 231 (1960).
282. Koch, M. S., Gyarfas, E. C., and Dwyer, F. P., *Aust. J. Biol. Sci.* **9**, 371 (1955).
283. Kovac, B., and Klasine, L., *Z. Naturforsch., A* **23A**, 247 (1978).
284. Kremer, S., Henke, W., and Reinen, D., *Inorg. Chem.* **21**, 3013 (1982).
285. Krohnke, F., *Synthesis*, p. 1 (1976).
286. Krumholz, P., *Inorg. Chem.* **4**, 612 (1965).
287. Kucharski, E. S., Skelton, B. W., and White, A. H., *Aust. J. Chem.* **31**, 47 (1978).
288. Kwik, W.-L., and Ang, K.-P., *Transition Met. Chem. (Weinheim, Ger.)* **10**, 50 (1985).
289. Landee, C. P., Wicholas, M., Willett, R. D., and Wolford, T., *Inorg. Chem.* **18**, 2317 (1979).
290. Lehn, J.-M., Sauvage, J.-P., Simon, J., Ziessel, R., Piccinni-Leopardi, C., Germain, G., Declercq, J.-P., and Van Meerssche, M., *Nouv. J. Chim.* **7**, 413 (1983).
291. Lindoy, L. F., and Livingstone, S. E., *Coord. Chem. Rev.* **2**, 173 (1967).
292. Lin, C.-T., Bottcher, W., Chou, M., Creutz, C., and Sutin, N., *J. Am. Chem. Soc.* **98**, 6536 (1976).
293. Lin, J.-L., Chang, L.-F., and Satake, M., *Bull. Chem. Soc. Jpn.* **56**, 2739 (1983).
294. Lions, F., Dance, I. G., and Lewis, J., *J. Chem. Soc. A*, p. 565 (1967).
295. Lip, H. C., and Plowman, R. A., *Aust. J. Chem.* **28**, 893 (1975).
296. Livingstone, S. E., Mayfield, J. H., and Moore, D. S., *Aust. J. Chem.* **29**, 1209 (1976).
297. Livingstone, S. E., and Mihkelson, A. E., *Inorg. Chem.* **9**, 2545 (1970).
298. Lovecchio, F. V., Pace, S. J., and Macero, D. J., *Inorg. Chim. Acta* **3**, 94 (1969).
299. Lutz, P. M., Long, G. J., and Baker, W. A., Jr., *Inorg. Chem.* **8**, 2528 (1969).
300. Lytle, F. E., Petrosky, L. M., and Carlson, L. R., *Anal. Chim. Acta* **57**, 239 (1971).
301. Macero, D. J., Lovecchio, F. V., and Pace, S. J., *Inorg. Chim. Acta* **3**, 65 (1969).
302. Maeda, Y., Ohshio, H., and Takashima, Y., *Chem. Lett.*, p. 1359 (1980).
303. Maeda, Y., Ohshio, H., and Takashima, Y., *Bull. Chem. Soc. Jpn.* **55**, 3500 (1982).
304. Maki, N., *Bull. Chem. Soc. Jpn.* **42**, 2274 (1969).
305. Marchese, A. L., and West, B. O., *J. Organomet. Chem.* **266**, 61 (1984).
306. Martin, R. B., and Lissfelt, J. A., *J. Am. Chem. Soc.* **78**, 938 (1956).
307. Maslen, E. N., Raston, C. L., and White, A. H., *J. Chem. Soc., Dalton Trans.*, p. 1803 (1974).
308. Maslen, E. N., Raston, C. L., and White, A. H., *J. Chem. Soc., Dalton Trans.*, p. 323 (1975).
309. Mathew, M., and Palenik, G. J., *J. Coord. Chem.* **1**, 243 (1971).
310. May, J. C., and Curran, C., *J. Organomet. Chem.* **39**, 289 (1972).
311. McArdle, J. V., Yocom, K., and Gray, H. B., *J. Am. Chem. Soc.* **99**, 4141 (1977).
312. McHatton, R. C., and Anson, F. C., *Inorg. Chem.* **23**, 3935 (1984).
313. McLendon, G., and Mooney, W. F., *Inorg. Chem.* **19**, 12 (1980).
314. McWhinnie, W. R., and Miller, J. D., *Adv. Inorg. Chem. Radiochem.* **12**, 135 (1969).
315. Mehrotra, R. N., and Wilkins, R. G., *Inorg. Chem.* **19**, 2177 (1980).
316. Melby, L. R., Rose, N. J., Abramson, E., and Caris, J. C., *J. Am. Chem. Soc.* **86**, 5117 (1964).
317. Melvin, W. S., Rablen, D. P., and Gordon, G., *Inorg. Chem.* **11**, 489 (1972).
318. Meyer, T. J., *Isr. J. Chem.* **15**, 200 (1976/7).
319. Miller, J. D., and Prince, R. H., *J. Chem. Soc. (A)*, p. 1049 (1966).
320. Miller, J. D., and Prince, R. H., *J. Chem. Soc. (A)*, p. 1371 (1966).

321. Mizuno, K., Imamura, S., and Lunsford, J. H., *Inorg. Chem.* **23**, 3510 (1984).
322. Mohr, R., Mietta, L. A., Ducommun, Y., and Van Eldick, R., *Inorg. Chem.* **24**, 757 (1985).
323. Morgan, G. T., and Burstall, F. H., *J. Indian Chem. Soc., Ray Commem.* **1**, (1933).
324. Morgan, G. T., and Burstall, F. H., *J. Chem. Soc.*, p. 1499 (1934).
325. Morgan, G. T., and Burstall, F. H., *J. Chem. Soc.*, p. 1675 (1938).
326. Morgan, G. T., and Burstall, F. H., *J. Chem. Soc.*, p. 20 (1932).
327. Morgan, G. T., and Burstall, F. H., *J. Chem. Soc.*, p. 1672 (1938).
328. Morgan, G. T., and Burstall, F. H., *J. Chem. Soc.*, p. 1649 (1937).
329. Morgan, G. T., and Davies, G. R., *J. Chem. Soc.*, p. 1858 (1938).
330. Morrison, M. M., and Sawyer, D. T., *Inorg. Chem.* **17**, 333 (1978).
331. Morris, R. L., *Anal. Chem.* **24**, 1376 (1952).
332. Morris, D. E., Hanck, K. W., and DeArmond, M. K., *J. Electroanal. Chem. Interfacial Electrochem.* **149**, 115 (1983).
333. Moss, M. L., and Mellon, M. G., *Ind. Eng. Chem., Anal. Ed.* **14**, 862 (1942).
334. Moss, M. L., and Mellon, M. G., *Ind. Eng. Chem., Anal. Ed.* **15**, 75 (1943).
335. Moyer, B. A., Thompson, M. S., and Meyer, T. J., *J. Am. Chem. Soc.* **102**, 2310 (1980).
336. Munukata, M., Nishibayashi, S., and Sakamoto, H., *J. Chem. Soc., Chem. Commun.*, p. 219 (1980).
337. Mureinik, R. J., and Bidani, M., *Inorg. Nucl. Chem. Lett.* **13**, 625 (1977).
338. Mureinik, R. J., and Bidani, M., *Inorg. Chim. Acta* **29**, 37 (1978).
339. Murphy jun., W. R., Takeuchi, K. J., and Meyer, T. J., *J. Am. Chem. Soc.* **104**, 5817 (1982).
340. Murtha, D. P., and Walton, R. A., *Inorg. Chem.* **12**, 1278 (1973).
341. Murtha, D. P., and Walton, R. A., *Inorg. Chem.* **12**, 368 (1973).
342. Murtha, D. P., and Walton, R. A., *Inorg. Nucl. Chem. Lett.* **9**, 819 (1973).
343. Murtha, D. P., and Walton, R. A., *Inorg. Chim. Acta* **8**, 279 (1974).
344. Musumeci, S., Rizzarelli, E., Sammartano, S., and Bonomo, R. P., *J. Inorg. Nucl. Chem.* **36**, 853 (1974).
345. Nagle, J. K., and Meyer, T. J., *Inorg. Chem.* **23**, 3663 (1984).
346. Naik, D. V., and Scheidt, W. R., *Inorg. Chem.* **12**, 272 (1973).
347. Naik, D. V., and Curran, C., *J. Organomet. Chem.* **81**, 177 (1974).
348. Nakamoto, K., *J. Phys. Chem.* **64**, 1421 (1960).
349. Nakamura, K., *Bull. Chem. Soc. Jpn.* **45**, 1943 (1971).
350. Newkome, G. R., and Lee, H.-W., *J. Am. Chem. Soc.* **105**, 5956 (1983).
351. Oepen, G., and Vogtle, F., *Liebigs Ann. Chem.*, p. 2114 (1979).
352. Ogura, K., Urabe, H., and Yosino, T., *Electrochim. Acta* **22**, 285 (1977).
353. Ohno, T., and Kato, S., *J. Phys. Chem.* **88**, 1670 (1984).
354. Ohno, T., and Lichtin, N. N., *J. Phys. Chem.* **84**, 3019 (1980).
355. Ohno, T., Kato, S., and Lichtin, N. N., *Bull. Chem. Soc. Jpn.* **55**, 2753 (1982).
356. Ortego, J. D., and Seymour, M., *Polyhedron* **1**, 21 (1982).
357. Otera, J., Hinoishi, T., and Okawara, R., *J. Organomet. Chem.* **202**, C93 (1980).
358. Pancheva, R., *Suvrem. Med.* **26**, 46 (1975).
359. Pankuch, B. J., Lacky, D. E., and Crosby, G. A., *J. Phys. Chem.* **84**, 2061 (1980).
360. Peloso, A., *J. Organomet. Chem.* **74**, 59 (1974).
361. Pflaum, R. T., and Brandt, W. W., *J. Am. Chem. Soc.* **76**, 6215 (1954).
362. Pflaum, R. T., and Brandt, W. W., *J. Am. Chem. Soc.* **77**, 2019 (1955).
363. Phillips, J., Langford, C. H., and Barradas, R. G., *Electrochim. Acta* **23**, 143 (1978).
364. Pierce, J., Busch, K. L., Walton, R. A., and Cooks, R. G., *J. Am. Chem. Soc.* **103**, 2581 (1981).
365. Pipes, D. W., and Meyer, T. J., *Inorg. Chem.* **23**, 2466 (1984).
366. Pipes, D. W., and Meyer, T. J., *J. Am. Chem. Soc.* **106**, 7653 (1984).
367. Pispisa, B., *NATO Adv. Study Inst. Ser., Ser. C* **100** (1983).

368. Pispisa, B., and Farinella, M., *Biopolymers* **23**, 1465 (1984).

369. Pispisa, B., and Paoletti, S., *J. Phys. Chem.* **84**, 24 (1980).

370. Pispisa, B., Barteri, M., and Farinella, M., *Inorg. Chem.* **22**, 3166 (1983).

371. Postmus, C., Ferraro, J. R., and Wozniak, W., *Inorg. Chem.* **6**, 2030 (1967).

372. Prasad, J., and Peterson, N. C., *Inorg. Chem.* **8**, 1622 (1969).

373. Prasad, J., and Scaife, D. B., *J. Electroanal. Chem.* **84**, 373 (1977).

374. Prow, W. F., Garmestani, S. K., and Farina, R. D., *Inorg. Chem.* **20**, 1297 (1981).

375. Rablen, D., and Gordon, G., *Inorg. Chem.* **8**, 395 (1969).

376. Radonovich, L. J., and Gluck, M. D., *Inorg. Chem.* **10**, 1463 (1971).

377. Rao, J. M., Hughes, M. C., and Macero, D. J., *Inorg. Chim. Acta* **18**, 127 (1976).

378. Rao, J. M., Hughes, M. C., and Macero, D. J., *Inorg. Chim. Acta* **16**, 231 (1976).

379. Rao, J. M., Macero, D. J., and Hughes, M. C., *Inorg. Chim. Acta* **41**, 221 (1980).

380. Raston, C. L., and White, A. H., *J. Chem. Soc., Dalton Trans.*, p. 7 (1976).

381. Raston, C. L., Rowbottom, G. L., and White, A. H., *J. Chem. Soc., Dalton Trans.*, p. 1383 (1981).

382. Reiff, W. M., *J. Am. Chem. Soc.* **96**, 3829 (1974).

383. Reiff, W. M., Long, G. J., and Baker, W. A., Jr., *J. Am. Chem. Soc.* **90**, 6347 (1968).

384. Reiff, W. M., Baker jun., W. A., and Erickson, N. E., *J. Am. Chem. Soc.* **90**, 4794 (1968).

385. Reiff, W. M., Erickson, N. E., and Baker, W. A., Jr., *Inorg. Chem.* **8**, 2019 (1969).

386. Renfrew, R. W., Osvath, P., and Weatherburn, D. C., *Aust. J. Chem.* **33**, 45 (1980).

387. Ridd, M. J., and Keene, F. R., *J. Am. Chem. Soc.* **103**, 5733 (1981).

388. Robinson, D. J., and Kennard, C. H. L., *Aust. J. Chem.* **19**, 1285 (1966).

389. Robinson, D. J., and Kennard, C. H. L., *J. Inorg. Nucl. Chem.* **31**, 3322 (1969).

390. Roche, T. S., and Wilkins, R. G., *J. Chem. Soc., Chem. Commun.*, p. 1681 (1970).

391. Roche, T. S., and Wilkins, R. G., *J. Am. Chem. Soc.* **96**, 5082 (1974).

392. Rojo, T., Darriet, J., Dance, J. M., and Beltran-Porter, D., *Inorg. Chim. Acta* **64**, L105 (1982).

393. Rojo, T., Vlasse, M., and Beltran-Porter, D., *Acta Crystallogr., Sect. C* **C39**, 194 (1983).

394. Root, M. J., and Deutsch, E., *Inorg. Chem.* **24**, 1464 (1985).

395. Root, M. J., Deutsch, E., Sullivan, J. C., and Meisel, D., *Chem. Phys. Lett.* **101**, 353 (1983).

396. Ruiz, F. J., Mesa, J. L., Rojo, T., and Arriortua, M. I., *J. Appl. Crystallogr.* **16**, 430 (1983).

397. Saji, T., and Aoyagui, S., *J. Electroanal. Chem. Interfacial Electrochem.* **58**, 401 (1975).

398. Saji, T., and Aoyagui, S., *J. Electroanal. Chem. Interfacial Electrochem.* **108**, 223 (1980).

399. Schilt, A. A., and Wong, S. W., *J. Coord. Chem.* **13**, 331 (1984).

400. Schilt, A. A., and Smith, G. F., *Anal. Chim. Acta* **15**, 567 (1956).

401. Schilt, A. A., "Analytical Applications of 1,10-Phenanthroline and Related Compounds." Pergamon, Oxford, 1969.

402. Schmidt, J. G., Brey, W. S., Jr., and Stoufer, R. C., *Inorg. Chem.* **6**, 268 (1967).

403. Schmidt, R., Weis, W., Klingmuller, V., and Staudinger, H., *Z. Klin. Chem. Klin. Biochem.* **5**, 304 (1967).

404. Schmidt, R., Weis, W., Klingmuller, V., and Staudinger, H.-J., *Z. Klin. Chem.* **5**, 308 (1967).

405. Serpone, N., Jamieson, M. A., Henry, M. S., Hoffman, M. Z., Bolletta, F., and Maestri, M., *J. Am. Chem. Soc.* **101**, 2907 (1979).

406. Serpone, N., Ponterini, G., Jamieson, M. A., Bolletta, F., and Maestri, M., *Coord. Chem. Rev.* **50**, 209 (1983).

407. Seto, H., and Tomaz, A., *Proc. Natl. Acad. Sci. U.S.A.* **74**, 296 (1977).

408. Silverman, L. D., Dewan, J. C., Giandomenico, C. M., and Lippard, S. J., *Inorg. Chem.* **19**, 3379 (1980).

409. Sinha, S. P., *Spectrochim. Acta* **22**, 57 (1966).

410. Sinha, S. P., *Z. Naturforsch. A* **20A**, 1661 (1965).

411. Sinha, S. P., Z. Naturforsch. A **20A**, 552 (1965).
412. Sinha, S. P., Z. Naturforsch. A **20A**, 835 (1965).
413. Sinha, S. P., Mehta, P. C., and Surana, S. S. L., Mol. Phys. **23**, 807 (1972).
414. Sinha, S. P., Z. Naturforsch. A **20A**, 164 (1965).
415. Smith, G. F., Anal. Chem. **26**, 1534 (1954).
416. Smith, G. F., and Banick, W. M., Anal. Chem. **18**, 269 (1958).
417. Spacu, P., and Antonescu, E., Rev. Roum. Chim. **16**, 373 (1971).
418. Spahni, W., and Calzaferri, G., Helv. Chim. Acta **67**, 450 (1984).
419. Spellane, P. J., and Watts, R. J., Inorg. Chem. **20**, 3561 (1981).
420. Sprouse, S., King, K. A., Spellane, P. J., and Watts, R. J., J. Am. Chem. Soc. **106**, 6647 (1984).
421. Srivastava, T. N., and Srivastava, P. C., J. Indian Chem. Soc. **53**, 365 (1976).
422. Stamm, D., Staudinger, H., and Weis, W., Z. Klin. Chem. **4**, 222 (1966).
423. Stanbury, D. M., and Lednicky, L. A., J. Am. Chem. Soc. **106**, 2847 (1984).
424. Stone, M. L., and Crosby, G. A., Chem. Phys. Lett. **79**, 169 (1981).
425. Stoufer, R. C., Smith, D. W., Clevenger, E. A., and Norris, T. E., Inorg. Chem. **5**, 1167 (1966).
426. Strothkamp, K. G., and Lippard, S. J., Proc. Natl. Acad. Sci. U.S.A. **73**, 2536 (1976).
427. Sudnick, D. R., and Horrocks, W. D., Jr., Biochim. Biophys. Acta **578**, 135 (1979).
428. Sullivan, P. B., Smythe, R. S., Kober, E. M., and Meyer, T. J., J. Am. Chem. Soc. **104**, 4701 (1982).
429. Sullivan, P. B., Calvert, J. M., and Meyer, T. J., Inorg. Chem. **19**, 1404 (1980).
430. Summers, M. F., and Marzilli, L. G., Inorg. Chem. **23**, 521 (1984).
431. Sutton, G. J., Aust. J. Chem. **20**, 1859 (1967).
432. Sutton, G. J., Aust. J. Scient. Res. Ser. A. **4**, 651 (1951).
433. Szalda, D. J., Eckstein, F., Sternbach, H., and Lippard, S. J., J. Inorg. Biochem. **11**, 279 (1979).
434. Takahashi, T., Fukushima Kogyo Koto Semmon Gakko Kiyo **15**, 33 (1979).
435. Takahashi, T., Fukushima Kogyo Koto Semmon Gakko Kiyo **15**, 39 (1979).
436. Takahashi, T., and Koiso, T., Bull. Chem. Soc. Jpn. **53**, 3400 (1980).
437. Takeuchi, K. J., Thompson, M. S., Pipes, D. W., and Meyer. T. J., Inorg. Chem. **23**, 1845 (1984).
438. Tandon, S. P., and Mehta, P. C., J. Chem. Phys. **52**, 5417 (1970).
439. Taylor, D. B., Callahan, K. P., and Shaikh, I., J. Med. Chem. **18**, 1088 (1975).
440. Taylor, C. G., and Waters, J., Anal. Chim. Acta **69**, 363 (1974).
441. Taylor, C. G., Waters, J., and Williams, P. V., Anal. Chim. Acta **69**, 373 (1974).
442. Taylor, P. J., and Schilt, A. A., Inorg. Chim. Acta **5**, 691 (1971).
443. Thompson, M. S., De Giovani, W. F., Moyer, B. A., and Meyer, T. J., J. Org. Chem. **49**, 4972 (1984).
444. Thompson, M. S., and Meyer, T. J., J. Am. Chem. Soc. **103**, 5577 (1981).
445. Thompson, M. S., and Meyer, T. J., J. Am. Chem. Soc. **104**, 4106 (1982).
446. Thompson, M. S., and Meyer, T. J., J. Am. Chem. Soc. **104**, 5070 (1982).
447. Tokel-Takvoryan, N. E., Hemingway, R. E., and Bard, A. J., J. Am. Chem. Soc. **95**, 6582 (1973).
448. Tomaja, D. L., and Zuckerman, J. J., Synth. React. Inorg. Met. Org. Chem. **6**, 323 (1976).
449. Toner, J. L., Tetrahedron Lett. **24**, 2707 (1983).
450. Truhaut, R., Ann. Biol. Clin. (Paris) **17**, 571 (1959).
451. Truhaut, R., Gosse, C., de Lesdain, N., and Paoletti, C., Ann. Biol. Clin. (Paris) **17**, 571 (1959).
452. Uhlig, E., and Dinjus, E., Z. Anorg. Allg. Chem. **418**, 45 (1975).
453. Usmani, M. A. A., and Scaife, D. B., Pak. J. Sci. Ind. Res. **19**, 4 (1976).

454. Vlasse, M., Rojo, T., and Beltran-Porter, D., *Acta Crystallogr., Sect. C* **C39**, 560 (1983).
455. Vogtle, F., Muller, W. M., and Rasshofer, W., *Isr. J. Chem.* **18**, 246 (1979).
456. Wagner-Jauregg, T., Hackley, B. E., Jr., Lies, T. A., Owens, O. O., and Proper, R., *J. Am. Chem. Soc.* **77**, 922 (1955).
457. Walters, W. S., Gillard, R. D., and Williams, P. A., *Aust. J. Chem.* **31**, 1959 (1978).
458. Walton, R. A., *Inorg. Chem.* **7**, 640 (1968).
459. Walton, R. A., *Inorg. Nucl. Chem. Lett.* **12**, 767 (1976).
460. Walton, R. A., *J. Chem. Soc. A*, p. 1485 (1967).
461. Walton, R. A., *J. Inorg. Nucl. Chem.* **32**, 2875 (1970).
462. Walton, R. A., *J. Inorg. Nucl. Chem.* **39**, 549 (1977).
463. Wang, A. H. J., Nathans, J., Van der Marel, G., Van Boom, J. H., and Rich, A., *Nature (London)* **276**, 471 (1978).
464. Wang, A. H. J., Quigley, G. J., and Kolpak, F. J., *Abstr. Am. Crystallogr. Assoc.* **6**, 50 (1979).
465. Wassef, M. A., Hegazi, W. S., and Ali, S. A., *Egypt. J. Chem.* **26**, 197 (1983).
466. Wassef, M. A., and Ahmed, S., *Egypt. J. Chem.* **25**, 505 (1982).
467. Wassef, M. A., and Gaber, M., *Egypt. J. Chem.* **24**, 165 (1981).
468. Wassef, M. A., and Hessin, S., *Egypt. J. Chem.* **24**, 97 (1981).
469. Wassef, M. A., and Hessin, S., *Commun. Fac. Sci. Univ. Ankara, Ser. B* **27**, 141 (1981).
470. Wassef, M. A., and Hessin, S., *Commun. Fac. Sci. Univ. Ankara, Ser. B* **27**, 153 (1981).
471. Wassef, M. A., and Hessin, S., *Commun. Fac. Sci. Univ. Ankara, Ser. B* **27**, 161 (1981).
472. Wassef, M. A., Hegazi, W. S., and Ali, S. A., *Acta Chim. Acad. Sci. Hung.* **111**, 199 (1982).
473. Wassef, M. A., Hegazi, W. S., and Ali, S. A., *J. Coord. Chem.* **12**, 97 (1983).
474. Wassef, M. A., and Ahmed, S., *Indian J. Chem., Sect. A* **21A**, 428 (1982).
475. Wassef, M. A., Hegazi, W. S., and Ali, S. A., *Commun. Fac. Sci. Univ. Ankara, Ser. B* **27**, 225 (1981).
476. Wassef, M. A., Hegazi, W. S., and Ali, S. A., *Commun. Fac. Sci. Univ. Ankara, Ser. B* **27**, 239 (1981).
477. Werber, M. M., and Shalitin, Y., *Bioorg. Chem.* **4**, 149 (1975).
478. Wickramasinghe, W. A., Bird, P. H., and Serpone, N., *J. Chem. Soc., Chem. Commun.*, p. 1284 (1981).
479. Wickramasinghe, W. A., Bird, P. H., and Serpone, N., *Inorg. Chem.* **21**, 2694 (1982).
480. Wickramasinghe, W. A., Bird, P. H., Jamieson, M. A., and Serpone, N., *J. Chem. Soc., Chem. Commun.*, p. 798 (1979).
481. Wieghardt, K., and Quiltzsch, U., *Z. Anorg. Allg. Chem.* **457**, 75 (1979).
482. Wieghardt, K., Holzbach, W., and Weiss, J., *Z. Naturforsch. B.* **37B**, 680 (1982).
483. Wieghardt, K., Holzbach, W., Nuber, B., and Weiss, J., *Chem. Ber.* **113**, 629 (1980).
484. Wieghardt, K., and Holzbach, W., *Angew. Chem., Int. Ed. Engl.* **18**, 549 (1979).
485. Wilkins, D. H., and Smith, G. F., *Anal. Chim. Acta* **9**, 338 (1953).
486. Willemen, H., Van de Vondel, D. F., and Van der Kelen, G. P., *Inorg. Chim. Acta* **34**, 175 (1979).
487. Williams, N. H., and Yandell, J. K., *Aust. J. Chem.* **35**, 1133 (1982).
488. Wink, D. J., and Cooper, N. J., *J. Chem. Soc., Dalton Trans.*, p. 1257 (1984).
489. Wilson, W. D., Heyl, B. L., Reddy, R., and Marzilli, L. G., *Inorg. Chem.* **21**, 2527 (1982).
490. Wong, Y.-S., and Lippard, S. J., *J. Chem. Soc., Chem. Commun.*, p. 825 (1977).
491. Wrighton, M. S., *Adv. Chem. Ser.* **168** (1978).
492. Yamauchi, O., Benno, H., and Nakahara, A., *Bull. Chem. Soc. Jpn.* **46**, 3458 (1973).
493. Yandell, J. K., and Yonetani, T., *Biochim. Biophys. Acta* **748**, 263 (1983).
494. Young, R. C., Nagle, J. K., Meyer, T. J., and Whitten, D. G., *J. Am. Chem. Soc.* **100**, 4773 (1978).
495. Young, R. C., Meyer, T. J., and Whitten, D. G., *J. Am. Chem. Soc.* **98**, 286 (1976).

496. Zak, B., Epstein, E., and Baginski, E. S., *Microchem. J.* **14**, 155 (1969).
497. Zak, B., Baginski, E. S., Epstein, E., and Weiner, L. M., *Clin. Chim. Acta* **29**, 77 (1970).
498. Zak, B., Baginski, E. S., Epstein, E., and Weiner, L. M., *Clin. Toxicol.* **4**, 621 (1971).
499. von Zelewsky, A., and Bemtgen, J.-M., *Inorg. Chem.* **21**, 1771 (1982).
500. Zhu, K. J., and Okamoto, Y., *Polym. Mater. Sci. Eng.* **49**, 78 (1983).
501. Zubairi, S. A., and Usmani, M. A. A., *Pak. J. Sci. Ind. Res.* **19**, 11 (1976).

HIGH-NUCLEARITY CARBONYL CLUSTERS: THEIR SYNTHESIS AND REACTIVITY

MARIA D. VARGAS and J. NICOLA NICHOLLS

University Chemical Laboratory, University of Cambridge, Cambridge CB2 1EW, England

I. Introduction

Over 20 years ago, the structure of the first high-nuclearity carbonyl cluster (HNCC), $Rh_6(CO)_{16}$, was elucidated (*164*). Since that time, the field has grown to a state where there is now an enormous number of such species. We have chosen to define HNCC as homo- or heteronuclear carbonyl clusters of transition and main group metals containing five or more metal atoms, each of which is linked to the metal core by at least one M—M bond.

A review of the area of HNCC by Chini *et al.* in 1976 (*196*) reflected the emphasis of the earlier studies. Efforts were concentrated on establishing new structural forms and rationalizing the bonding patterns rather than understanding the reactivity of these compounds. To some extent this has been corrected over the past few years. With the development of more efficient synthetic routes, a large number of HNCC have become available in reasonable amounts and more effort has been devoted to the investigation of their chemistry.

Several reviews have included aspects of the synthesis and reactivity of HNCC (*77, 456–469*), but the vast amount of information now accumulated in this area justifies a separate discussion of these topics. There is often no clear dividing line between synthesis and reactivity. For convenience, under the heading "synthesis" have been gathered all reactions involving the formation of HNCC through a change in nuclearity. We have chosen to discuss the chemistry of the HNCC in terms of reaction type rather than metal by metal. We hope that this approach will provide a better understanding of what may be viewed as a highly complicated subject. We must emphasize that our expertise is as synthetic chemists, and we apologize for any error that might occur within our discussions of the more theoretical aspects of the field.

All the HNCC that have been characterized to date by X-ray crystallography are listed in Table I together with the methods used for their synthesis and references to their spectroscopic data.

123

Copyright © 1986 by Academic Press, Inc.
All rights of reproduction in any form reserved.

TABLE I

COMPENDIUM OF HNCC WHOSE STRUCTURES HAVE BEEN DETERMINED CRYSTALLOGRAPHICALLY[a]

Cluster[b]	References		Reagents and conditions	Yield (%)	Spectroscopic and theoretical studies[d]
	X-Ray[c]	Synthesis[c]			
$Re_6(CO)_{18}(PMe_3)$	1	1	$Re_4(CO)_{15}MePP(Me)PMeCl_2 + Re_2(CO)_{10}/\Delta$, Ar	—	—
$Re_6(CO)_{19}P(PMe)(PMe_2)$	1	1	$Re_6(CO)_{15}MePP(Me)PMeCl_2 + Re_2(CO)_{10}/\Delta$, Ar	—	—
$[Re_6C(CO)_{19}]^{2-}$	376	376	$[Re_7C(CO)_{21}]^{3-} + I_2$/MeCN, CO	—	IR (416)
$[Re_6C(CO)_{18}H_2]^{2-}$	2	2	$[Re(CO)_4H_2]^-$/n-tetradecane, decaline, Δ	35	^{13}C NMR, ^{1}H NMR
$[Re_7C(CO)_{21}]^{3-}$	3	3	$[Re(CO)_4H_2]^-$/n-tetradecane, Δ	70	^{1}H NMR, IR (416)
$[Re_8C(CO)_{24}]^{2-}$	4	4	$[Re(CO)_4H_2]^-$/n-tetradecane, Δ	30	^{1}H NMR
$Fe_5(CO)_{13}(CS)S_2$	5	5	$Fe_3(CO)_{12} + CS_2 + CO$/Ar, hexane	2	
$Fe_5C(CO)_{15}$	6	7	$[Fe_6C(CO)_{16}]^{2-} + [C_7H_7][BF_4]$/MeOH	100	^{13}C NMR (8), IR (9), MO calc. (6), ^{57}Fe Mössbauer (10, 11), ESCA (10)
$[Fe_5C(CO)_{14}]^{2-}$	13	14	$Fe_5C(CO)_{15} + NaOH$, $NaBH_4$ or Na/Hg, THF	80	^{57}Fe Mössbauer (10–12), ESCA (10)
$Fe_5C(CO)_{12}Br_2$	7	7	$[Fe_6C(CO)_{16}]^{2-} + C_7H_7Br$/MeOH	—	^{57}Fe Mössbauer (11)
$[Fe_5N(CO)_{14}]^-$	13	15	$NOBF_4 + [Fe_2(CO)_8]^{2-} + Fe(CO)_5$/diglyme, Δ	66	^{1}H, ^{13}C NMR (15), ^{57}Fe Mössbauer (11)
$Fe_5N(CO)_{14}H$	15	15	$[Fe_5N(CO)_{14}]^- + H_2SO_4$/toluene	—	
$[Fe_6C(CO)_{16}]^{2-}$	16	16	$Fe(CO)_5 + [Mn(CO)_5]^-$/diglyme, Δ	—	^{13}C NMR (8), ^{57}Fe Mössbauer (10, 11), ESCA (10)
$[Fe_6C(CO)_{15}(NO)]^-$	17	17	$[Fe_6C(CO)_{16}]^{2-} + NOBF_4/CH_2Cl_2$	50	MS, ^{1}H, ^{13}C NMR
$Fe_6C(CO)_{11}(NO)_4$	17	17	$[Fe_6C(CO)_{15}(NO)]^- + NOBF_4/CH_2Cl_2$	10	
$Ru_5C(CO)_{14}(CNBu^t)_2$	18	18	$Ru_5C(CO)_{11}(CNBu^t)/\Delta$, N_2	—	^{1}H NMR
$Ru_5C(CO)_{13}H(CCHPPh_2)(PPh_2)$	19	19	$Ru_5C(CO)_{13}(CCPh_2)(PPh_2) + H_2$/cyclohexane, Δ	30	^{31}P NMR
$Ru_5C(CO)_{13}(CCPh)(PPh_2)$	19	19	$Ru_5C(CO)_{11}(Ph_2PCCPh)$/heptane,	100	^{31}P NMR
$Ru_5C(CO)_{14}(CCPh)(PPh_2)$	20	20	$Ru_5C(CO)_{13}(CCPh)(PPh_3) + CO$		
$Ru_5C(CO)_{13}(CCPh)(PPh_2)$	20	20			
$Ru_5C(CO)_{13}(CCPPh_2)(PPh_2)$	21	21	$Ru_5C(CO)_{12} + $ (i) $Na(Ph_2CO) + C_2(PPh_2)_2$ + (ii) toluene, Δ	62	

124

No.	Compound	Reaction	Yield (%)	Characterization	Ref.
22	$Ru_5(CO)_{15}(CCPPh_2)(PPh_2)$ (two isomers)	$Ru_5(CO)_{13}(CCPPh_2)(PPh_2)$ + CO/cyclohexane (isomer I) + CO/cyclohexane, Δ (isomer II)	80–85		22
23	$Ru_5(CO)_{15}(PR)$ (R = Ph, Et)	$Ru_5(CO)_{12}$ + [Mn(CO)$_2$CpPRCl$_2$]/toluene, Δ	35	MS, ^{31}P NMR	23
24	$[Ru_5N(CO)_{14}]^-$	$Ru_3(CO)_{12}$ + [N(PPh$_3$)$_2$][N$_3$]/THF, Δ → $[Ru_6N(CO)_{16}]^-$, $[Ru_6N(CO)_{16}]^-$ + CO/THF overall yield	5; 78	^{15}N NMR	24
25	$Ru_5C(CO)_{15}$	$Ru_6C(CO)_{17}$ + CO	97	IR (9)	25
26	$Ru_5C(CO)_{13}H(PPh_3)(SEt)$	$Ru_5C(CO)_{14}H(SEt)$ + PPh$_3$	66	MS, ^{1}H NMR	26
26	$Ru_5C(CO)_{12}H(PPh_3)(SEt)$	$Ru_5C(CO)_{13}H(PPh_3)(SEt)$/cyclohexane, Δ	71	MS, ^{1}H NMR	26
26	$Ru_5C(CO)_{14}H(SEt)$	$Ru_5C(CO)_{15}$ + EtSH	90	MS, ^{1}H NMR	26
25	$Ru_5C(CO)_{12}H_2\{Ph_2P(CH_2)_2PPh_2\}$	$Ru_5C(CO)_{15}$ + (i) Ph$_2$P(CH$_2$)$_2$PPh$_2$ + (ii) H$_2$	70	^{1}H NMR	25
19	$Ru_5C(CO)_{11}H_3(PPh_2)(PMePh_2)$	$Ru_5C(CO)_{13}(CCPPh_2)(PPh_2)$ + H$_2$/cyclohexane, Δ	—	^{1}H NMR	19
25	$Ru_5C(CO)_{15}(MeCN)$	$Ru_5C(CO)_{15}$ + MeCN	—	IR (27)	25
25	$Ru_5C(CO)_{14}(PPh_3)$	$Ru_5C(CO)_{15}$ + PPh$_3$	—		25
25	$Ru_5C(CO)_{13}(PPh_3)_2$	$Ru_5C(CO)_{14}(PPh_3)$ + PPh$_3$	100		25
28	$Ru_5C(CO)_{13}\{Ph_2P(CH_2)_4PPh_2\}$	$Ru_5C(CO)_{15}$ + Ph$_2$P(CH$_2$)$_4$PPh$_2$/hexane	—	^{31}P NMR	28
29	$Ru_5P(CO)_{16}(PPh_2)$	$Ru_3(CO)_9H(PPh_2)$/heptane, Δ	20	^{31}P NMR	29
30	$[Ru_6C(CO)_{18}]^{2-}$	$Ru_3(CO)_{12}$ + KOH/H$_2$O	80		31
31	$[Ru_6C(CO)_{18}H]^-$	$[Ru_3(CO)_{11}H]^-$ + H$_2$SO$_4$/THF	50	^{1}H, ^{13}C NMR, IR (33), neutron analysis (32)	31
34	$Ru_6(CO)_{18}H(OCNMe_2)_2$	$Ru_3(CO)_{10}H(OCNMe_2)$ + KOH/THF	45	No IR reported	34
35	$Ru_6(CO)_{18}H_2$	$[Ru_6(CO)_{18}]^{2-}$ + H$_2$SO$_4$/CH$_2$Cl$_2$	90	MS (36), IR (37)	31
38	$Ru_6C(CO)_{17}$	$Ru_3(CO)_{12}$ + CH$_2$Cl$_2$/Δ	70	MS (40)	39
39, 42	$[Ru_6C(CO)_{16}]^{2-}$	$Ru_3(CO)_{12}$ + NaMn(CO)$_5$/diglyme, Δ	60	^{13}C NMR (42)	43
44	$Ru_6C(CO)_{15}H(NO)$	$Ru_6C(CO)_{17}$ + (i) [N(PPh$_3$)$_3$]NO$_2$/CH$_2$Cl$_2$ + (ii) H$_2$SO$_4$	—	^{1}H NMR	44
45	$Ru_6C(CO)_{17}H(SEt)_3$	$Ru_6C(CO)_{17}$ + EtSH	45	No IR reported	45
46	$Ru_6C(CO)_{16}(CNBu^i)$	$Ru_5C(CO)_{14}(CNBu^i)_2$/nonane, Δ	13	MS, ^{13}C-labeling experiments	46
47	$Ru_6C(CO)_{15}(MeHC(CH)_2CHMe)$	$Ru_3(CO)_{12}$/CH$_2$CH$_2$, Δ	—	^{1}H NMR, MS, no IR reported	47
48	$Ru_6C(CO)_{14}(C_6H_3Me_3)$	$Ru_3(CO)_{12}$/C$_6$H$_3$Me$_3$, Δ	—	^{1}H NMR, MS (139)	139

(*continued*)

TABLE I (*continued*)

Cluster[b]	References X-Ray	References Synthesis[c]	Reagents and conditions	Yield (%)	Spectroscopic and theoretical studies[a]
$Ru_6C(CO)_{11}(C_6H_6)_2$	49	49	$[Ru_5C(CO)_{14}]^{2-}$ + (i) $[Ru(C_6H_6)(PhCN)_3][ClO_4]_2/CH_2Cl_2$, Δ + (ii) $Na_2CO_3/MeOH$ + (iii) $[Ru(C_6H_6)(PhCN)_3][ClO_4]_2/CH_2Cl_2$	—	^1H NMR
$Ru_6C(CO)_{14}(C_{14}H_{14})$	50	43	$[Ru_6C(CO)_{16}]^{2-}$ + $[C_7H_7]Br$	—	No IR reported
$Ru_6C(CO)_{14}(NO)_2$	51	51	$Ru_6C(CO)_{17}$ + (i) $[N(PPh_3)_2]NO_2$ + (ii) $NOBF_4$	—	
$Ru_6C(CO)_{16}(PPh_2Et)$	52	52	$Ru_6C(CO)_{17}$ + PPh_2Et	85	^{13}C NMR
$[Ru_{10}C_2(CO)_{24}]^{2-}$	53	53	$[Ru_6C(CO)_{16}]^{2-}$/tetraglyme, Δ	35	FABMS, ^{13}C NMR
$Os_5C(CO)_{16}$	54	41	Vacuum pyrolysis $Os_3(CO)_{12}$	7	MS, ΔH M-M, M-L (55)
$Os_5C(CO)_{19}$	56	56	$Os_6(CO)_{18}$ + CO (90 atm), heptane, Δ	80	MS
$[Os_5(CO)_{15}H]^-$	57	58	$[Os_5(CO)_{15}]^{2-}$ + $H_2SO_4/MeCN$	—	^{13}C and ^1H NMR (57)
$Os_5(CO)_{13}H(PhNC_6H_5N)$	59	59	$Os_3(CO)_{10}H_2$ + azobenzene, Δ	20–30	^1H NMR, MS
$Os_5(CO)_{13}H(PhNC_6H_4N)(PEt_3)$	60	60	$Os_5(CO)_{13}H(PhNC_6H_4N)$ + PEt_3	60	^1H NMR, MS
$Os_5(CO)_{16}H_2$	61	62	$Os_5(CO)_{15}H_2$ + CO	40	^1H NMR, MS (63)
$Os_5(CO)_{15}H_2(CCPh)$	64	64	$Os_5(CO)_{15}H_2$ + HCCPh/Δ	—	^1H NMR, MS
$Os_5(CO)_{14}H_2(C_5H_4N)$	177	177	$Os_5(CO)_{15}H_2$ + C_5H_5N/octane, Δ	23	MS
$Os_5(CO)_{15}H_2\{P(OMe)_3\}$	65	65	$Os_5(CO)_{15}H_2$ + PR_3 (R = OMe, Et, Ph)	60	^1H NMR, MS
$Os_5(CO)_{14}H_2(PEt_3)$	66	66	$Os_5(CO)_{15}H_2(PR_3)$/Δ (R = OMe, Et, Ph)	—	MS
$Os_5(CO)_{13}H_2(PEt_3)\{P(OMe)_3\}$	66	66	$Os_5(CO)_{14}H_2(PEt_3)$ + $P(OMe)_3$/Δ	—	MS
$Os_5(CO)_{14}H_2S_2$	67	67	$Os_5(CO)_{15}S$ + H_2S/Δ	7	
$[Os_5(CO)_{15}H_2I]^-$	65	65	$Os_5(CO)_{15}H_2$ + Bu_4NI	80	^1H NMR
$Os_5(CO)_{14}H_3(C_5H_4N)$	177	177	$Os_5(CO)_{15}H_2$ + C_5H_5N/octane, Δ	9	
$Os_5(CO)_{17}(HCCH)$	68	68	$Os_5(CO)_{19}$ + HCCH	15	MS
$Os_5(CO)_{14}(PhCCPh)$	69	69	$Os_5(CO)_{15}(MeCN)$ + $RCCR/CH_2Cl_2$ (R = Ph, Et, Me)	—	MS
$Os_6(CO)_{20}(HCCPh)$	70	70	$Os_6(CO)_{20}(HCCPh)/CH_2Cl_2$, Δ	—	^1H NMR, MS
$Os_5(CO)_{13}(HCCPh)(PPh_3)$	377	377	$Os_5(CO)_{15}(HCCPh)$ + PPh_3/toluene, Δ	25	^1H NMR, MS
$Os_5(CO)_{13}(PhCCPh)_2$	64	64	$Os_5(CO)_{15}H_2$ + PhCCPh/Δ	—	

126

Complex			Synthesis	Yield (%)	Methods of characterization (ref.)
Os$_5$(CO)$_{15}$(POMe)	71	72	Vacuum pyrolysis Os$_3$(CO)$_{11}$P(OMe)$_3$	—	^{1}H NMR, MS (72)
Os$_5$(CO)$_{16}$\{P(OMe)$_3$\}$_3$	56	56	Os$_5$(CO)$_{19}$ + P(OMe)$_3$	80	MS
Os$_5$(CO)$_{15}$S	73	73	Os$_3$(CO)$_{10}$(MeCN)$_2$ + Os$_3$(CO)$_{10}$S/Δ	18	^{1}H NMR, MS
Os$_5$(CO)$_{14}$S(PPh$_3$)	377	377	Os$_6$(CO)$_{19}$(PPh$_3$)$_2$ + S$_8$/toluene, Δ	65	No IR quoted
[Os$_5$(CO)$_{15}$I]$^-$	74	75	Os$_5$(CO)$_{16}$ + Bu$_4$NI	—	MS, IR (9)
Os$_5$C(CO)$_{15}$	76	54, 77	Vacuum pyrolysis Os$_3$(CO)$_{12}$	5	^{1}H NMR, MS
Os$_5$C(CO)$_{16}$	78	78	Os$_5$C(CO)$_{15}$ + CO (50 atm), Δ	80	^{1}H NMR, MS
[Os$_5$C(CO)$_{14}$]$^{2-}$	78	78	Os$_5$C(CO)$_{15}$ + NaCO$_3$/MeOH	75	^{1}H NMR, MS (81)
Os$_5$C(CO)$_{14}$H(CO$_2$Et)	79	79	Os$_5$C(CO)$_{15}$ + ROH/Δ (R = Me, Et, Bu^i)	60–80	^{1}H NMR, MS (82)
Os$_5$C(CO)$_{14}$H(C$_5$H$_4$N)	80	80	Vacuum pyrolysis Os$_3$(CO)$_{11}$(C$_5$H$_4$N)	4	^{1}H NMR, MS
Os$_5$C(CO)$_{14}$H\{OP(OMe)$_2$\}	81	72	Vacuum pyrolysis Os$_3$(CO)$_{11}$P(OMe)$_3$	—	MS, ^{31}P NMR
Os$_5$C(CO)$_{13}$H\{OP(OMe)OP(OMe)$_2$\}	82	72	Vacuum pyrolysis Os$_3$(CO)$_{11}$P(OMe)$_3$	—	^{1}H NMR, MS (67)
Os$_5$C(CO)$_{13}$H\{OP(OMe)$_2$\}\{P(OMe)$_3$\}	72	72	Vacuum pyrolysis Os$_3$(CO)$_{11}$P(OMe)$_3$	—	
Os$_5$C(CO)$_{15}$\{Ph$_2$P(CH$_2$)$_2$PPh$_2$\}	83	83	Os$_5$C(CO)$_{15}$ + Ph$_2$P(CH$_2$)$_2$PPh$_2$/CH$_2$Cl$_2$	95	
Os$_5$C(CO)$_{14}$(CO$_2$Me)I	79	79	Os$_5$C(CO)$_{15}$I$_2$ + MeOH	80–90	
[Os$_5$C(CO)$_{15}$I]$^-$	76	76	Os$_5$C(CO)$_{15}$ + Bu$_4$NI/CH$_2$Cl$_2$	—	
Os$_6$(CO)$_{18}$	84	41	Vacuum pyrolysis Os$_3$(CO)$_{12}$	80	^{13}C NMR (VT) (85), MS (41), ΔH M–M, M–L (55), XAFS (86)
[Os$_6$(CO)$_{18}$]$^{2-}$	87	88	Os$_6$(CO)$_{18}$ + Bu$_4$NI/CH$_2$Cl$_2$	—	^{13}C NMR (VT) (88)
[Os$_6$(CO)$_{18}$H]$^-$	87	88	Os$_6$(CO)$_{18}$ + NaBH$_4$,THF	—	^{1}H, ^{13}C NMR (88), IR (37)
Os$_6$(CO)$_{17}$H(CCEt)	89	89	Os$_6$(CO)$_{17}$(MeCN) + RCCH/toluene, 70°C (R = Et, Ph, Me)	—	MS, ^{1}H NMR
Os$_6$(CO)$_{16}$H(C$_5$H$_4$N)	177	177	Os$_6$(CO)$_{18}$ + C$_5$H$_5$N/octane, Δ	40	^{1}H NMR, MS
Os$_6$(CO)$_{16}$H\{C$_5$H$_3$(Me)N\}	177	177	Os$_6$(CO)$_{18}$ + C$_5$H$_4$(Me)N/octane, Δ	—	
Os$_6$(CO)$_{19}$H(SEt)	90	90	Os$_6$(CO)$_{19}$(MeCN)$_2$ + (i) HSEt + NEt$_3$/CH$_2$Cl$_2$ + (ii) HBF$_4$/CH$_2$Cl$_2$	—	
Os$_6$(CO)$_{18}$H$_2$	87, 91	88	[Os$_6$(CO)$_{18}$]$^{2-}$ + H$_2$SO$_4$/MeCN	100	^{1}H, ^{13}C NMR (88), MS (63)
Os$_6$(CO)$_{19}$H$_2$	92	92	Os$_5$(CO)$_{16}$ + (i) Me$_3$NO + MeCN/CH$_2$Cl$_2$ + (ii) H$_2$OsCO$_4$/CH$_2$Cl$_2$	52	^{1}H NMR, MS
Os$_6$(CO)$_{18}$H$_2$(PPh)	93	93	Os$_6$(CO)$_{20}$(MeCN) + (i) PPhH$_2$/CH$_2$Cl$_2$ + (ii) toluene, Δ	80	^{1}H NMR

(continued)

TABLE I (*continued*)

Cluster[b]	References		Reagents and conditions	Yield (%)	Spectroscopic and theoretical studies[d]
	X-Ray	Synthesis[c]			
Os$_6$(CO)$_{20}$H$_2$(PH)	94	94	Os$_3$(CO)$_{10}$H(PH$_2$) + Os$_3$(CO)$_{10}$(MeCN)$_2$	90	^{1}H NMR
Os$_6$(CO)$_{17}$H$_2$\{P(OMe)$_3$\}	313	313	Os$_6$(CO)$_{18}$ + (i) Me$_3$NO/CH$_2$Cl$_2$, + (ii) HBF$_4$/CH$_2$Cl$_2$, + (iii) POMe$_3$	72	^{1}H NMR
Os$_6$(CO)$_{18}$H$_2$S$_2$	95	95	Os$_3$(CO)$_{10}$H(SCH$_2$C$_6$H$_5$), hv/N$_2$/hexane	8	^{1}H NMR
Os$_6$(CO)$_{17}$H$_2$S$_2$ (two isomers)	95	95	Os$_3$(CO)$_{10}$H(SCH$_2$C$_6$H$_5$)/nonane, Δ isomer I isomer II	2.5 10	^{1}H NMR ^{1}H NMR
Os$_6$(CO)$_{15}$H$_4$S$_2$(HCNC$_6$H$_5$)$_2$ and Os$_6$(CO)$_{14}$H$_6$S$_2$(HCNC$_6$H$_5$)$_2$	96	96	Os$_3$(CO)$_9$HS(HCNC$_6$H$_5$) + H$_2$/octane, Δ	47 12	^{1}H NMR ^{1}H NMR
Os$_6$(CO)$_{17}$(PPh$_3$)	97	97	Os$_6$(CO)$_{17}$(MeCN) + PR$_3$/CH$_2$Cl$_2$ (R = Ph, OMe)	69	^{1}H NMR, MS
Os$_6$(CO)$_{16}$(PPh$_3$)$_2$	97	97	Os$_6$(CO)$_{16}$(MeCN)$_2$ + PPh$_3$/MeCN, CH$_2$Cl$_2$	25	
Os$_6$(CO)$_{20}$\{P(OMe)$_3$\}	98	99	Os$_3$(CO)$_{10}$(CH$_3$CN)$_2$ + PdCl$_2$/CH$_2$Cl$_2$ → Os$_6$CO$_{20}$(MeCN) (two isomers) + Os$_6$CO$_{19}$(MeCN)$_2$ (two isomers)	46 30	
Os$_6$(CO)$_{19}$\{P(OMe)$_3$\}$_2$	98	99	Os$_6$CO$_{20}$(MeCN) + P(OMe)$_3$/CH$_2$Cl$_2$	70	^{1}H NMR, MS
Os$_6$(CO)$_{17}$\{P(OMe)$_3$\}$_4$	100	100	Os$_6$(CO)$_{19}$(MeCN)$_2$ + P(OMe)$_3$/CH$_2$Cl$_2$	80	^{1}H, ^{31}P\{^{1}H\} NMR
Os$_6$(CO)$_{17}$P(OMe)$_3$ (MeCN)$_2$	101	101	Os$_6$(CO)$_{21}$ + P(OMe)$_3$ (excess)/benzene, Δ	—	
Os$_6$(CO)$_{17}$\{P(OMe)$_3$\} + MeCN		101	Os$_6$(CO)$_{17}$\{P(OMe)$_3$\} + MeCN	—	
Os$_6$(CO)$_{18}$(PPh$_3$)O	102	102	Os$_6$(CO)$_{20}$(PPh$_3$) + O$_2$/toluene, Δ	80	^{1}H NMR, MS
Os$_6$(CO)$_{20}$PhCCH	70	70	Os$_6$(CO)$_{20}$(MeCN) + RCCH, (R = Ph, Me)/CH$_2$Cl$_2$	90	^{1}H NMR, MS
Os$_6$(CO)$_{17}$(EtCCH)	89	89	Os$_6$(CO)$_{17}$(MeCN) + RCCH/toluene, 45°C (R = Et, Ph)	—	MS, ^{1}H NMR
Os$_6$(CO)$_{16}$(CMe)$_2$	103	103	Os$_6$(CO)$_{18}$ + CH$_2$CH$_2$/decane, Δ	10	^{1}H NMR, MS
Os$_6$(CO)$_{16}$(MeCCEt)	89	89	Os$_6$(CO)$_{17}$MeCN + RCCR'/toluene, Δ (R = R' = Me, Et, Ph; R = Me, R' = Et)	50–60	MS, ^{1}H NMR
Os$_6$(CO)$_{16}$(CPh)$_2$	104	104	Os$_6$(CO)$_{18}$ + PhCCPh, hv	—	No spectral data reported
Os$_6$(CO)$_{16}$(CMe)(CEt)	89	89	Os$_6$(CO)$_{16}$RCCR'/octane, Δ (R = R' = Me, Ph; R = Me, R' = Et)	40–50	MS, ^{1}H NMR

$Os_6(CO)_{15}(MeCCMe)_2$ (isomer I)	$Os_6(CO)_{16}(MeCCMe)$ + (i) Me_3NO + $MeCN/CH_2Cl_2$ + (ii) $MeCCMe$/toluene, Δ	*106*	*106*	45	MS, ^{1}H NMR
(isomer II)	Isomer I/toluene, 45°C	*106*	*106*	100	MS, ^{1}H NMR
$Os_6(CO)_{16}(C_4H_6)$	$Os_6(CO)_{16}(MeCN)_2$ + C_4H_6/toluene, Δ	*105*	*105*	50	MS, ^{1}H NMR
$Os_6(CO)_{16}(CNCMe_3)_2$	Pyrolysis of $Os_3(CO)_{11}(CNCMe_3)$	*107*	*108*	—	MS, ^{1}H NMR
$Os_6(CO)_{18}(CNC_6H_4\text{-}p\text{-Me})_2$	$Os_6(CO)_{18}$ + CNC_6H_4-p-Me	*109*	*110*	1	^{1}H NMR *(110)*
$Os_6(CO)_{15}S_2(HCNC_6H_5)_2$	$Os_3(CO)_9HS(HCNC_6H_5)_2$/octane, Δ	*111*	*111*	10	^{1}H NMR
$Os_6(CO)_{17}(NC_5H_5)_2$	$Os_6(CO)_{18}$ + C_5H_5N/CH_2Cl_2	*112*	*112*	—	^{1}H NMR
$Os_6(CO)_{19}O$	$Os_6(CO)_{21}$ + O_2/toluene, Δ	*113*	*113*	31	MS
$Os_6(CO)_{19}S$	$Os_3(CO)_{10}(MeCN)_2$ + $Os_3(CO)_{10}S$/benzene, Δ	*114*	*114*	23	
$Os_6(CO)_{17}S$	$Os_6(CO)_{19}S$/toluene, Δ	*114*	*114*	27	
$Os_6(CO)_{17}S_2$	$Os_3(CO)_{10}(MeCN)_2$ + $Os_3(CO)_9S_2$/hexane, Δ	*115*	*115*	100	
$Os_6(CO)_{16}S_2$	$Os_6(CO)_{17}S_2$/octane, Δ	*115*	*115*	34	
$Os_6(CO)_{16}S_4$	$Os_3(CO)_9S_2/N_2/h\nu$	*116*	*116*	5	^{1}H NMR, MS
$Os_6C(CO)_{16}(MeCCMe)$	$Os_6(CO)_{18}$ + CH_2CH_2/decane, Δ	*103*	*103*		
$Os_7(CO)_{21}$	Vacuum pyrolysis $Os_3(CO)_{12}$	*117*	*41*	max. 20	^{13}C NMR *(117)*, MS *(41)*
$Os_7(CO)_{20}H_2$	$Os_6(CO)_{16}(MeCN)_2$ + $Os(CO)_4H_2$	*118*	*119*	64	^{1}H NMR *(118)*, MS *(118)*
$Os_7(CO)_{21}H_2$	$Os_6(CO)_{18}$ + $Os(CO)_4H_2$ + 1.5Me_3NO/CH_2Cl_2	*119*	*119*	41	MS, ^{1}H NMR
$Os_7(CO)_{22}H_2$	$Os_6(CO)_{18}$ + $Os(CO)_4H_2$ + 1.5Me_3NO/CH_2Cl_2	*119*	*119*	13	MS, ^{1}H NMR
$Os_7(CO)_{19}S$	$Os_4(CO)_{12}S$ + $Os_3(CO)_{10}(MeCN)_2$/octane, Δ	*120*	*120*	23	
$Os_7(CO)_{20}S_2$	$Os_4(CO)_{12}S_2$ + $Os_3(CO)_{10}(MeCN)_2$/octane, Δ	*121*	*121*	16	
$Os_8(CO)_{23}$	Vacuum pyrolysis $Os_3(CO)_{12}$	*122*	*41*	max. 5	MS *(41)*
$[Os_8(CO)_{22}]^{2-}$	$Os_8(CO)_{23}$ + Bu_4N/THF	*122*	*123*	30	^{1}H NMR, MS
$[Os_8(CO)_{22}H]^-$	$Os_8(CO)_{22}/Bu^tOH$, Δ	*124*	*124*	95	^{1}H NMR, MS
$Os_8(CO)_{22}HI$	$[Os_8(CO)_{22}H]^-$ + I_2/CH_2Cl_2	*124*	*124*	—	^{1}H NMR, FABMS
$[Os_9(CO)_{21}\{CHC(R)CH\}]^-$ (R = Me, Et)	$Os_3(CO)_{12}/Bu^tOH$, Δ	*125*	*125*	79	^{13}C NMR (VT) *(127)*
$[Os_3(CO)_{10}HO_2Co_6(CO)_{17}]^-$	$[Os_3(CO)_{11}H]^-$ + $Os_6(CO)_{18}$	*126*	*127*	—	^{1}H NMR (VT)
$[Os_{10}(CO)_{24}H_4]^{2-}$	$Os_3(CO)_{12}/Bu^tOH$, Δ	*128*	*128*	65	FABMS
$Os_{10}(CO)_{23}S_2$	Vacuum pyrolysis $Os_3(CO)_{12}/S_8$	*129*	*129*	95	XAFS *(86)*, IR *(424)*
$[Os_{10}C(CO)_{24}]^{2-}$	Vacuum pyrolysis $Os_3(CO)_{11}(C_5H_5N)$	*80*	*80*	85	^{1}H NMR *(130, 131)*, ^{13}C NMR *(131)*
$[Os_{10}C(CO)_{24}H]^-$	$[Os_{10}C(CO)_{24}]^{2-}$ + H_2SO_4/CH_2Cl_2	*130*	*130*	—	FABMS
$[Os_{10}C(CO)_{24}(NO)]^-$	$[Os_{10}C(CO)_{24}]^{2-}$ + $NOBF_4$/MeCN	*132*	*132*	85	^{13}C NMR *(131)*
$[Os_{10}C(CO)_{23}(NO)]^-$	$[Os_{10}C(CO)_{24}(NO)]^-/CH_2Cl_2$	*132*	*132*	—	

(continued)

TABLE I (continued)

Cluster[b]	References		Reagents and conditions	Yield (%)	Spectroscopic and theoretical studies[d]
	X-Ray	Synthesis[c]			
$[Os_{10}C(CO)_{22}(NO)I]^{2-}$	133	133	$Os_{10}C(CO)_{24}I_2 + [N(PPh_3)_2]NO_2/CH_2Cl_2$	64	
$Os_{10}C(CO)_{23}\{P(OMe)_3\}I_2$	133	133	$Os_{10}C(CO)_{24}I_2 + P(OMe)_3/xylene, \Delta$	—	
$Os_{10}C(CO)_{21}\{P(OMe)_3\}_4$	133	133	$Os_{10}C(CO)_{24}I_2 + P(OMe)_3/xylene, \Delta$	—	
$[Os_{10}C(CO)_{24}I]^-$	133	133	$[Os_{10}C(CO)_{24}]^{2-} + 2I_2/CH_2Cl_2$	71	
$Os_{10}C(CO)_{24}I_2$	133	133	$[Os_{10}C(CO)_{24}]^{2-} + 4I_2/CH_2Cl_2$	82	FABMS
$[Os_{11}C(CO)_{27}]^{2-}$	134	134	Vacuum pyrolysis $Os_3(CO)_{12}$	1	
$Co_5(CO)_{11}(PMe_2)_3$	135	136	$Co(CO)_3(C_3H_5) + Co(CO)_2(NO)(PMe_2H)$	7	$^1H, ^{31}P$ NMR (135)
$Co_6(CO)_{16}$	137	138	$[Co_6(CO)_{15}]^{2-} + Na_2[HgCl_4]$	8	
$[Co_6(CO)_{15}]^{2-}$	137	140	$Co_2(CO)_8 +$ (i) EtOH, (ii) Δ, vacuum, + (iii) H_2O/KBr	80	^{13}C NMR (156)
$[Co_6(CO)_{14}]^{4-}$	141	142	$Co_4(CO)_{12} + Na$ or Li/THF	46	MO calc. (143)
$[Co_6(CO)_{15}(H)]^-$	144	144	$[Co_6(CO)_{15}]^{2-} + HCl/H_2O$	—	^1H NMR
$[Co_6C(CO)_{15}]^{2-}$	145	145	$[Co(CO)_4]^- +$ (i) $Co_3(CO)_9CCl/diisopropyl$ ether, Δ + (ii) KBr	70–80	^{13}C NMR (156), IR (146)
$[Co_6C(CO)_{14}]^-$	147	147	$[Co_6C(CO)_{15}]^{2-} + FeCl_3/acetone$	75–85	ESR (148)
$Co_6C(CO)_{12}S_2$	149	150	$Co_2(CO)_8 + CS_2/hexane$	20–35	IR (149)
$Co_6(C_2)(CO)_{14}S$	151	151	$Co_2(CO)_8 + CS_2/petroleum$ ether, Δ	5	IR
$[Co_6N(CO)_{15}]^-$	152	152	$[Co_6(CO)_{15}]^{2-} + NOBF_4/THF$	40–50	$^{13}C, ^{15}N, ^{14}N$ NMR
		234	$Co_4(CO)_{12} + [N(PPh_3)_2][NO_2]/THF$	40–50	IR (146)
$[Co_6P(CO)_{16}]^-$	153	154	$[Co(CO)_4]^- + PCl_3/THF$	5–10	^{31}P NMR (154)
$[Co_8C(CO)_{18}]^{2-}$	155	145	$[Co(CO)_4]^- +$ (i) $Co_3(CO)_9CCl + Co_4(CO)_{12}/diisopropyl$ ether + (ii) KBr	75–85	
$[Co_9Si(CO)_{21}]^{2-}$	157	157	$Si[Co_3(CO)_7]_2 + [Co(CO)_4]^-/CH_2Cl_2, \Delta$	12	
$[Co_{11}(C_2)(CO)_{22}]^{3-}$	158	158	$[Co_6(CO)_{15}]^{2-}/diglyme, \Delta$	50	
$[Co_{13}C_2(CO)_{24}]^{4-}$	159	159	$[Co_6(CO)_{15}]^{2-}/diglyme, \Delta$	40–50	ESR (148)
$[Co_{13}(C)_2(CO)_{24}]^{3-}$	160	160	$Co_{13}(C)_2(CO)_{24}]^{4-} + \frac{1}{2}I_2/MeCN$	75–85	
$Co_{16}As_2(CO)_{32}(AsPh)_4$	161	161	$Co_2(CO)_8 + (AsPh)_6/toluene, \Delta$	21	
$[Rh_5(CO)_{15}]^-$	162	162	$Rh_4(CO)_{12} + [Rh(CO)_4]^-$	80	^{13}C NMR
$[Rh_5(CO)_{14}I]^{2-}$	163	163	$Rh_4(CO)_{12} + Bu^n_4NI$	—	

Compound			Preparation	Yield (%)	Characterization
$Rh_6(CO)_{16}$	164	165	$RhCl_3 \cdot 3H_2O$ + CO (40 atm)/MeOH, Δ	80–90	MO calc. (167)
		166	$Rh_2(O_2CMe)_4$ + HBF_4 + CO/Pr^iOH, Δ	87	^{13}C NMR (168), MS (165)
$[Rh_6(CO)_{14}]^{4-}$	169	170	$Rh_6(CO)_{16}$ + KOH/H_2O	70–80	
$[Rh_6(CO)_{15}(COEt)]^-$	171	172	$Rh_4(CO)_{12}$ + C_2H_4 + H_2O	—	^1H NMR
$[Rh_6(CO)_{15}CO(OMe)]^-$	171	173	$Rh_4(CO)_{12}$ + Na_2CO_3/MeOH	80	
$[Rh_6(CO)_{15}]^{2-}$	174	174	$[Rh_6(CO)_{15}]^{2-}$ + C_3H_5Cl	—	
$Rh_6(CO)_{10}(C_7H_8)_3$	175	175	$Rh_6(CO)_{16}$ + C_7H_8/methylcyclohexane	65	
$Rh_6(CO)_{12}\{P(OPh)_3\}_4$	176	176	$Rh_6(CO)_{16}$ + $P(OPh)_3$/CH_2Cl_2, Δ	87	
$Rh_6(CO)_{10}\{Ph_2P(CH_2)PPh_2\}_3$	178	179	$Rh_6(CO)_{16}$ + $Ph_2P(CH_2)PPh_2$/CH_2Cl_2	—	$^{13}C,\{^{103}Rh\}$ NMR (178), ^{31}P NMR (179)
$Rh_6(CO)_9(Bu^t_2As)_2(Bu^tAs)$	180	180	$[Rh(CO)_2Cl]_2$ + $Li(Bu^t)_2As$/THF, −78°C	35	^1H NMR
$[Rh_6(CO)_{15}I]^-$	181	173	$[Rh_6(CO)_{15}]^{2-}$ + I_2	10	^{13}C NMR (182)
$[Rh_6C(CO)_{15}]^{2-}$	183	183	$Rh_4(CO)_{12}$ + (i) NaOH/MeOH + (ii) $CHCl_3$	70	^{13}C NMR (184), IR (146),
		187	$[RhCl_6]^{3-}$ + KOH + CO + $CHCl_3$/MeOH	70	$^{13}C\{^{103}Rh\}$ NMR (185), ^{103}Rh NMR (186)
$[Rh_6C(CO)_{13}]^{2-}$	188	188	$K_2[Rh_6C(CO)_{15}] \cdot MeO(CH_2)_4OMe$/Δ	—	$^{13}C\{^{103}Rh\}$, ^{13}C NMR (185)
$[Rh_6N(CO)_{15}]^-$	189	234	$Rh_6(CO)_{16}$ + $[N(PPh_3)_2][NO_2]$/THF	77	^{13}C, ^{15}N NMR (190), IR (146)
$[Rh_7(CO)_{16}]^{3-}$	191	192	$Rh_6(CO)_{16}$ + KOH/MeOH	90	$^{13}C\{^{103}Rh\}$ (193), ^{13}C NMR (169)
$[Rh_7(CO)_{16}I]^{2-}$	194	195	$Rh_6(CO)_{16}$ + R_4NI	80	^{13}C NMR (196)
$[Rh_7N(CO)_{15}]^{2-}$	197	197	$[Rh_6N(CO)_{15}]^{2-}$ + $[Rh(CO)_4]^-$	80	
$Rh_8C(CO)_{19}$	198	199	$K_2[Rh_6C(CO)_{15}]$ + (i) $Fe[NH_4][SO_4]_2$ + CO/H_2O + (ii) CO/CH_2Cl_2	—	
$[Rh_9(CO)_{19}]^{3-}$	200	200	$[Rh_5(CO)_{15}]^-$ + $[Rh_4(CO)_{11}]^{2-}$	—	
$[Rh_9P(CO)_{21}]^{2-}$	201	201	$Rh(CO)_2(acac)$ + CO/H_2 (400 atm) + $Cs[C_6H_5COO]$ + PPh_3/tetraglyme, Δ	80	^{31}P NMR, ^{103}Rh NMR (186)
$[Rh_{10}S(CO)_{22}]^{2-}$	202	203	$Rh_4(CO)_{12}$ + SCN^-	85	^{103}Rh, ^{13}C, $^{13}C\{^{103}Rh\}$ NMR
$[Rh_{10}P(CO)_{22}]^{2-}$	204	204	$Rh(CO)_2(acac)$ + CO/H_2 (400 atm) + $Cs[C_6H_5COO]$ + $Cs[BH_4]$ + PPh_3/diglyme, Δ	98	^{13}C NMR (205)
$[Rh_{10}As(CO)_{22}]^{2-}$	205	205	$Rh(CO)_2(acac)$ + CO/H_2 (400 atm) + $Cs[C_6H_5COO]$ + $AsPh_3$/diglyme, Δ	98	^{13}C NMR
$[Rh_{11}(CO)_{23}]^{3-}$	206	206	$[Rh_7(CO)_{16}]^{3-}$ + $FeCl_3$	—	

(continued)

131

TABLE I (*continued*)

Cluster[b]	References: X-Ray	References: Synthesis[c]	Reagents and conditions	Yield (%)	Spectroscopic and theoretical studies[d]
$[Rh_{12}(CO)_{30}]^{2-}$	*207*	*208*	$Rh_4(CO)_{12}$ + (i) Na[MeCOO] + H_2O/acetone + (ii) NaCl/H_2O	85	^{13}C NMR (*209*)
$Rh_{12}(C_2)(CO)_{25}$	*210*	*210*	$[Rh_6C(CO)_{15}]^{2-}$ + (i) $Fe[NH_4][SO_4]_2/H_2O$ + (ii) CH_2Cl_2	79	
$[Rh_{12}(C)_2(CO)_{24}]^{2-}$	*211*	*211*	$[Rh_6C(CO)_{15}]^{2-}$ + H_2SO_4/Pr^{i}OH	90	
$[Rh_{12}(C)_2(CO)_{23}]^{3-}$	*212*	*212*	$[Rh_{12}(C)_2(CO)_{23}]^{2-}$ + NaOH/MeOH	40	ESR (*148*)
$[Rh_{12}(C)_2(CO)_{23}]^{4-}$	*213*	*213*	$[Rh_{12}(C)_2(CO)_{24}]^{2-}$ + KOH/Pr^{i}OH, Δ	65–75	
$[Rh_{13}(CO)_{24}H]^{4-}$	*214*	*214*	$[Rh_{13}(CO)_{24}H_3]^{2-}$ + KOBu^t/THF	—	^1H NMR
$[Rh_{13}(CO)_{24}(H)_2]^{3-}$	*215*	*216*	$Rh(CO)_2$(acac) + CO/H_2 (12 atm) + $Cs[C_6H_5COO)]/(CH_2OH)_2$, tetraglyme, Δ	91	^1H, ^{13}C, ^1H {^{103}Rh} NMR (*217*)
$[Rh_{13}(CO)_{24}H_3]^{2-}$	*218*	*218*	$[Rh_{12}(CO)_{30}]^{2-}$ + H_2	50	^1H, ^{13}C, ^1H {^{103}Rh} NMR (*217*)
$[Rh_{14}(CO)_{26}]^{2-}$	*219*	*216*	$[Rh_7(CO)_{16}]^{3-}$ + CF_3SO_3H	86	^{13}C NMR (*216*)
$[Rh_{14}(CO)_{25}]^{4-}$	*220*	*221*	$Rh(CO)_2$(acac) + CO/H_2 (200 atm) + $Cs[C_6H_5COO]$ + $(CH_2OH)_2$ + *n*-methyl morpholine/tetraglyme, Δ	60	
$[Rh_{14}(CO)_{25}H]^{3-}$	*223*	*223*	$[Rh_{14}(CO)_{25}]^{4-}$ + H_3PO_4, $MeCN/H_2O$	—	^1H NMR (*233*), ^{13}C, ^{103}Rh NMR (*222*)
$[Rh_{14}(C)_2(CO)_{33}]^{2-}$	*224*	*224*	$[Rh_6C(CO)_{15}]^{2-}$ + $[Rh(CO)_2(MeCN)_2]^+$	30	
$[Rh_{15}(CO)_{27}]^{3-}$	*225*	*226*	$Rh(CO)_2$(acac) + CO/H_2 (19 atm) + $Cs[C_6H_5COO]$ + $(CH_2OH)_2$ + *n*-methylmorpholine/tetraglyme, Δ	87	^{13}C NMR (*226*)
$[Rh_{15}(CO)_{30}]^{2-}$	*227*	*227*	$Rh(CO)_2$(acac) + (i) CO/H_2 (15 atm) + $Cs[C_6H_5COO]$ + 18-crown-6 + $(CH_2OH)_2$ + *n*-methylmorpholine, Δ + (ii) MeOH + $(CH_2OH)_2$	—	
$[Rh_{15}(C)_2(CO)_{28}]^-$	*228*	*199*	$[Rh_6C(CO)_{15}]^{2-}$ + $Fe[NH_4][SO_4]_2/N_2$	—	
$[Rh_{17}(CO)_{30}]^{3-}$	*229*	*229*	$Rh_4(CO)_{12}$ + NaOH/Pr^{i}OH	20	
$[Rh_{17}(S)_2(CO)_{32}]^{3-}$	*230*	*230*	$Rh(CO)_2$(acac) + CO/H_2 (300 atm) + $Cs[C_6H_5COO]$ + H_2O + H_2S or SO_2/tetraglyme, Δ	73	^1H NMR (*230*), ^{13}C NMR (*231*), ^{103}Rh NMR (*186*)

Compound	Reaction	Ref.	Ref.	Yield (%)	Characterization
[Rh$_{22}$(CO)$_{37}$]$^{4-}$	Rh$_4$(CO)$_{12}$ + NaOH/Pr^{i}OH	232	232	1–10	^{13}C, ^{1}H NMR, MO calc.
[Rh$_{22}$(CO)$_{35}$H$_x$]$^{5-}$ and [Rh$_{22}$(CO)$_{35}$H$_{(x+1)}$]$^{4-}$	Rh(CO)$_2$(acac) + 18-crown-6 + CO + Cs[C$_6$H$_5$COO] + H$_2$O	233	233	25	^{1}H, ^{13}C NMR
Ir$_6$(CO)$_{16}$ (red)	[Ir$_6$(CO)$_{15}$]$^{2-}$ + MeCOOH/CO	235	236	87	
Ir$_6$(CO)$_{16}$ (black)	[Ir$_6$(CO)$_{15}$]$^{2-}$ + MeCOOH/CH$_2$Cl$_2$	235	235	10	
	[Ir$_6$(CO)$_{15}$]$^{2-}$ + CF$_3$SO$_3$H/CH$_2$Cl$_2$		234^e	94	
[Ir$_6$(CO)$_{15}$]$^{2-}$	Ir$_4$(CO)$_{12}$ + (i) Na sand + CO/THF (ii) Ir$_4$(CO)$_{12}$ + CO	234	237	84	FABMS (234)
[Ir$_6$(CO)$_{15}$(COEt)]$^-$	Ir$_6$(CO)$_{16}$ + CO/H$_2$ + H$_2$CCH$_2$ + H$_2$O/THF	238	238	—	^{1}H NMR
Ir$_6$(CO)$_{12}${P(OPh)$_3$}$_4$	Ir$_6$(CO)$_{16}$ + P(OPh)$_3$/toluene, Δ	239	239	—	^{13}C NMR
Ir$_6$(CO)$_{11}${P(OMe)$_3$}$_5$	Ir$_6$(CO)$_{16}$ + P(OMe)$_3$/toluene, Δ	240	240	—	^{1}H NMR
Ir$_7$(CO)$_{12}$(C$_8$H$_{12}$)(C$_8$H$_{11}$)(C$_8$H$_{10}$)	Ir$_4$(CO)$_{12}$ + 1,5-cyclooctadiene/chlorobenzene, Δ	241	242	3	MS, ^{1}H NMR (242)
[Ir$_8$(CO)$_{22}$]$^{2-}$	Ir$_4$(CO)$_{12}$ + (i) Na/THF/CO + (ii) Et$_4$NCl/H$_2$O	243	244	42	
[Ni$_5$(CO)$_{12}$]$^{2-}$	Ni(CO)$_4$ + Na/THF	245	246		^{13}C NMR (247)
Ni$_5$(CO)$_6${(Me$_3$Si)$_2$HCPPCH(SiMe$_3$)$_2$}Cl	[Ni$_6$(CO)$_{12}$]$^{2-}$ + P{CH(SiMe$_3$)$_2$}Cl$_2$/ether	248	248	15	^{31}P NMR
[Ni$_6$(CO)$_{12}$]$^{2-}$	Ni(CO)$_4$ + NaBH$_4$/THF	249	250	68	^{13}C NMR (247)
[Ni$_8$(CO)$_{18}$(PPh)$_6$]	[Ni$_6$(CO)$_{12}$]$^{2-}$ + PhPCl$_2$/THF	251	251	20	^{1}H NMR
[Ni$_6$C(CO)$_{16}$]$^{2-}$	[Ni$_6$(CO)$_{12}$]$^{2-}$ + CO/THF	252	252	69	IR (416)
[Ni$_9$(CO)$_{18}$]$^{2-}$	Ni(CO)$_4$ + [N(PPh$_3$)$_2$][BH$_4$]/CH$_2$Cl$_2$	253	253	80	^{13}C NMR (247)
[Ni$_9$C(CO)$_{17}$]$^{2-}$	[Ni$_6$(CO)$_{12}$]$^{2-}$ + CCl$_4$/THF	252	252	80	IR (416)
[Ni$_{10}$(C$_2$)(CO)$_{16}$]$^{2-}$	[Ni$_6$(CO)$_{12}$]$^{2-}$ + CCl$_4$ or C$_2$Cl$_6$/MeCN	254	254	60	
[Ni$_{12}$(CO)$_{21}$H]$^{4-}$	Ni(CO)$_4$ + Na/THF	255	255	—	
[Ni$_{12}$(CO)$_{21}$H]$^{3-}$	Ni(CO)$_4$ + (i) NaOH/DMSO/N$_2$ + (ii) H$_2$O + NH$_4$Cl	256	256		^{1}H NMR (257)
[Ni$_{12}$(CO)$_{21}$H$_2$]$^{2-}$	Ni(CO)$_4$ + (i) NaOH/DMSO/N$_2$ + (ii) H$_2$O + H$_3$PO$_4$	257	257		^{1}H NMR (257)
[Ni$_{16}$(C$_2$)$_2$(CO)$_{23}$]$^{2-}$	[Ni$_{10}$(C$_2$)(CO)$_{16}$]$^{2-}$ + 8PPh$_3$/THF	380	380	100	
Pd$_7$(CO)$_7$(PMe$_3$)$_7$	Pd(C$_8$H$_{12}$)(PMe$_3$) + CO/toluene	258	258	—	
Pd$_{10}$(CO)$_{14}$(PBu$_3$)$_4$	Pd(OAc)$_2$ + PBu$^n_3$ + CF$_3$COOH/dioxane/acetone	259	259	17	^{31}P NMR (260)
Pd$_{10}$(CO)$_{12}$(PBu$_3$)$_6$	Pd(OAc)$_2$ + PBu$^n_3$ + CF$_3$COOH/dioxane/acetone	261	261	—	^{31}P NMR (260)
Pt$_5$(CO)$_6$(PPh$_3$)$_4$	Pt(PPh$_3$)$_2$Cl$_2$ + [Fe(CO)$_3$NO]$^-$	262	263	75	
Pt$_5$(CO)$_3$(PPh$_3$)$_4$(SO$_2$)$_3$	Pt$_5$(CO)$_6$(PPh$_3$)$_4$ + SO$_2$	264	264	60	^{31}P, ^{1}H NMR, powder diffraction study

(continued)

133

TABLE I (continued)

Cluster[b]	References		Reagents and conditions	Yield (%)	Spectroscopic and theoretical studies[a]
	X-Ray	Synthesis[c]			
$[Pt_6(CO)_{12}]^{2-}$	265	265	$[PtCl_6]^{2-}$ + CO + NaOH/MeOH	—	^{195}Pt NMR (266)
$[Pt_9(CO)_{18}]^{2-}$	265	265	$[PtCl_6]^{2-}$ + CO + NaOH/MeOH	—	^{195}Pt NMR (266)
$[Pt_{15}(CO)_{30}]^{3-}$	265	265	$[PtCl_6]^{2-}$ + CO + NaOH/MeOH	—	
$[Pt_{19}(CO)_{22}]^{4-}$	267	267	$[Pt_9(CO)_{18}]^{2-}$/MeCN, Δ	50	^{13}C NMR
$[MoFe_2C(CO)_{17}]^{2-}$	268	269	$[Fe_2C(CO)_{14}]^{2-}$ + Mo(CO)$_3$(THF)	51	^{13}C NMR (269)
$MoRu_3Hg(CCBu^t)(CO)_9(C_5H_5)$	270	270	Ru$_3$(CO)$_{12}$ + (i) Bu^{t}CCH + (ii) KOH/EtOH + (iii) HgI$_2$ + (iv) MoC$_5$H$_5$)(CO)$_3$	14	^{1}H NMR, MS
$[Mo_2Ni_3(CO)_{16}]^{2-}$	271	271	$[Mo_2(CO)_{10}]^{2-}$ + Ni(CO)$_4$/THF, Δ	60	
$[Mo_2Ni_4(CO)_{14}]^{2-}$	271	271	$[Mo_2(CO)_{10}]^{2-}$ + Ni(CO)$_4$	—	
$WRu_5Au_2C(CO)_{17}(PEt_3)_2$	272	272	$[Ru_5C(CO)_{14}]^{2-}$ + (i) W(CO)$_3$(MeCN)$_3$ + (ii) Au(PEt$_3$)Cl	—	^{31}P, ^{1}H NMR
$W_2Os_3(CO)_{14}(PMe_2Ph)_2(S)_2$	273	273	Os$_3$(CO)$_9$(S)$_2$ + W(CO)$_5$(PMe$_2$Ph)/hv → WOs$_3$(CO)$_{12}$(PMe$_2$Ph)(S)$_2$ + W$_2$Os$_3$(CO)$_{14}$(PMe$_2$Ph)$_2$(S)$_2$; WOs$_3$(CO)$_{12}$(PMe$_2$Ph)(S)$_2$ + W(CO)$_5$(PMe$_2$Ph)/hv	27	
$[W_2Ni_3(CO)_{16}]^{2-}$	271	271	$[W_2(CO)_{10}]^{2-}$ + Ni(CO)$_4$/THF, Δ	68	
$W_2Pt_3(CO)_4(CR)(C_5H_5)_2(COD)_2$	274	274	W(CO)$_2$(CR)(C$_5$H$_5$) + Pt(COD)$_2$, 2 steps	33	^{13}C{^{1}H}, ^{195}Pt{^{1}H} NMR
$W_3Pt_2(CO)_6(CR)(C_5H_5)_3$	274	274	W(CO)$_2$(CR)(C$_5$H$_5$) + Pt(COD)$_2$, 2 steps	30	^{13}C{^{1}H}, ^{195}Pt{^{1}H} NMR
$Mn_4Hg_4(CO)_8(MeC_5H_4)_4$	275	275	$[MnGe(CO)_2(H)_3(MeC_5H_4)]^-$ + HgCl$_2$/H$_2$O	55	
$Re_2Os_3(CO)_{20}H_2$	276	277	Os$_3$(CO)$_{10}$(C$_8$H$_{14}$)$_2$ + Re(CO)$_5$H/benzene	90	^{1}H, ^{13}C, ^{1}H NMR, MS (277)
$Re_8In_4(CO)_{32}$	278	279	Re$_2$(CO)$_{10}$ + In/Δ	41	
$FeOs_3Au(CO)_{13}H(PPh_3)$	378	326	$[FeOs_3CO_{13}H]^-$ + Au(PPh$_3$)Cl + TlPF$_6$	60–65	
$FeCo_3Au(CO)_{12}(PPh_3)$	280	280	$[FeCo_3(CO)_{12}]^-$ + Au(PPh$_3$)NO$_3$	—	
$[FeRh_4(CO)_{15}]^{2-}$	281	281	Rh$_2$(CO)$_4$Cl$_2$ + (i) [Fe(CO)$_4$H]$^-$, N$_2$ + (ii) CO	—	^{103}Rh, ^{13}C{^{1}H} NMR
$[FeRh_5(CO)_{16}]^-$	282	281	Rh$_2$(CO)$_4$Cl$_2$ + (i) [Fe(CO)$_4$H]$^-$, N$_2$ + (ii) Rh$_2$(CO)$_4$Cl$_2$	—	^{103}Rh, ^{13}C{^{103}Rh} NMR (281)

134

Compound	No.	Reaction	Ref.	Yield (%)	Data
$[Fe_2Rh_4(CO)_{16}]^{2-}$	281	$Fe(CO)_5 + RhCl_3 + Na/Hg/diglyme$, Δ	282	—	^{103}Rh, $^{13}C\{^{103}Rh\}$ NMR
		$Rh_2(CO)_4Cl_2$ + (i) $[Fe(CO)_4H]^-$, N_2 + (ii) Pr^iOH, N_2	281	—	
$[Fe_3Pt_3(CO)_{15}]^{2-}$	283	$[Pt_3(CO)_6]^{2-} + Fe(CO)_5/MeCN$, Δ	283	70	ESR (284)
$[Fe_3Pt_3(CO)_{15}]^-$	283	$[Fe_3Pt_3(CO)_{15}]^{2-} + H^+$ or I_2	283	100	
$[Fe_3Cu_3(CO)_{12}]^{3-}$	379	$[Fe(CO)_4]^{2-} + CuBr/THF$	379	100	^{63}Cu, ^{13}C, ^{17}O NMR (470)
$Fe_3Ag_6(CO)_{12}\{(Ph_2P)_3CH\}$	285	$Na_2[Fe(CO)_4](dioxane)_{3/2} + Ag_3Cl_3\{(Ph_2P)_3CH\}$	285	50	^{31}P NMR
$Fe_3Au_2(CO)_9(PPh_3)_2S$	286	$Fe_3(CO)_9H_2S$ + (i) KH/THF + (ii) $Au(PPh_3)Cl$	286	47	
$Fe_3Bi_2(CO)_9$	287	$[Fe(CO)_4H]^-$ + (i) $NaBiO_3$ + (ii) H_2SO_4	287	7	
$[Fe_4CoC(CO)_{14}]^-$	288	$[Fe_4C(CO)_{12}]^{2-} + Co_2(CO)_8$, CH_2Cl_2	288	65	^{13}C NMR
$Fe_4CoRhC(CO)_{14}$	381	$[Fe_6C(CO)_{16}]^{2-}$ + (i) $CoCl_2$/diglyme, Δ + (ii) $[Rh(CO)_2Cl]_2/CH_2Cl_2$	381	28	
$[Fe_4Rh(CO)_{14}]^-$	268	$[Fe_4C(CO)_{12}]^{2-} + [Rh(C_8H_{12})Cl]_2$	269	28	^{13}C NMR (269)
$[Fe_4Pd(CO)_{16}]^{2-}$	289	$[Fe_3(CO)_{11}]^{2-} + 0.7PdCl_2/MeCN$	289	70–80	
$[Fe_4Pt(CO)_{16}]^{2-}$	289	$[Fe_3(CO)_{11}]^{2-} + 0.7PtCl_2/MeCN$	289	70–80	
$[Fe_4Pt_3(CO)_{22}]^{2-}$	283	$[Fe_3Pt_3(CO)_{15}]^-/MeCN$, Δ	283	—	
$[Fe_4Cu_5(CO)_{16}]^{3-}$	379	$CuBr + [Fe_3Cu_3(CO)_{12}]^{3-}$	379	26	^{63}Cu, ^{13}C, ^{17}O NMR (470)
$Fe_4AuC(CO)_{12}H(PPh_3)$	290	$[Fe_4C(CO)_{13}]^-$ + (i) $Au(PR_3)Cl + TlPF_6$ + (ii) $HBF_4 \cdot Et_2O$	290	47	^1H NMR, MS
$Fe_4Au_2C(CO)_{12}(PEt_3)_2$	290	$Fe_4AuC(CO)_{12}H(PEt_3)$ + (i) NEt_3 + (ii) $Au(PEt_3)Cl + TlPF_6$	290	90	MS
$Fe_4Cd_4(CO)_{16}$	291	$Fe(CO)_5$ + (i) $Cd(OAc)_2 + NH_3 + H_2O$ + (ii) Δ vacuum + (iii) acetone	291	—	
$[Fe_4Bi_4(CO)_{13}]^{2-}$	292	$[BiFe_3(CO)_{10}]^-$ + CO (35–50 atm)$/CH_2Cl_2$	292	—	
$[Fe_5RhC(CO)_{16}]^-$	282	$[Fe_5C(CO)_{14}]^{2-} + [Rh(CO)_2Cl]_2/MeOH$	269	—	^{13}C NMR (269)
$Fe_5Au_2C(CO)_{14}(PEt_3)_2$	293	$[Fe_5C(CO)_{14}]^{2-} + Au(PEt_3)Cl + TlPF_6$	293	80	
$[Fe_5Sn(CO)_4(C_5H_5)]$	294	$[FeSn(CO)_4(C_5H_5)]_2/toluene$, Δ	294	33	
$[Fe_6Pd_6(CO)_{24}H]^{3-}$	289	$[Fe_4(CO)_{13}]^{2-} + [PdCl]^{2-}$	289	5–10	
$RuCo_3Au(CO)_{12}(PPh_3)$	295	$[Co(CO)_4]^-$ + (i) $RuCl_3 \cdot xH_2O + RuCl_3L_3$ ($L = o\text{-}MePhCN$, PhSMe) + (ii) $Au(PPh_3)Cl$	295	60–70	
$RuCo_3Hg(CO)_{16}$	296	$[RuCo_3(CO)_{12}]^-$ + (i) $HgBr_2/toluene$ + (ii) $[Co(CO)_4]^-$	296	32	^{13}C NMR
$[RuRh_4(CO)_{15}]^{2-}$	297	$Ru_3(CO)_{12} + Rh_4(CO)_{12} + CO + NaOH/MeOH$	297	78	^{13}C NMR
$[RuIr_4(CO)_{15}]^{2-}$	298	$Ir_4(CO)_{12} + RuCl_3 \cdot xH_2O + NaOH/MeOH + CO$	298	30	

(continued)

TABLE I (*continued*)

| Cluster[b] | References | | Reagents and conditions | Yield (%) | Spectroscopic and theoretical studies[d] |
	X-Ray	Synthesis[c]			
$Ru_2Co_2Au_2(CO)_{12}(PPh_3)_2$	286	286	$Ru_2Co_2(CO)_{12}H_2$ + (i) KH/THF + (ii) $Au(PPh_3)Cl$	38	^1H NMR
$Ru_3CoAu(CO)_{13}(PPh_3)$	299	299	$Ru_3Co(CO)_{12}$ + (i) $[Co(CO)_4]^-$ + (ii) $Au(PPh_3)Cl$	45	^1H NMR
$Ru_3CoAu_2(CO)_{12}H(PPh_3)_2$	299	299	$Ru_3Co(CO)_{13}H + [(Ph_3PAu)_3O][BF_4]$	9	^1H NMR
$Ru_3CoAu_3(CO)_{12}(PPh_3)_3$	299	299	$[Ru_3Co(CO)_{13}]^- + [(Ph_3PAu)_3O][BF_4]$	55	
$Ru_3Rh_2(CO)_{13}(PEt_3)(PPh)$	300	300	$Ru_3(CO)_9H_2(PPh)$ + (i) KOH/MeOH + (ii) $[Rh(CO)_3(PEt_3)_2]^+$	5	^1H NMR
$Ru_3Ni_2(CO)_8(C_5H_5)_2(PhCCPh)$	301	301	$Ru_3(CO)_{12} + Ni_2(C_5H_5)_2(PhCCPh) + H_2$/octane	6–10	^1H NMR
$Ru_3Au_2(CO)_8(PPh_3)_3S$	302	302	$Ru_3(CO)_9H_2S + Au(PPh_3)Me$/toluene, Δ	18	^1H, $^{31}P\{^1H\}$ NMR, FABMS (303)
$Ru_3Au_2(CO)_9(PPh_3)_2S$	304	304	$Ru_3(CO)_9H_2S$ + (i) $K(HBBu^s_3)$/THF + (ii) $[(Ph_3PAu)_3O][BF_4]$	14	^1H NMR, FABMS
$Ru_3Au_2(CO)_9(CCHBu^t)(PPh_3)_2$	305	305	$Ru_3(CO)_9(CCHBu^t)H$ + (i) $K[HBBu^s_3]$ + (ii) $[(Ph_3PAu)_3O][BF_4]$	16	^1H NMR
$Ru_3Au_2(CO)_9(COMe)H(PPh_3)_2$	306	306	$Ru_3(CO)_9H_3(COMe) + AuMe(PPh_3)$	21	^1H, $^{31}P\{^1H\}$, $^{13}C\{^1H\}$ NMR
$Ru_3Au_3(CO)_9(COMe)(PPh_3)_3$	306	306	$Ru_3(CO)_9H_3(COMe) + AuMe(PPh_3)$	12	$^{31}P\{^1H\}$, $^{13}C\{^1H\}$ NMR
$Ru_3Au_3(CO)_8(C_{12}H_{15})(PPh_3)_3$	307	307	$Ru_3(CO)_9H(C_{12}H_{15})$ + (i) $K(HBBu^s_3)$/THF + (ii) $[(Ph_3PAu)_3O][BF_4]$	29	FABMS (303)
$Ru_4Ni(CO)_9(CCPr^i)_2(PPh_2)_2$	308	308	$Ru_3(CO)_9(CCPr^i)(PPh_2) + [(C_5H_5)Ni(CO)]_2$	—	^1H, $^{31}P\{^1H\}$ NMR
$Ru_4CuAg(CO)_{12}H_2(PPh_3)_2$	309	309	$[Ru_4(CO)_{12}H_2]^{2-} + Cu(PPh_3)Cl + Ag(PPh_3)I$ + TlPF$_6$	40	
$Ru_4Cu_2(CO)_{12}H_2(PPh_3)_2$	309	309	$[Ru_4(CO)_{12}H_2]^{2-} + Cu(PPh_3)Cl$	40	^1H, $^{31}P\{^1H\}$ NMR
$Ru_4Ag_2(CO)_{12}H_2(PPh_3)_2$	309	309	$[Ru_4(CO)_{12}H_2]^{2-} + Ag(PPh_3)I$ + TlPF$_6$	50	^1H, $^{31}P\{^1H\}$, $^{13}C\{^1H\}$ NMR
$Ru_5RhC(CO)_{12}(C_5Me_5)$	310	310	$Ru_5RhC(CO)_{14}(C_5Me_5)$ + CO/heptane	35	MS
$Ru_4RhC(CO)_{11}H(C_5Me_5)I$	310	310	$Ru_4RhC(CO)_{12}(C_5Me_5)$ + (i) NEt_4BH_4/THF + (ii) I_2/CH_2Cl_2	10	

Compound	Ref	Synthesis	Yield	Methods
$Ru_4AuC(CO)_{12}H(PPh_3)$	311	$Ru_5Au_2C(CO)_{14}(PPh_3)_2 + HI$	—	^{31}P NMR, MS
$Ru_4AuC(CO)_{12}(PEt_3)I$	311	$Ru_5Au_2C(CO)_{14}(PEt_3)_2 + I_2$	—	^{31}P NMR, ^1H NMR, MS
$Ru_4Au_2C(CO)_{12}(PMe_2Ph)_2$	311	$Ru_5AuC(CO)_{14}(PMe_2Ph)_2 + CO$ (80 atm) toluene, Δ	80	MS, ^{31}P NMR
$Ru_4Au_3(CO)_{12}H(PPh_3)_3$	312	$Ru_4(CO)_{12}H_4 + Au(PPh_3)Me$	59	^1H, $^{31}P\{^1H\}$, $^{13}C\{^1H\}$ NMR, FABMS
$Ru_5RhC(CO)_{14}(C_5Me_5)$	310	$[Ru_5C(CO)_{14}]^{2-} + [Rh(C_5Me_5)(MeCN)_2][SbF_6]_2$	70	MS
$Ru_5RhAuC(CO)_{14}(COD)PPh_3$	310	$[Ru_5C(CO)_{14}]^{2-} +$ (i) $[Rh(COD)_2][SbF_6]/CH_2Cl_2$ + (ii) $Au(PPh_3)Cl/CH_2Cl_2$	70	
$Ru_5RhAuC(CO)_{16}PPh_3$	310	$Ru_5RhAuC(CO)_{14}(COD)PPh_3 + CO$/heptane	30	
$Ru_5AuC(CO)_{14}[C(MeO)](PPh_3)$	314	$Ru_5C(CO)_{15} + Au(PPh_3)Me$	64	^1H NMR (314)
$Ru_5AuC(CO)_{13}(C_5H_5)(PPh_3)$	314	$Ru_5C(CO)_{15} +$ (i) $Na[C_5H_5]$ + (ii) $[Au(PPh_3)][ClO_4]$	—	MS
$Ru_5AuC(CO)_{13}(PPh_3)_2I$	26	$Ru_5Au(CO)_{15}(PPh_3)I +$ (i) heptane, Δ + (ii) PPh_3	75	
$Ru_5AuC(CO)_{15}(PPh_3)Cl$	315	$Ru_5C(CO)_{15} + Au(PPh_3)Cl$	—	
$Ru_5AuC(CO)_{14}(PPh_3)Br$	315	$Ru_5C(CO)_{15} + Au(PPh_3)Br/CH_2Cl_2$, N_2	—	
$Ru_5Au_2C(CO)_{14}(PEt_3)_2$	473	$[Ru_5C(CO)_{14}]^{2-} + [Au(PEt_3)][ClO_4]$	95	$^{31}P\{^1H\}$ NMR
$Ru_6Cu_2C(CO)_{16}(MeCN)_2$	316	$[Ru_6C(CO)_{16}]^{2-} + [Cu(MeCN)_4][BF_4]$	77	^1H, ^{13}C NMR
$Ru_6Cu_2C(CO)_{18}(C_6H_4\text{-}Me)_2$	317	$[Ru_6C(CO)_{18}]^{2-} +$ (i) $[Cu(MeCN)_4][BF_4]$/acetone + (ii) toluene, Δ	—	
$Ru_6AuC(CO)_{15}(NO)(PPh_3)$	51	$[Ru_6C(CO)_{15}(NO)]^- + Au(PPh_3)Cl + TlPF_6$	—	^{31}P NMR
$Ru_6Au_2C(CO)_{16}(PMePh_2)_2$	272	$[Ru_6C(CO)_{16}]^{2-} + Au(PMePh_2)Cl$	72	^1H NMR, MS
$Ru_6Hg(CO)_{18}(CCBu^t)_2$	270	$Ru_3(CO)_9(CCBu^t) +$ (i) Bu^tCCH + (ii) $KOH/EtOH + HgI_2$ + (iii) $[Ru_3(CO)_9(CCBu^t)]^-$		
$Ru_6Hg(CO)_{20}(NO)_2$	318	$[Ru_3(CO)_{10}(NO)]^- + HgCl_2$	—	MS
$[Ru_{12}Tl(C)_2(CO)_{32}]^-$	319	$[Ru_6C(CO)_{16}]^{2-} + Tl(NO_3)_3/MeOH$	—	
$Os_3NiAu(CO)_9(C_5H_5)H_3(PPh_3)$	321	$Os_3Ni(CO)_9(C_5H_5)H_3 +$ (i) NaH/THF + (ii) $Au(PPh_3)Cl$/toluene	36	^1H NMR
$Os_3Ni_3(CO)_9(C_5H_5)_3$	322	$Os_3(CO)_{12} + [Ni(CO)(C_5H_5)]_2$/octane, Δ	40	^1H NMR, MS, ESR (323)
$Os_3Au_2(CO)_{11}H_2(PPh_3)_2$	324	Synthesis not reported		
$Os_3Au_2(CO)_{10}(PEt_3)_2$	325	$[Os_3(CO)_{11}H]^- + Au(PEt_3)Cl + TlPF_6/CHCl_3$, Δ	80	^1H NMR
$Os_4AuC(CO)_{13}H(PEt_3)$	326	$[Os_4(CO)_{13}H]^- + Au(PEt_3)Cl + TlPF_6/CHCl_3$	65	^1H NMR
$Os_4Au(CO)_{12}H_3(PEt_3)$	326	$[Os_4(CO)_{12}H_3]^- + Au(PEt_3)Cl + TlPF_6/CHCl_3$	50	^1H NMR

(continued)

TABLE I (continued)

Cluster[b]	References X-Ray	References Synthesis[c]	Reagents and conditions	Yield (%)	Spectroscopic and theoretical studies[d]
$Os_4Au_2(CO)_{12}H_2(PPh_3)_2$	327	327	$Os_4Au(CO)_{12}H_3(PPh_3) + Au(PPh_3)Cl$ (two isomers)	—	MS
$Os_4Au_2(CO)_{13}(PEt_3)_2$	328	328	$[Os_3(CO)_{13}H]^- + 2Au(PEt_3)Cl + TlBF_4$	—	^{31}P NMR
$Os_4Au_2(CO)_{12}(PMePh_2)_2$	328	328	$Os_4Au_2(CO)_{13}(PMePh_2)_2/CH_2Cl_2$, Δ	—	^{31}P NMR
$Os_4Bi_2(CO)_{12}$	329	329	$Os_3(CO)_{12} + NaBiO_3$, Δ	—	
$Os_5AuH(CO)_{15}(PEt_3)$	69	69	$[Os_5(CO)_{15}H]^- + Au(PEt_3)Cl + TlPF_6/CH_2Cl_2$	80	
$Os_5Au_2C(CO)_{14}(PPh_3)_2$	78	78	$[Os_5C(CO)_{14}]^{2-} + Au(PPh_3)Cl + TlPF_6/CH_2Cl_2$	50	^{31}P, ^1H NMR
$Os_6Au_2(CO)_{17}(PMe_3)_2$	330	330	$Os_6(CO)_{18}$ + (i) Me_3NO/CH_2Cl_2 + (ii) $Au(PMe_3)Cl/TlPF_6/CH_2Cl_2$	90	
$Os_6Au_2(CO)_{18}(PMe_3)_2$	330	330	$Os_6Au_2(CO)_{17}(PMe_3)_2 + CO/CH_2Cl_2$	90	
$Os_6Pt_2(CO)_{17}(C_8H_{12})_2$	331	331	$Os_6(CO)_{20} + Pt(C_8H_{12})_2$, toluene/$CH_3CN/C_2H_4$	15	
$Os_6Pt_2(CO)_{16}(C_8H_{12})_2$	322	332	$Os_6(CO)_{16}(MeCN)_2 + Pt(C_8H_{12})_2$, CH_2Cl_2	—	
$[Os_6Ag(CO)_{20}H_2]^-$	333	333	$[Os_3(CO)_{11}H]^- + AgPF_6/THF$, Δ	90	^{13}C, ^1H NMR
$[Os_6Au(CO)_{20}H_2]^-$	334	334	$Os_3Au(CO)_{10}H(PR_3) + [N(PPh_3)_2]Cl/CH_2Cl_2$ (R = Ph, Et)	90	^1H NMR
$Os_7Au_2(CO)_{20}(PEt_3)_2$	335	335	$[Os_7(CO)_{20}]^{2-} + Au(PEt_3)Cl + TlPF_6/CH_2Cl_2$	90	
$Os_8Au_2(CO)_{22}(PPh_3)_2$	339	339	$[Os_8(CO)_{22}]^{2-} + Au(PPh_3)Cl + TlPF_6/CH_2Cl_2$	65	
$Os_9Hg_3(CO)_{33}$	336	336	$[Os_3(CO)_{11}H]^- + Hg^{2+}$	60	
$[Os_{10}CuC(CO)_{24}(MeCN)]^-$	337	337	$[Os_{10}C(CO)_{24}]^{2-} + [Cu(MeCN)_4][BF_4]/CH_2Cl_2$	—	
$[Os_{10}AuC(CO)_{24}(PPh_3)]^-$	337	337	$[Os_{10}C(CO)_{24}]^{2-} + Au(PPh_3)Cl + TlPF_6/CH_2Cl_2$	—	
$[Os_{10}AuC(CO)_{24}Br]^-$	338	338	$[Os_{10}C(CO)_{24}]^{2-} + AuPPh_3Br + AgClO_4$	—	
$[Os_{11}CuC(CO)_{27}(MeCN)]^-$	134	134	$[Os_{11}C(CO)_{27}]^{2-} + [Cu(MeCN)_4][BF_4]/CH_2Cl_2$	—	
$[Os_{20}AuC_2(CO)_{48}]^{2-}$	338	338	$[Os_{10}C(CO)_{24}]^{2-} + M(PPh_3)X + AgClO_4$ (excess) (M = Cu, Ag, Au; X = Cl, Br)	—	
$[Os_{20}Hg(C_2)(CO)_{48}]^{2-}$	338	338	$[Os_{10}C(CO)_{24}]^{2-} + Hg^{2+}$ salts or $Hg/AgBF_4/CH_2Cl_2$	—	ESR
$[CoRh_6N(CO)_{15}]^{2-}$	197	197	$[Rh_6N(CO)_{15}]^{2-} + [Co(CO)_4]^-/THF$, Δ	70–80	

138

Compound	Ref.	Synthesis	Ref.	Yield (%)	Characterization
$Co_2Rh_2(CO)_{16}$	340	$Co_2Rh_2(CO)_{12}$/n-heptane, Δ	340	—	MS
$Co_2P_3(CO)_9(PEt_3)_3$	341	$[Co(CO)_4]^- + Pt(PEt_3)_2Cl_2$/THF	341	66	
$Co_2Au_6(CO)_8(PPh_3)_4$	342	$[Au_8(PPh_3)_7]^{2+} + [Co(CO)_4]^-$/THF	342	—	
$[Co_3Ni_7C_2(CO)_{15}]^{3-}$	343	$Co_3(CO)_9CCl + [Ni_6(CO)_{12}]^{2-}$/THF	343	—	
$[Co_3Ni_9C(CO)_{20}]^{3-}$	344	$Co_3(CO)_9CCl + [Ni_6(CO)_{12}]^{2-}$/THF	344	—	
$[Co_4Ni_2(CO)_{14}]^{2-}$	345	$NiCl_2 + [Co(EtOH)_2][{Co(CO)_4}_2 + $ (i) Δ, vac. + (ii) KBr/H_2O	346	60	
$Co_4Cu_4(CO)_{16}$	347	$Na[Co(CO)_4]$/THF + CuCl/HCl	347	—	
$Co_4Ag_4(CO)_{16}$	348	Synthesis not reported			
$Co_4Zn_2(CO)_{15}$	349	$Hg[Co(CO)_4]_2 + $ (i) Zn + (ii) octane/Δ	349	35	MS
$Co_4Ge(CO)_{16}$	350	Synthesis not reported	350		MS
$Co_4Ge(CO)_{16}$	351	$Na[Co(CO)_4] + GeI_4$/benzene, hexane	351	52	MS
$Co_4Ge(CO)_{14}$	352	$Na[Co(CO)_4] + GeI_4$/benzene, pet. ether	352	62	MS
$Co_4Ge_2(CO)_{11}(Me)_2$	353	$MeGe + Co_4Ge(CO)_{14}$/hexane, sealed tube, dark	353	—	MS
$[Co_5HgGe(CO)_{17}]^-$	354	$Co_2(CO)_8 + $ (i) NaHg/THF + (ii) GeI_4/benzene, hexane	354	5	—
$[Co_5Ge(CO)_{16}]^-$	355	$[NEt_4][Co_2(CO)_4] + Co_4Ge(CO)_{14}/CH_2Cl_2$, Δ	355	64	—
$[Co_5Sn_2(CO)_{19}Cl_2]^-$	356	$ClSn{Co(CO)_4}_3 + [HB(pyrazole)_3]^-$/THF	356	35	—
$[Co_6Ni_2(C)_2(CO)_{16}]^{2-}$	357	$[NEt_4]_3[Co_3Ni_9C(CO)_{20}] + [Co_3(CO)_9CCl]$/acetone	357	70	^{1}H NMR
$Co_6Ge_2(CO)_{19}$	358	$GeH_2 + Co_2(CO)_8$/hexane	358		
$Rh_4Pt(CO)_4(C_5Me_5)_4$	359	$Pt(C_2H_4)_3 + [Rh(CO)(C_5Me_5)]_2$/toluene	359	90	^{13}C, ^{195}Pt NMR
$[Rh_4Pt(CO)_{14}]^{2-}$	360	$RhCl_3 \cdot xH_2O + $ (i) $[PtCl_6]^{2-}$/MeOH + (ii) CO + NaOH	361	53	^{13}C, ^{13}C{^{103}Rh}, ^{103}Rh NMR
$[Rh_4Pt(CO)_{12}]^{2-}$	361	$[Rh_4Pt(CO)_{14}]^{2-} + $ THF/N_2	361	92	^{195}Pt NMR (361)
$[Rh_5Pt(CO)_{15}]^-$	362	$[PtCl_6]^{2-} + RhCl_3 + Na_2CO_3 + $ CO/MeOH	361	87	^{195}Pt NMR (250), ^{13}C, ^{13}C{^{103}Rh}, ^{103}Rh, ^{195}Pt NMR (361)
$[Rh_6Ni(CO)_{16}]^{2-}$	363	$[Rh_6(CO)_{15}]^{2-} + Ni(CO)_4$/THF	363	90	^{13}C, ^{13}C{^{103}Rh}, ^{103}Rh NMR (364)
$[Rh_6IrN(CO)_{15}]^{2-}$	197	$[Rh_6N(CO)_{15}]^{2-} + [Ir(CO)_4]^-$/THF, Δ	197	79–80	
$Rh_6Cu_3C(CO)_{15}(MeCN)_2$	365	$[Rh_6C(CO)_{15}]^{2-} + [Cu(MeCN)_4][BF_4]$/MeOH	365	70	
$[Rh_9Pt_2(CO)_{22}]^{3-}$	366	$[PtRh_5(CO)_{15}]^-$/MeOH, Δ	366	—	
$[Rh_{10}PtN(CO)_{21}]^{3-}$	367	$[Rh_6N(CO)_{15}]^- + [Rh_4Pt(CO)_{12}]^{2-}$/acetone, Δ	367	15	
$[Rh_{11}Pt_2(CO)_{24}]^{3-}$	368	$[Rh_5Pt(CO)_{15}]^- + NaHCO_3$/MeOH, Δ	368	15	^{1}H NMR, ESR

(continued)

TABLE 1 (continued)

Cluster[b]	References		Reagents and conditions	Yield (%)	Spectroscopic and theoretical studies[d]
	X-Ray	Synthesis[c]			
[Rh$_{12}$Pt(CO)$_{24}$]$^{4-}$	368	368	[Rh$_5$Pt(CO)$_{15}$]$^-$ + NaHCO$_3$/MeOH, Δ	—	
[Rh$_{12}$AgC(CO)$_{30}$]$^{3-}$	369	369	[Rh$_6$C(CO)$_{15}$]$^{2-}$ + AgBF$_4$/acetone	75	
[Rh$_{12}$Sb(CO)$_{27}$]$^{3-}$	370	370	Rh(CO)$_2$(acac) + CO/H$_2$ (400 atm) + Cs[C$_6$H$_5$COO] + SbPh$_3$/tetraglyme, dimethyl ether, Δ	66	^{13}C NMR
Ir$_3$Pt$_3$(CO)$_6$(C$_5$Me$_5$)$_3$	371	371	Ir(CO)$_2$(C$_5$Me$_5$) + Pt(C$_2$H$_4$)$_3$/diethyl ether	100	^{13}C{^{1}H}, ^{195}Pt NMR
Ir$_8$Au$_2$(CO)$_{20}$(PhPPPh)(PEt$_3$)$_2$	372	372	Ir$_4$CO$_{11}$PPhH$_2$ + (i) DBU/THF + (ii) AgClO$_4$ + (iii) Au(PEt$_3$)ClO$_4$	30	
[Ni$_{38}$Pt$_6$(CO)$_{48}$H$_2$]$^{4-}$	373	373	[Ni$_6$(CO)$_{12}$]$^{2-}$ + 6PtCl$_2$/MeCN	80	^{195}Pt NMR (374)
[Ni$_{38}$Pt$_6$(CO)$_{48}$H]$^{5-}$	373	373	[Ni$_{38}$Pt$_6$(CO)$_{48}$H$_2$]$^{4-}$ + NaCO$_3$/MeCN		
Pd$_4$Hg$_2$(CO)$_4$(PEt$_3$)$_4$(Br)$_2$	375	375	Pd(CO)$_5$(PEt$_3$)$_4$ + C$_9$H$_8$NCH(Me)HgBr + CO/benzene	24	

[a] Table is comprehensive up to September 1985.

[b] Clusters are listed under the earliest transition metal contained within their framework. Homometallic clusters are grouped together, followed by heterometallic clusters. Groups of the periodic table are listed in succession and within each group metals are arranged by period. The entries in each metal listing are further categorized according to the size of the clusters, which appear in order of increasing nuclearity. It must be noted that in the text cluster formulas are written in a more descriptive fashion.

[c] Where several synthetic routes have been reported, the synthesis giving the highest yield is given.

[d] References are given for spectroscopic studies, other $\nu_{(C-O)}$ IR. When these studies are reported in the synthetic/structural reference no further reference number is given. MS, mass spectrum; FABMS, fast atom bombardment mass spectrum; VT, variable-temperature.

[e] The author does not specify which isomer is produced by this method.

II. Synthesis of High-Nuclearity Carbonyl Clusters

The logical synthesis of HNCC presents a major problem. The transition metals exhibit a complex array of bonding modes and a far more extensive range of coordination numbers than their main group counterparts, which have dependable valences and, in the main, strongly directional bonds. This, combined with the relative similarities of M—M, M—CO, and M—H bond energies (454), makes the designed synthesis of HNCC difficult.

Initial studies seem to have been based upon chance discoveries. Early work by Chini *et al.* on the elements of the cobalt triad led to the discovery of useful but in general unsystematic routes (196). These methods, primarily concerned with redox-condensation reactions, have now been improved and extended to embrace the elements of adjacent triads. Similarly, pyrolytic syntheses, which were originally used to prepare osmium HNCC, have been shown to have a greater potential than initially anticipated and are now used to synthesize HNCC of other heavy metals. Recently, however, elegant work by Stone, Vahrenkamp, and others has shown that designed synthesis of clusters is possible (118, 458, 471, 472). Use of the isolobal analogy in particular (455) has been most beneficial in devising routes for the preparation of many homo- and heteronuclear clusters. The synthetic strategies they have devised for small clusters have been applied with increasing frequency to the synthesis of HNCC.

Synthetic methods used for the preparation of HNCC may be classified in to two broad categories, depending on whether or not they involve the use of redox conditions; they are discussed below accordingly.

A. Syntheses Not Involving Redox Conditions

1. Pyrolytic or Photolytic Generation of Reactive Fragments

Pyrolysis reactions of mononuclear carbonyls and low-nuclearity cluster compounds have been used extensively in the syntheses of HNCC of osmium (54, 72, 80, 95, 108), ruthenium (18, 20, 29), and, more recently, rhenium (2–4). The reactions have been carried out either in inert solvents or, to facilitate the ejection of CO or other volatile ligands, in the solid state under vacuum. Condensation processes under pyrolytic conditions are rarely specific and, as such, lead to the formation of a wide range of products. In order to obtain optimum yields of a particular HNCC, the reaction conditions must be carefully screened. Solution reactions offer advantages such as the ability to monitor the progress of the reaction using IR spectroscopy. As they often give products or product distributions which differ from those under vacuum, both methods have been widely used.

A large number of HNCC of osmium have been produced from the vacuum pyrolysis of $Os_3(CO)_{12}$ and some of its derivatives (Scheme 1). The reaction of $Os_3(CO)_{12}$ has been shown to be extremely sensitive to temperature, time, and moisture, giving a series of products whose nuclearities range from 5 to 11 and possibly higher (41, 54, 80, 134). By optimizing the conditions, yields up to 80% of $Os_6(CO)_{18}$ have been obtained (41). High specificity has also been observed in the vacuum pyrolysis of $Os_3(CO)_{11}(C_5H_5N)$, from which $[Os_{10}C(CO)_{24\,5}]^{2-}$ has been obtained in 65% yield (80).

The HNCC of rhenium, $[Re_6C(CO)_{18}H_2]^{2-}$, $[Re_7C(CO)_{21}]^{3-}$, and $[Re_8C(CO)_{24}]^{2-}$, have been produced in high yields from the solution thermolysis of $[Re(CO)_4H_2]^-$ (2-4). In contrast, vacuum pyrolysis of $Re_2(CO)_{10}$ (182–300°C) has been shown to yield only rhenium metal (41), unless pyrolyzed in the presence of metallic indium, in which case $Re_4(CO)_{12}\{InRe(CO)_5\}_4$ was obtained in good yield (279).

The formation of carbido clusters under pyrolytic conditions is a common process. In a number of cases the carbon atom has been shown to be derived from a CO ligand, probably by disproportionation of CO to CO_2 and "C". In the case of the vacuum pyrolysis of $Ru_3(CO)_{12}$, from which $Ru_6C(CO)_{17}$ is formed in 65% yield, carbon dioxide was detected as a product of the reaction (41). Pyrolysis of the ^{13}CO-enriched species $Os_3(CO)_{11}(C_5H_5N)$ (80) and $[Re(CO)_4H_2]^-$ (4) as described above yields $[Os_{10}C(CO)_{24}]^{2-}$ and $[Re_8C(CO)_{24}]^{2-}$ with labeled carbide atoms and CO groups. The other possible sources of carbide atoms, the pyridine ligand in the osmium cluster and the n-tetradecane solvent used in the pyrolysis of $[Re(CO)_4H_2]^-$, can therefore be ruled out. This is in contrast with the pyrolysis of $Ru_5(CO)_{14}(CNBu^t)(\mu_5\text{-}CNBu^t)$ carried out in refluxing nonane, which yields the carbido cluster $Ru_6C(CO)_{16}(CNBu^t)$ (46). In this case the use of isotopically labeled $^{13}CNBu^t$ revealed that abstraction of a carbon atom from an isocyanide ligand had occurred. The fate of the NR portion of the fragmented isocyanide has not been established with any certainty.

Incorporation of other main group heteroatoms in pyrolysis products, as a consequence of the splitting of organic ligands under the reaction conditions, has also been observed. For example, the solution thermolysis of $Os_3(CO)_{10}H(\mu\text{-SR})$ [R = C_6H_5 (73) and $CH_2C_6H_5$ (445)] (Scheme 2) results in the cleavage of a C—S bond in the thiolato ligands. For R = $CH_2C_6H_5$, the reaction was shown to occur by two competing processes: cleavage followed by elimination of toluene, and formation of sulfido clusters which do not contain hydrides, or homolysis of the C—S bond and formation of dibenzyl. The latter route yields two isomers of $Os_6(CO)_{17}H_2S_2$, which do not interconvert.

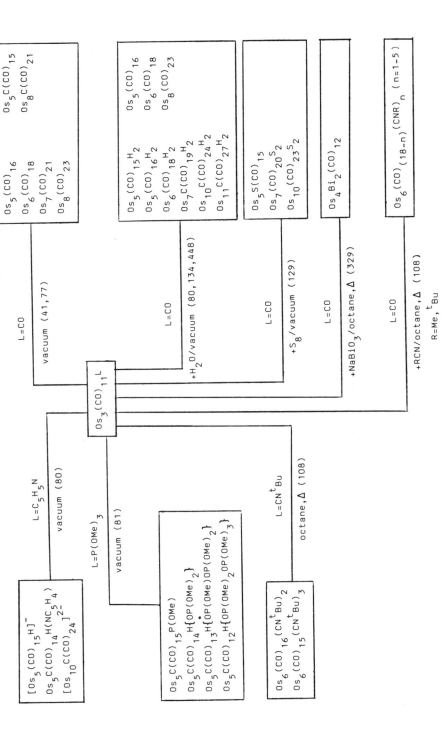

SCHEME 1. The pyrolysis reactions of $Os_3(CO)_{12}$ and some of its derivatives (see Table I for yields).

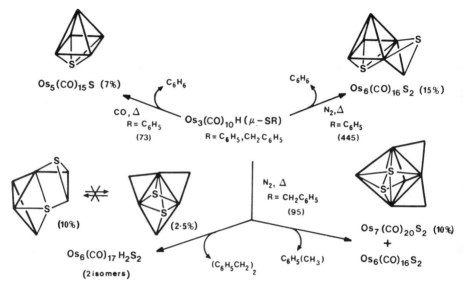

SCHEME 2. Pyrolysis reactions of $Os_3(CO)_{10}H(\mu\text{-SR})$ ($R = C_6H_5$ and $CH_2C_6H_5$).

2. Chemical Activation of Carbonyl Clusters for Condensation under Pyrolytic Conditions

The mechanism by which cluster expansion occurs in pyrolysis reactions is thought to involve formation of coordinatively unsaturated species by dissociation of ligands or cleavage of M—M bonds. These coordinatively unsaturated fragments are apparently the key intermediates that condense to give HNCC products (456). Major developments in HNCC synthesis have been achieved by using cluster precursors that contain a ligand such as acetonitrile, which is less strongly bound than CO. In this way milder conditions, which are less likely to lead to cluster fragmentation, can be employed.

Condensation of $Os_6(CO)_{16}(MeCN)_2$ with $Pt(COD)_2$ ($COD =$ 1,2-cyclooctadiene), for example, has been found to yield $Os_6(CO)_{16}(PtCOD)_2$ (332). As shown in Scheme 3, this process is accompanied by a reorganization of the Os_6 polyhedron. Displacement of the MeCN ligands in $Os_6(CO)_{(18-n)}(MeCN)_n$ ($n = 1, 2$) has also been achieved by $Os(CO)_4H_2$ to give the heptanuclear hydrido clusters $Os_7(CO)_nH_2$ ($n = 20$, 21, and 22). The bicapped "bow tie" cluster $Os_7(CO)_{22}H_2$ loses CO under mild conditions to produce $Os_7(CO)_{21}H_2$ and $Os_7(CO)_{20}H_2$ with a sequential closing of the polyhedral framework (Scheme 3) (119). Similarly, substitution of the acetonitrile ligand in $Os_5(CO)_{15}(MeCN)$ by $Os(CO)_4H_2$ has

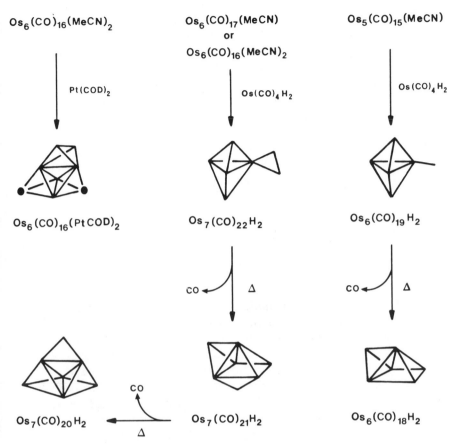

SCHEME 3. Synthesis of HNCC from $Os_6(CO)_{18-n}(MeCN)_n$ ($n = 1$, 2) and $Os_5(CO)_{15}(MeCN)$.

been performed successfully to give the spiked pentagonal-bipyramidal species $Os_6(CO)_{19}H_2$. This cluster also undergoes CO loss under mild conditions to give the capped square-based pyramidal cluster $Os_6(CO)_{18}H_2$ (92). The fortuitous condensation of two $Os_3(CO)_{10}(MeCN)_2$ clusters upon reaction with $PdCl_2$ has been reported to give the "raft" clusters $Os_6(CO)_{(21-n)}(MeCN)_n$ ($n = 0$, 1, 2) (99), while condensation of $Os_3(CO)_{10}(MeCN)_2$ with excess $HRe(CO)_5$ produces $Re_2Os_3H_2(CO)_{20}$ (277).

Sequential build-up of clusters, utilizing as coupling agents ligands that contain uncoordinated pairs of electrons, has been widely used by Vahrenkamp's group (450) and others in the synthesis of trinuclear and tetranuclear systems. This approach has been extended by Adams et al. to the synthesis of sulfur-containing HNCC of osmium. The formation of $Os_6(CO)_{19}S$ (*114*), $Os_6(CO)_{17}S_2$ (*115*), $Os_7(CO)_{19}S$ (*120*), and $Os_7(CO)_{20}S_2$ (*121*) by condensation of $Os_3(CO)_{10}(MeCN)_2$ with trinuclear and tetranuclear μ_3-S-containing clusters is shown in Scheme 4. The mechanisms involved in these condensation reactions are not understood. An initial link has been suggested to occur via formation of a coordination bond between the sulfido atoms in the trinuclear and tetranuclear clusters and a metal atom in $Os_3(CO)_{10}(MeCN)_2$ upon displacement of an acetonitrile ligand. The sulfur atoms seem to play a key role not only in the initial linking to the acetonitrile clusters but also in preventing cluster breakdown during the reorganization of the hexanuclear compounds $Os_6(CO)_{17}S_2$ and $Os_6(CO)_{19}S$ upon loss of CO.

B. SYNTHESES REQUIRING REDUCING OR OXIDIZING CONDITIONS

1. Reduction of Mononuclear Carbonyl Complexes or Small Carbonyl Clusters

Synthesis of the great majority of HNCC of the cobalt and nickel subgroups involves the reduction of mononuclear carbonyls or small carbonyl clusters (*196*). In general, the reactive fragments for the condensation processes have been generated by the action of reducing agents such as alkali metals, alkali hydroxides, or carbonates. Soft nucleophiles such as halides, thiocyanate, and nitride, as well as phosphines and carbon monoxide, have also been used to this end. Alternatively, reactive fragments have been produced from the thermolysis reactions, particularly of small carbonyl cluster compounds, in coordinating high-boiling solvents.

a. Reactions Leading to Cluster Build-up. A range of HNCC of nickel has been prepared via the reduction of $Ni(CO)_4$ (*245, 250, 257, 401*). As shown in Scheme 5, the nuclearity of the products is critically dependent on the experimental conditions. For example, when alkali metals in tetrahydrofuran (THF) are used as reducing agent, $[Ni_5(CO)_{12}]^{2-}$ or $[Ni_{12}(CO)_{21}]^{4-}$ is produced. Alternatively, sodium borohydride yields $[Ni_6(CO)_{12}]^{2-}$ (68%) if the reaction is performed in tetrahydrofuran, but $[Ni_9(CO)_{18}]^{2-}$ (80%) if dichloromethane is used instead. Scheme 5 illustrates the lability of these clusters, which results in their facile interconversion under mild conditions.

Reduction of the carbonyls $M_4(CO)_{12}$ (M = Co, Rh, Ir) is the basis for the synthesis of many of the HNCC of the cobalt triad. Thus $Co_4(CO)_{12}$ is converted by the action of alkali metals in tetrahydrofuran to the hexanuclear dianion $[Co_6(CO)_{15}]^{2-}$. The dianion may be reduced further under these conditions to give $[Co_6(CO)_{14}]^{4-}$ (*142*). Reduction of $Ir_4(CO)_{12}$ under the

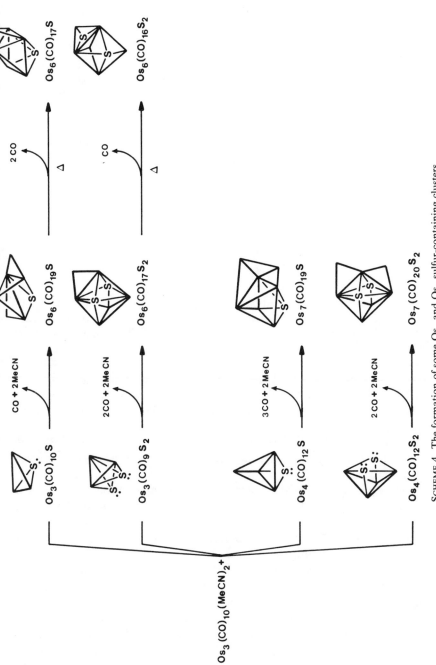

SCHEME 4. The formation of some Os_6 and Os_7 sulfur-containing clusters.

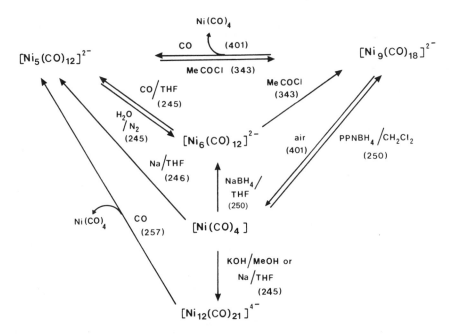

SCHEME 5. Synthesis of HNCC of nickel by reduction of $Ni(CO)_4$ and some redox interconversions between these clusters.

same conditions has also been shown to yield a hexanuclear dianion, $[Ir_6(CO)_{15}]^{2-}$ (244). In this case, however, the conversion has been shown to occur via the sequential formation of the octanuclear species $[Ir_8(CO)_{22}]^{2-}$ and $[Ir_8(CO)_{20}]^{2-}$. When the reduction of $Ir_4(CO)_{12}$ is carried out with potassium hydroxide in ethanol a mixture of brown anions possessing a high Ir/CO ratio is formed, whose components have not yet been characterized (244).

A large number of HNCC of rhodium have been produced from the reduction reactions of $Rh_4(CO)_{12}$ (Scheme 6). The reaction of this cluster with alkali hydroxides in isopropanol under thermolytic conditions has afforded binary carbonyl anions containing up to 22 rhodium atoms (225, 229, 232). The difficulties encountered in obtaining high yields of each product and in the separation process have led to the development of alternative synthetic routes to these clusters. Very selective, high-yield preparations of a series of rhodium HNCC have been achieved by high-pressure reduction of $Rh(CO)_2(acac)$ (acac, acetylacetone) with CO/H_2 mixtures in glyme solvents in the presence of bases (Scheme 7). In addition to the clusters previously prepared from the reduction of $Rh_4(CO)_{12}$, new species with encapsulated main-group atoms, $[Rh_9Y(CO)_{21}]^{2-}$

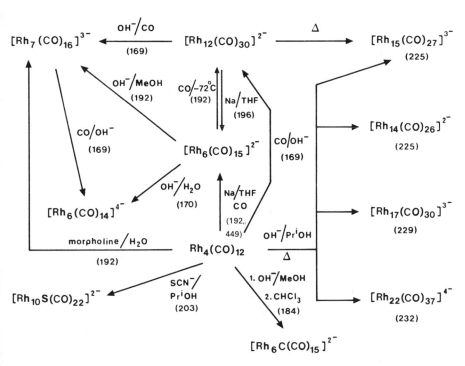

SCHEME 6. Synthesis of HNCC of rhodium by reduction of $Rh_4(CO)_{12}$ and some redox interconversions between these clusters.

(201), $[Rh_{10}Y(CO)_{22}]^{3-}$ (204, 205) $[Rh_{12}Sb(CO)_{27}]^{3-}$ (370), and $[Rh_{17}S_2(CO)_{32}]^{3-}$ (230) (Y = P, As) have been obtained when an external source of the heteroatom is added to the reaction mixture.

Halocarbons have been used successfully as a source of carbide atoms in the synthesis of carbido clusters of rhodium and nickel. For example, $[Rh_6C(CO)_{15}]^{2-}$ may be synthesized by the reduction of $Rh_4(CO)_{12}$ with alkali hydroxide in methanol followed by reaction with chloroform (184), or by the reaction of $[Rh(CO)_4]^-$ with CCl_4 according to Eq. (1).

$$6[Rh(CO)_4]^- + CCl_4 \xrightarrow[25°C]{Pr^iOH} [Rh_6C(CO)_{15}]^{2-} + 4Cl^- + 9CO \qquad (1)$$

In the latter case, the source of the carbide atom was established unequivocally by using $^{13}CCl_4$ in the reaction (184).

Carbide clusters of nickel have been obtained from the reaction of $[Ni_6(CO)_{18}]^{2-}$ with carbon tetrachloride, which yields $[Ni_9C(CO)_{17}]^{2-}$ (252) or $[Ni_{10}(C_2)(CO)_{16}]^{2-}$ (254), depending on the reaction conditions.

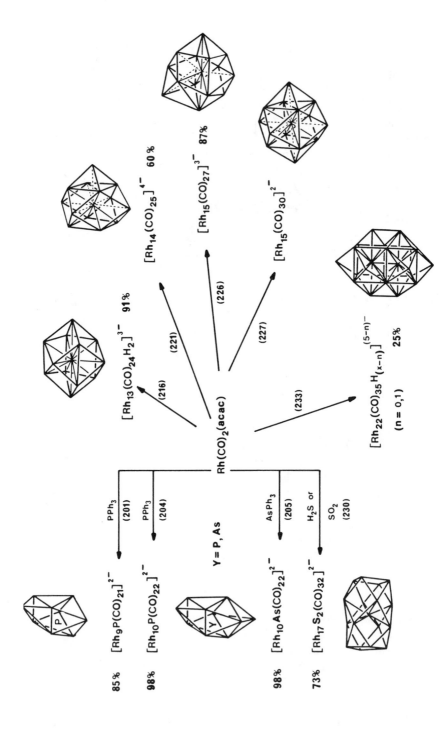

SCHEME 7. Synthesis of some HNCC of rhodium from $Rh(CO)_2(acac)$ (acac, acetylacetone).

$[Rh_9P(CO)_{21}]^{2-}$ PPh_3 (201)   85%

$[Rh_{10}P(CO)_{22}]^{2-}$ PPh_3 (204)   98%

$Y = P, As$

$[Rh_{10}As(CO)_{22}]^{2-}$ $AsPh_3$ (205)   98%

$[Rh_{17}S_2(CO)_{32}]^{2-}$ H_2S or SO_2 (230)   73%

$Rh(CO)_2(acac)$

$[Rh_{13}(CO)_{24}H_2]^{3-}$ (216)   91%

$[Rh_{14}(CO)_{25}]^{4-}$ (221)   60%

$[Rh_{15}(CO)_{27}]^{3-}$ (226)   87%

$[Rh_{15}(CO)_{30}]^{2-}$ (227)

$[Rh_{22}(CO)_{35}H_{(x-n)}]^{(5-n)-}$ (233) $(n = 0,1)$ 25%

The dicarbide is also formed if hexachloroethane is used as a source of carbide atoms (*254*).

The incorporation of a main group atom derived from the reducing agent itself has also been noted in a few cases. For example, reduction of $Rh_4(CO)_{12}$ with SCN^- in isopropanol has been found to give the sulfido cluster $[Rh_{10}S(CO)_{22}]^{2-}$ in high yield (*203*), while $[Ru_6N(CO)_{16}]^-$ was produced from the reduction of $Ru_3(CO)_{12}$ with N_3^- in refluxing tetrahydrofuran (*24*).

Reduction of $Ru_3(CO)_{12}$ and $Os_3(CO)_{12}$ with sodium in ether solvents at various reaction temperatures has been found to provide an effective route to HNCC of ruthenium and osmium (*382*). Although most of the dianions produced by this method had been isolated previously, the study of these reactions has allowed a better understanding of the build-up processes of these clusters.

As shown in Scheme 8 reduction of $M_3(CO)_{12}$ initially gives $[M_6(CO)_{18}]^{2-}$ (M = Ru, Os) in good yield (*382*). When heated in diglyme, $[Ru_6(CO)_{18}]^{2-}$ is quantitatively converted to the monocarbide $[Ru_6C(CO)_{16}]^{2-}$. Further thermolysis of this cluster in tetraglyme gives the dicarbide dianion $[Ru_{10}(C)_2(CO)_{24}]^{2-}$, whose structure has been shown to consist of two edge-sharing octahedra (*53*). In contrast, upon heating $[Os_6(CO)_{18}]^{2-}$ in triglyme, the anticipated carbido cluster $[Os_6C(CO)_{16}]^{2-}$ was only isolated in yields around 2% (*382*). Instead the reaction gave the tetracapped octahedral dianion $[Os_{10}C(CO)_{24}]^{2-}$ as the major product.

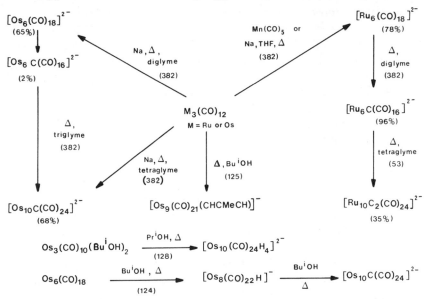

SCHEME 8. Synthesis of some HNCC of ruthenium and osmium under reducing conditions.

This suggests that $[Os_6C(CO)_{16}]^{2-}$ is probably generated as a reaction intermediate but under the reaction conditions it is converted to the thermo-dynamically more stable $[Os_{10}C(CO)_{24}]^{2-}$.

Prolonged thermolysis of $Os_3(CO)_{10}(OBu^i)_2$ in isobutanol has been found to produce a mixture of $[Os_{10}C(CO)_{24}]^{2-}$ and $[Os_{10}C(CO)_{24}H]^-$. However, when the reaction was carried out in isopropanol, the tetrahydrido dianion $[Os_{10}(CO)_{24}H_4]^{2-}$ was formed instead (128). The importance of the reaction temperature in the formation of the carbido species $[Os_{10}C(CO)_{24}]^{2-}$ is also evidenced by the thermolysis reactions of $Os_6(CO)_{18}$. When carried out in refluxing n-hexanol the only HNCC formed from the reaction is $[Os_{10}C(CO)_{24}H]^-$, while $[Os_8(CO)_{22}H]^-$ is produced in 30% yield when isobutanol is used (124).

Mixed-metal clusters have also been shown to undergo cluster build-up under reducing conditions. For instance, prolonged thermolysis of $[Rh_5Pt(CO)_{15}]^-$ in methanol in the presence of $NaHCO_3$ gives $[Rh_{11}Pt_2(CO)_{24}]^{3-}$ and $[Rh_{12}Pt(CO)_{24}]^{4-}$ (368). In the absence of base, $[Rh_9Pt_2(CO)_{22}]^{3-}$ is the major product of the reaction (366).

b. Syntheses Involving a Reduction in Cluster Nuclearity.   The syntheses of HNCC under reducing conditions discussed above all result in cluster buildup. There are certain HNCC, however, which are best prepared by degradation of preformed clusters with nucleophiles such as CO, PR_3, or halides. The capped square-antiprismatic dianion $[Ni_9C(CO)_{17}]^{2-}$, for instance, is "decapped" by CO in tetrahydrofuran to give $[Ni_8C(CO)_{16}]^{2-}$ in 69% yield (252). Similarly, carbonylation of the octahedral species $Ru_6C(CO)_{17}$ and $[Ru_6N(CO)_{16}]^-$ is the method of choice for the prep-aration of the square-based pyramidal clusters $Ru_5C(CO)_{15}$ and $[Ru_5N(CO)_{14}]^-$, respectively (24, 25). The synthesis of $Os_5(CO)_{19}$ is also achieved by carbonylation of $Os_6(CO)_{18}$. By using high-pressure infrared techniques to monitor the progress of the reaction, this cluster has been obtained in up to 80% yield (56).

2. Redox Condensations

The synthesis of HNCC by redox condensation involves the reaction of an anionic mononuclear or polynuclear carbonyl species with a neutral, cationic, or even anionic fragment.

a. Reactions of Carbonyl Metallates with Neutral Metal Complexes.   Re-dox condensation reactions of mononuclear and cluster carbonyl metallates with neutral carbonyl compounds provide a selective synthetic route to homo- and heteronuclear HNCC anions. For example, reaction of

$[Rh(CO)_4]^-$ with the neutral cluster $Rh_4(CO)_{12}$ affords the trigonal-bipyramidal anion $[Rh_5(CO)_{15}]^-$ under mild conditions (162). Similarly, the octahedral species $[Fe_6C(CO)_{16}]^{2-}$ has been prepared by the condensation of $[Fe(CO)_4]^{2-}$ with stoichiometric amounts of $Fe(CO)_5$ under thermolytic conditions, which are necessary for the generation of the carbide atom [Eq. (2)] (269).

$$5Fe(CO)_5 + [Fe(CO)_4]^{2-} \xrightarrow[150-160°C]{diglyme} [Fe_6C(CO)_{16}]^{2-} \qquad (2)$$

Further examples of this synthetic technique are shown in Eqs. (3) (363) and (4) (283).

$$[Rh_6(CO)_{15}]^{2-} + Ni(CO)_4 \longrightarrow [Rh_6Ni(CO)_{16}]^{2-} + 3CO \qquad (3)$$

$$[Pt_3(CO)_6]^{2-} + 3Fe(CO)_5 \longrightarrow [Fe_3Pt_3(CO)_{15}]^{2-} + 6CO \qquad (4)$$

The condensation reactions of carbonyl metallates with neutral species, which are either coordinatively unsaturated or which will readily generate coordinatively unsaturated fragments, have also yielded a variety of mixed-metal clusters. The square-based pyramidal dianion $[Fe_5C(CO)_{14}]^{2-}$, for example, has been shown to react with a number of such species to yield octahedral Fe_5MC cluster compounds (Scheme 9) (269, 381). In some cases,

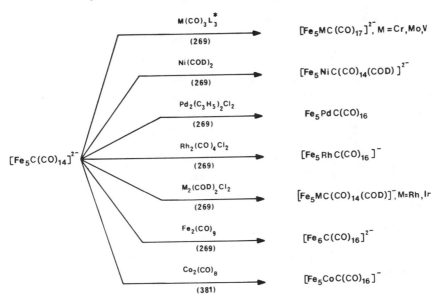

*$M=Cr, L=C_5H_5N; M=Mo, L=THF; M=W, L=MeCN.$

SCHEME 9. Some reactions of $[Fe_5C(CO)_{14}]^{2-}$ with unsaturated metal fragments.

the unsaturation is achieved by the displacement of halides in halogenated metal complexes. Reaction of the dimer $Rh_2(CO)_2Cl_2$ with $[Fe(CO)_4H]^-$ affords $[FeRh_4(CO)_{15}]^{2-}$, $[FeRh_5(CO)_{16}]^-$, or $[Fe_2Rh_4(CO)_{16}]^{2-}$, depending on the reaction conditions employed (281). Interestingly, both hexanuclear clusters were found to be isostructural with the octahedral $Rh_6(CO)_{16}$ species, while $[FeRh_4(CO)_{15}]^{2-}$ possesses a different carbonyl distribution to that found for $[Rh_5(CO)_{15}]^-$ both in solution and in the solid state (281).

The chloromethynyl derivative $Co_3(CO)_9CCl$ has been used as a source of "$Co_3(CO)_9C$" fragments for the synthesis of large cobalt and mixed-metal carbido clusters. The chlorine atom in this species has been found to be easily displaced by mononuclear and cluster carbonyl metallates. Reaction of $Co_3(CO)_9CCl$ with $[Co(CO)_4]^-$ yields $[Co_6C(CO)_{15}]^{2-}$ (145), while $[Co_3Ni_9C(CO)_{20}]^{3-}$ is one of the products from the reaction of $Co_3(CO)_9CCl$ with $[Ni_6(CO)_{12}]^{2-}$ (344). The monocarbido cluster $[Co_3Ni_9C(CO)_{20}]^{3-}$ reacts further with $Co_3(CO)_9CCl$ to yield $[Co_6Ni_2(C)_2(CO)_{16}]^{2-}$, according to Eq. (5) (357).

$$2[Co_3Ni_9C(CO)_{20}]^{3-} + 4Co_3(CO)_9CCl \longrightarrow$$

$$3[Co_6Ni_2(C)_2(CO)_{16}]^{2-} + 2NiCl_2 + 7Ni(CO)_4 + 3Ni \quad (5)$$

Even simple metal halides have been used as sources of metal atoms for the condensation reactions of carbonyl metallates [e.g., Eq. (6)] (379).

$$3[Fe(CO)_4]^{2-} + 3CuBr \longrightarrow [Fe_3Cu_3(CO)_{12}]^{3-} + 3Br^- \quad (6)$$

Another example of this synthetic method is the reaction of $[Ni_6(CO)_{12}]^{2-}$ with $PtCl_2$ in acetonitrile, which yields $[Ni_{38}Pt_6(CO)_{48}H_2]^{4-}$, the largest metal carbonyl cluster characterized to date (373).

Clusters of the heavier elements undergo coupling reactions with salts of Ag(I), Au(I), Hg(I), Hg(II), and Tl(III), in which the basic metal cluster geometry is preserved. Even in these cases it is difficult to predict the nuclearity of the mixed species formed. These reactions may result in the simple coupling of two cluster units, as in the formation of $[\{Ru_6C(CO)_{16}\}_2Tl]^-$ (Fig. 1a) shown in Eq. (7) (320).

$$2[Ru_6C(CO)_{16}]^{2-} + Tl(NO_3)_3 \longrightarrow [\{Ru_6C(CO)_{16}\}_2Tl]^- + 3NO_3^- \quad (7)$$

Dissociation of CO ligands may occur concomitantly with the condensation process as examplified by the formation of $\{RuCo_3(CO)_9\}_2Hg$ (Fig. 1b) from the reaction shown in Eq. (8) (453).

$$2[RuCo_3(CO)_{12}]^- + HgBr_2 \longrightarrow \{RuCo_3(CO)_9\}_2Hg + 6CO + 2Br^- \quad (8)$$

The formation of the dianions $[Os_{20}(C)_2(CO)_{48}M]^{2-}$ (M = Au and Hg) (Fig. 1c) from the reactions of $[Os_{10}C(CO)_{24}]^{2-}$ with $Au(PPh_3)Br$ in the

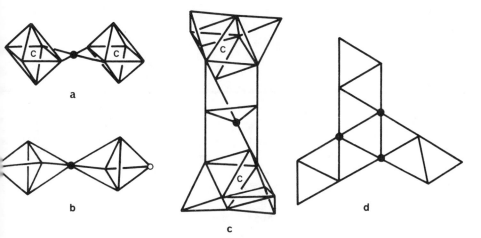

FIG. 1. Metal core geometry of (a) $[\{Ru_6C(CO)_{16}\}_2Tl]^-$, (b) $\{RuCo_3(CO)_9\}_2Hg$, (c) $[Os_{20}M(C)_2(CO)_{48}]^{2-}$ (M = Au, Hg), and (d) $Os_9Hg_3(CO)_{33}$.

presence of $AgClO_4$ and RHgY (R = C_6F_5, C_6H_5, Cl, CF_3COO; Y = CF_3COO or Cl), respectively, has been shown to occur via the formation of the intermediates $[Os_{10}C(CO)_{24}(AuBr)]^-$ and $[Os_{10}C(CO)_{24}(HgR)]^-$, which undergo ligand disproportionation and condensation on standing in solution (338).

Condensation of more than two metal clusters has been found to occur in some cases. For example, $[Os_3(CO)_{11}H]^-$ reacts with $HgBr_2$ to yield the "raft" $Os_9Hg_3(CO)_{33}$ (Fig. 1d) (336). Reaction of the trigonal-prismatic dianion $[Rh_6C(CO)_{15}]^{2-}$ with $AgBF_4$ yields a series of oligomers containing Rh_6 prismatic units bridged by Ag(I) atoms. The nature of these species was found to be critically dependent upon the ratio Ag(I): $[Rh_6C(CO)_{15}]^{2-}$. The condensation process was studied by monitoring the ^{13}C and $^{13}C\{^{103}Rh\}$ NMR spectra of both $[Rh_6C(^{13}CO)_{15}]^{2-}$ and $[Rh_6\,^{13}C(CO)_{15}]^{2-}$ after sequential addition of $AgBF_4$, and is summarized in Fig. 2 (369).

b. Reactions of Anionic Carbonyl Clusters with Cationic Mononuclear Complexes. A number of heterometallic HNCC have been synthesized by the condensation reactions of anionic carbonyl clusters with the cationic fragments $[M(MeCN)]^+$ and $[M(PR_3)]^+$, generated from $[M(MeCN)_4][Y]$ (M = Cu, Ag, Y = BF_4, PF_6) and $M(PR_3)X$ (M = Cu, Ag, Au; X = Cl, Br, I; R = alkyl or aryl), respectively. While formation of the coordinatively unsaturated fragments $[M(MeCN)]^+$ occurs readily in solution, the $[M(PR_3)]^+$ species have either been prepared with silver salts or, in the case of the chloride complexes, generated *in situ* by reaction with $TlPF_6$.

charge   2⁻   3⁻   4⁻   n⁻   2⁻   1⁻   0

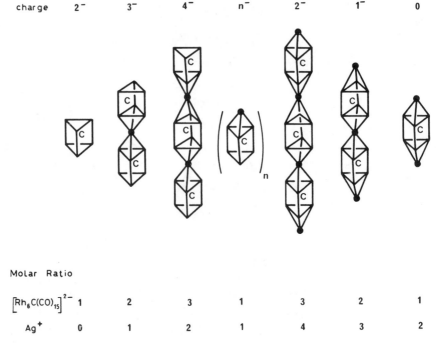

Molar Ratio

| $[Rh_6C(CO)_{15}]^{2-}$ | 1 | 2 | 3 | 1 | 3 | 2 | 1 |
| Ag^+ | 0 | 1 | 2 | 1 | 4 | 3 | 2 |

FIG. 2. $[Rh_6C(CO)_{15}]^{2-}/Ag^+$ adducts formed on progressive addition of $AgBF_4$ (369).

The condensation reactions of the dianion $[Ru_4(CO)_{12}H_2]^{2-}$ with $[M(MeCN)]^+$ and $[M(PPh_3)]^+$ (M = Cu and Ag) are shown in Scheme 10 and demonstrate the flexibility of this synthetic method (309, 430). The number of metal fragments incorporated in the anionic clusters can often be controlled by using stoichiometric amounts of the metal cations. In the case of the dianion $[Os_{10}C(CO)_{24}]^{2-}$, however, only $[Os_{10}C(CO)_{24}(AuPPh_3)]^-$ is formed, even in the presence of excess $[Au(PPh_3)]^+$ (337). In contrast, both the mono- and dicopper species $[Os_{10}C(CO)_{24}(CuMeCN)]^-$ and $Os_{10}C(CO)_{24}(CuMeCN)_2$ are produced from the reaction of the dianion with $[Cu(MeCN)_4]^+$ (337). In fact, $[Au(PR_3)]^+$ and $[Cu(MeCN)]^+$ have often been found to bind differently to the same HNCC due to their different electronic and steric requirements (451). Examples are $Ru_6C(CO)_{16}(AuPMePh_2)_2$ (272) and $Ru_6C(CO)_{16}(CuMeCN)_2$ (316), whose structures are shown in Fig. 3. More surprising perhaps is the difference in the bonding modes of the $AuPR_3$ groups in the series of closely related compounds $M_5C(CO)_{14}(AuPR_3)_2$ (M = Fe, (293) Ru (473), Os (78)] (Fig. 4).

According to electron counting rules, the electrophilic addition of $[M(MeCN)]^+$ and $[M(PPh_3)]^+$ fragments to anionic carbonyl clusters

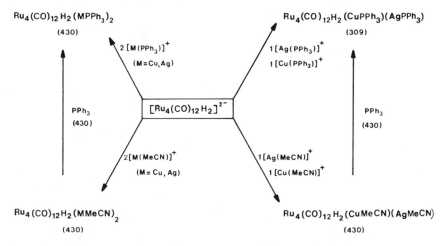

SCHEME 10. Condensation reactions of $[Ru_4(CO)_{12}H_2]^{2-}$ with $[M(MeCN)]^+$ and $[M(PPh_3)]^+$ (M = Cu, Ag).

FIG. 3. Metal core geometry of (a) $Ru_6C(CO)_{16}(AuPMePh_2)_2$ and (b) $Ru_6C(CO)_{16}(CuMeCN)_2$.

FIG. 4. Metal core geometry of (a) $M_5C(CO)_{14}(AuPEt_3)_2$ (M = Fe, Ru) and (b) $Os_5C(CO)_{14}(AuPR_3)_2$ (R = Et, Ph).

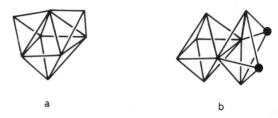

a   b

FIG. 5. Metal arrangements in (a) $[Os_8(CO)_{22}]^{2-}$ and (b) $Os_8(CO)_{22}(AuPPh_3)_2$.

should not lead to any metal polyhedral changes as these species donate no electrons to the cluster. Clearly, these rules do not always hold true for HNCC. The addition of $[Au(PPh_3)]^+$ to $[Os_8(CO)_{22}]^{2-}$ gives $Os_8(CO)_{22}(AuPPh_3)_2$, which exhibits an Os_8 polyhedral geometry different from that of the dianion (339) (Fig. 5).

The reagent $[(R_3PAu)_3O][BF_4]$ has also been used for the synthesis of clusters containing "$AuPR_3$" fragments, e.g. (452),

$$[H_3Ru_4(CO)_{12}]^- + [(Ph_3PAu)_3O]^+ \longrightarrow$$

$$H_3Ru_4(CO)_{12}AuPPh_3 + H_2Ru_4(CO)_{12}(AuPPh_3)_2 + HRu_4(CO)_{12}(AuPPh_3)_3 \quad (9)$$

In this case, however, the reaction does not always involve a simple addition of the gold moiety; often a carbonyl ligand is replaced by two or three $AuPR_3$ fragments [e.g., Eq. (10)] (299).

$$[Ru_3Co(CO)_{13}]^- + [(Ph_3PAu)_3O]^+ \longrightarrow Ru_3(CO)_{12}(AuPPh_3)_3 \quad (10)$$

A few homometallic HNCC have also been produced by the condensation reactions of anionic carbonyl clusters with cationic complexes. For instance, sequential buildup of rhodium HNCC via incorporation of "$Rh(CO)_2$" fragments has been achieved by reacting $[Rh(CO)_2(MeCN)_2]^+$ in acetonitrile with a series of anionic rhodium clusters as illustrated by the reactions shown in Eqs. (11) and (12) (223).

$$[Rh(CO)_2(MeCN)_2]^+ + [Rh_{13}(CO)_{24}H]^{4-} \longrightarrow [Rh_{14}(CO)_{25}H]^{3-} \quad (11)$$

$$[Rh(CO)_2(MeCN)_2]^+ + [Rh_{14}(CO)_{25}]^{4-} \longrightarrow [Rh_{15}(CO)_{27}]^{3-} \quad (12)$$

In spite of the relatively mild reaction conditions employed, further condensation of unstable products has been shown to occur. For example, reaction of $[Rh_6C(CO)_{15}]^{2-}$ with $[Rh(CO)_2(MeCN)_2]^+$ yields $[Rh_{14}(C)_2(CO)_{23}]^{2-}$ rather than the anticipated Rh_7C cluster (224). This contrasts with the condensation reaction of $[Rh_6(CO)_{15}]^{2-}$ with the same rhodium complex, which yields $[Rh_7(CO)_{16}]^{3-}$ (170).

3. Oxidation of Carbonyl Clusters

Only a relatively small number of HNCC have been synthesized by oxidation of other carbonyl clusters.

a. Oxidation Reactions Resulting in Expansion of Cluster Size. The oxidation reactions of anionic HNCC of rhodium have often been found to result in cluster expansion. For instance, reaction of $[Rh_6C(CO)_{15}]^{2-}$ with ferric ammonium alum gives $Rh_8C(CO)_{19}$, $[Rh_{15}(C)_2(CO)_{28}]^-$, or $Rh_{12}(C_2)(CO)_{25}$, depending on the reaction conditions, while oxidation of $[Rh_6C(CO)_{15}]^{2-}$ with sulfuric acid gives $[Rh_{12}(C)_2(CO)_{24}]^{2-}$ (*211*) (Scheme 11).

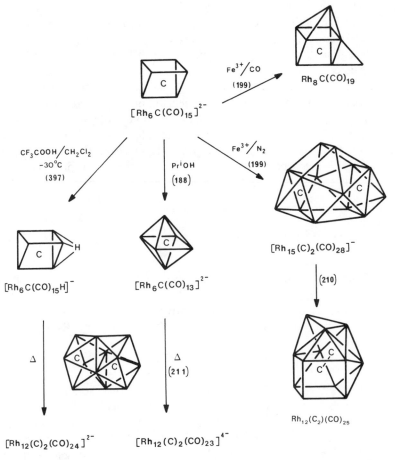

SCHEME 11. Synthesis of large carbido clusters of rhodium from $[Rh_6C(CO)_{15}]^{2-}$.

Oxidative aggregation of two Rh_6 clusters was also observed in the reaction of $[Rh_6(CO)_{15}]^{2-}$ with acids giving $[Rh_{12}(CO)_{30}]^{2-}$. It was established by variable-temperature multinuclear NMR studies that the protonation of both $[Rh_6(CO)_{15}]^{2-}$ and $[Rh_6C(CO)_{15}]^{2-}$ proceeds first with formation of the intermediate species $[Rh_6(CO)_{15}H]^-$ and $[Rh_6C(CO)_{15}H]^-$, which were fully characterized at low temperatures. Warming the solution was shown to result in the loss of hydrogen and formation of $[Rh_{12}(CO)_{30}]^{2-}$ and $[Rh_{12}(C)_2(CO)_{24}]^{2-}$, respectively (397). It is interesting to note that the reaction of the hexacobalt dianion $[Co_6(CO)_{15}]^{2-}$ with acids does not lead to an increase in nuclearity following formation of $[Co_6(CO)_{15}H]^-$ (144), and the large carbido clusters $[Co_{11}(C_2)(CO)_{22}]^{3-}$ and $[Co_{13}(C)_2(CO)_{24}]^{4-}$ have only been produced by the thermolysis of $[Co_6C(CO)_{15}]^{2-}$ in diglyme (158, 159).

The reaction of $[Rh_7(CO)_{16}]^{3-}$ with trifluorosulfonic acid also results in selective oxidative coupling of two Rh_7 clusters to give $[Rh_{14}(CO)_{26}]^{2-}$. Even though neither of the protonated species $[Rh_7(CO)_{16}H_{(3-x)}]^{x-}$ ($x = 1$, 2) were detected by ^1H NMR spectroscopy over a range of temperatures during the course of the reaction (216), the reaction probably follows a similar pathway to that of the hexanuclear dianions.

Oxidative aggregation is not restricted to rhodium HNCC. The dianions $[Ni_{12}(CO)_{21}H_{(4-x)}]^{x-}$ ($x = 2, 3$), for example, have been produced from the reaction of $[Ni_6(CO)_{18}]^{2-}$ with acids (257). In this case it was established by IR and ^1H NMR spectroscopy that the condensation processes occur through the sequence of reactions given by Eqs. (13)–(15).

$$3[Ni_6(CO)_{12}]^{2-} + 2H^+ \longrightarrow 2[Ni_9(CO)_{18}]^{2-} + H_2 \qquad (13)$$

$$2[Ni_9(CO)_{18}]^{2-} + H^+ \longrightarrow [Ni_{12}(CO)_{21}H]^{3-} + 3Ni(CO)_4 + 3Ni + 3CO \qquad (14)$$

$$[Ni_{12}(CO)_{21}H]^{3-} + H^+ \rightleftharpoons [Ni_{12}(CO)_{21}H_2]^{2-} \qquad (15)$$

Interestingly, the only example of oxidative cluster aggregation in the iron triad is the formation of $[Ru_6(CO)_{18}H]^-$ in the reaction of $[Ru_3(CO)_{11}H]^-$ with mineral acids (31).

b. *Selective Oxidative Degradation.*   The oxidation reactions of a number of large carbide clusters have been found to provide relatively selective synthetic routes to clusters of reduced nuclearity. For example, polyhedral contraction of the octahedral species $[Fe_5MC(CO)_x]^{y-}$ to the square-based pyramidal clusters $[Fe_4MC(CO)_z]^{w-}$ has been achieved selectively with ferric chloride as shown in Eqs. (16)–(20) (269).

$$[Fe_5CrC(CO)_{17}]^{2-} \xrightarrow{\ FeCl_3\ } Fe_4CrC(CO)_{16} \tag{16}$$

$$[Fe_5MoC(CO)_{17}]^{2-} \xrightarrow{\ FeCl_3\ } Fe_4MoC(CO)_{16} \tag{17}$$

$$[Fe_5WC(CO)_{17}]^{2-} \xrightarrow{\ FeCl_3\ } Fe_4WC(CO)_{16} \tag{18}$$

$$[Fe_5RhC(CO)_{16}]^{2-} \xrightarrow{\ FeCl_3\ } [Fe_4RhC(CO)_{14}]^- \tag{19}$$

$$[Fe_6C(CO)_{16}]^{2-} \xrightarrow{\ FeCl_3\ } Fe_5C(CO)_{15} \tag{20}$$

Similarly, the octahedral cluster $[Re_6C(CO)_{19}]^{2-}$ has been synthesized by the oxidation of $[Re_7C(CO)_{21}]^{3-}$ with iodine, which leads to abstraction of the capping rhenium unit (376).

III. Reactions of High-Nuclearity Carbonyl Clusters

In attempting to rationalize the reactivity of HNCC one faces an extremely complex problem. Compared with the boranes, which have been the subject of detailed electron-density calculations, our knowledge of the electronic properties of HNCC is still insufficient for correlations with reactivity patterns and prediction of the nature and structure of the products.

A full understanding of HNCC reactivity is hampered by several factors. First, HNCC exhibit a large number of polyhedral shapes, and within each given structure transition metals are found with coordination numbers ranging from five to nine and on occasion even more. It is therefore difficult to be sure of the site of attack of a substrate, whether it is a nucleophile or an electrophile. There is some evidence to suggest that nucleophilic attack occurs at sites of lowest coordination number, but this has not been established with certainty outside a few systems (387, 388). It has also been suggested that electrophilic attack occurs at sites of high electron density, viz, M—M bonds (408). The problem then is that of understanding the precise nature of the M—M bond in a metal cluster. To be more precise, one cannot overemphasize that polyhedral edges do not necessarily correspond to chemical bonds. Secondly, it is often difficult to predict the nature of the products formed. A number of pathways are always available for these reactions, and, depending on the type of metal and on the cluster nuclearity, the course of the reaction may vary considerably. This is complicated by the fact that in various reactions a large number of products are formed. In many instances it is clear that cluster degradation followed by recombination of the fragments occurs during the course of the reaction, leading to the formation of both lower and higher nuclearity compounds. This is particularly the case for HNCC of cobalt, nickel, iron, and rhodium. For clusters of the heavier metals, which normally require vigorous conditions to induce chemical reactivity, subsequent transformation of the products is often a problem. Thirdly, HNCC have a tendency to undergo polyhedral rearrangements. These often result from the addition of electron density to or its removal from the cluster as charge or as ligand. However, even more subtle changes resulting, for example, from the substitution of carbonyls by related ligands with different electronic and steric properties have been found to lead to changes in geometry.

For all these reasons it is difficult to suggest mechanisms for reactions of HNCC on the basis of product distribution and structure. These problems cannot be resolved readily. At present there is a paucity of kinetic data available and it may be that because of the complexity of these reactions such data will not be forthcoming. Nevertheless this is clearly an area where further studies are necessary.

Because of the difficulties discussed above, our classification of the reactions of HNCC is based on the products obtained from those reactions. They are discussed in six separate sections: A. Oxidation Reactions, B. Protonation and Deprotonation Reactions, C. Reduction Reactions, D. Electron-Transfer Reactions and Electrochemical Studies, E. Reactions with Soft Nucleophiles, and F. Oxidative Addition of the Small Molecules H_2, I_2, and HX.

A. OXIDATION REACTIONS

Oxidation of HNCC often results in cluster fragmentation or, as a consequence of redox condensation, cluster expansion as discussed in Section II,B,3. The systematic formation of oxidized species of unchanged nuclearity has been investigated only recently. Depending on the reagent, the products of oxidation reactions may or may not incorporate the oxidant and these reactions are classified below accordingly.

1. Oxidation of HNCC Anions without Incorporation of the Oxidant

Simple two-electron oxidation of anionic HNCC has been shown to proceed with or without concomitant addition of CO. For example, while oxidation of $[Os_5(CO)_{15}]^{2-}$ with $FeCl_3$ under CO gives $Os_5(CO)_{16}$ (383), $Os_6(CO)_{18}$ is the product of the oxidation of $[Os_6(CO)_{18}]^{2-}$ under the same reaction conditions (383). In the first case the cluster electron count is not altered in the process and the accommodation of an extra CO ligand results in only a slight distortion of the metal polyhedron in the dianion in comparison to the neutral species. Oxidation of the $[Os_6(CO)_{18}]^{2-}$ dianion though brings about reorganization of the metal polyhedron from an octahedral to a bicapped tetrahedral geometry. Cluster rearrangement is also observed in the oxidation of the 90-electron trigonal-prismatic $[Co_6C(CO)_{15}]^{2-}$ to give the 87-electron octahedral monoanion $[Co_6C(CO)_{14}]^-$ [Eq. (21)] (145, 147).

$$[Co_6C(CO)_{15}]^{2-} + FeCl_3 \longrightarrow [Co_6C(CO)_{14}]^- + FeCl_2 + Cl^- + CO \quad (21)$$

As oxidation reactions may often be complicated by further attack on the product by the oxidant, the choice of both oxidant and reaction conditions is critical. Ferric chloride under CO is the oxidant of choice, but even this sometimes causes cluster breakdown or buildup (see Section II,B.3). Mercuric salts have been used in the oxidation of $[Co_6(CO)_{14}]^{4-}$ and $[Co_6(CO)_{15}]^{2-}$ to give $Co_6(CO)_{16}$ (138, 389), but condensation of two cluster species incorporating atomic Hg may occur, as in the oxidation of $[Os_{10}C(CO)_{24}]^{2-}$ to $[Os_{20}Hg(C)_2(CO)_{48}]^{2-}$ (338). Iodine has been used successfully in a few cases, for example in the one-electron oxidation of $[Fe_3Pt_3(CO)_{15}]^{2-}$ to the monoanion $[Fe_3Pt_3(CO)_{15}]^-$ (283), as well as in the two-electron oxidation of $[Os_6(CO)_{18}]^{2-}$ to $Os_6(CO)_{18}$ (88). As discussed in Section III,A,2, however, the use of iodine as oxidant often results in the addition of I^+ to the anionic cluster.

Oxidation via intermediate formation of hydrido species has been achieved but only with clusters of the cobalt triad. At least two isomers of $Ir_6(CO)_{16}$ (235) have been isolated from the oxidation of $[Ir_6(CO)_{15}]^{2-}$ by varying the conditions employed, [Eqs. (22) (235) and (23) (236)].

$$[Ir_6(CO)_{15}]^{2-} + MeCOOH + CO \xrightarrow{CH_2Cl_2} Ir_6(CO)_{16} \text{ (black)} \quad (22)$$

$$[Ir_6(CO)_{15}]^{2-} + CO \xrightarrow{MeCOOH} Ir_6(CO)_{16} \text{ (red)} \quad (23)$$

Both isomers have an octahedral arrangement of metal atoms but with different CO distributions: the four COs that are not terminally bound in these species are edge-bridging in the black isomer but face-capping in the red one.

The nature of the oxidation products of $[Rh_6(CO)_{14}]^{4-}$ with acids shows a greater dependence on the H^+ to tetraanion ratio and on the strength of the acid. p-Toluenesulfonic acid in a 1 to 2 ratio with respect to $[Rh_6(CO)_{14}]^{4-}$ in acetonitrile gives $[Rh_6(CO)_{15}]^{2-}$, while an excess of the acid results in the formation of $Rh_6(CO)_{16}$; the use of an excess of the weaker acetic acid, however, gives mainly $[Rh_{12}(CO)_{30}]^{2-}$ (170). This illustrates the extremely sensitive balance between formation of M—M and M—CO bonds as well as the kinetic dependence of the oxidation reactions.

2. Addition of I^+ and NO^+ to Anionic HNCC

The reactions of anionic HNCC with I_2 and $NOBF_4$ resulting in the oxidative addition of I^+ and NO^+, respectively, have been studied in only a few systems. Even so it is becoming increasingly possible to see a pattern emerging in these processes. Because of a lack of kinetic data for these reactions, only a simplified discussion based on their final products is possible.

The nature of the products may well depend on the site of attack of the electrophile, among other factors. The metal atoms in HNCC are effectively shielded by a close-packed CO ligand envelope (384), which renders direct attack upon them difficult; however, oxygen atoms in compounds containing bridging COs provide electron-rich sites for attack. Kinetic studies have suggested that the formation of transition adducts of the type μ-{COE} takes place during the initial stages of the reaction of a number of polynuclear metal carbonyl derivatives and halogens (437).

This site of attack has also been postulated in the reaction of $[Co_6(CO)_{15}]^{2-}$ with acetyl chloride [Eq. (24)] (390).

$$3[Co_6(CO)_{15}]^{2-} + 2MeCOCl \xrightarrow{\;\;CO\;\;}$$
$$2[Co_6C(CO)_{14}]^- + 4[Co(CO)_4]^- + 2Co^{2+} + 2Cl^- + 2MeCOO^- \quad (24)$$

Addition of the acylium ion is believed to give $[Co_6(CO)_{14}$-{CO—C(=O)—Me}]$^-$ as an intermediate (Scheme 12,B). There is a precedent for such a ligand: In $Fe_3(CO)_{10}\{\mu_2$-CO—C(=O)—Me}]$^-$ (438), the acyl group is indeed attached to the oxygen atom of a bridging CO. Upon scission of this CO, the carbido cluster $[Co_6C(CO)_{14}]^-$ (Scheme 12,C) is formed and acetate is liberated. The site of I_2 attack in the reaction with $[Rh_6(CO)_{15}]^{2-}$ might also be at an oxygen of a bridging CO. Loss of I^- and formation of $[Rh_6(CO)_{15}I]^-$ with a terminally bound iodine atom (Scheme 12,D) occur in this case [Eq. (25)] (173).

$$[Rh_6(CO)_{15}]^{2-} + I_2 \xrightarrow{\hspace{2cm}} [Rh_6(CO)_{15}I]^- + I^- \quad (25)$$

The formation of the nitrido cluster $[Co_6N(CO)_{15}]^-$ (Scheme 12,E) from the reaction of $[Co_6(CO)_{15}]^{2-}$ and NO^+ (152) may similarly involve initial formation of $[Co_6(CO)_{15}(NO)]^-$ related to $[Rh_6(CO)_{15}I]^-$, followed by reduction of the coordinated NO to nitrogen by the cluster or by CO.

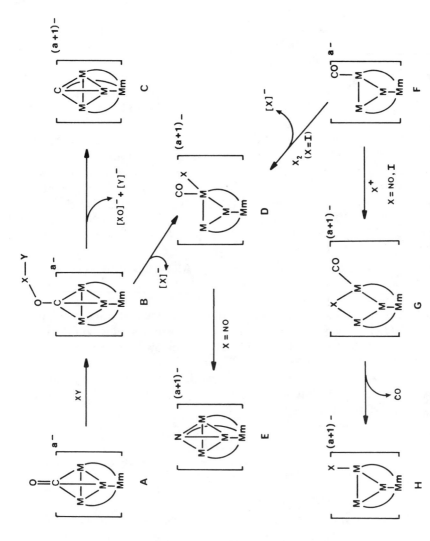

SCHEME 12. Pathways for the reactions of electrophiles with anionic HNCC (see text).

For anionic HNCC that do not contain bridging COs, as is often the case with compounds of osmium and ruthenium (Scheme 12,F), one could envisage the negatively charged oxygen layer orientating the attacking molecule XY for the tunneling process through the CO barrier (439). Dissociation of Y^- and attack of X^+ at a center of high electron density on the metal core might be followed by migration of X and metal polyhedral rearrangements. The nature of the addition product may depend not only on the electrophile but probably also on the type of cluster and charge distribution within it. For example, addition of I^+ to the "electron precise" trigonal-bipyramidal dianion $[Os_5(CO)_{15}]^{2-}$ gives $[Os_5(CO)_{15}I]^-$ (Scheme 12,D). The metal polyhedron in the dianion is slightly distorted in $[Os_5(CO)_{15}I]^-$. The iodine atom that is terminally bound and formally acts as a two-electron donor bears some negative charge and is easily abstracted by Ag^+ (75). In contrast, addition of I^+ to the tetracapped octahedral dianion $[Os_{10}C(CO)_{24}]^{2-}$ involves a sequential opening up of two capping tetrahedra to give $[Os_{10}C(CO)_{24}(\mu\text{-}I)]^-$ and $Os_{10}C(CO)_{24}$ $(\mu\text{-}I)_2$, which contain Os_4 butterflies bridged by iodine atoms acting as three-electron donors (Scheme 12,G) (133). For all HNCC studied to date, dissociation of bridging iodine ligands is effected by I^-, even when major structural rearrangements have occurred upon I^+ addition, e.g., in $Os_8(CO)_{22}H(\mu\text{-}I)$ (124). Addition of NO^+ to the dianion $[Os_{10}C(CO)_{24}]^{2-}$ initially gives the NO-bridged species $[Os_{10}C(CO)_{24}(\mu\text{-}NO)]^-$ with a structure similar to that of the monoiodine derivative. The final product of the reaction, though, is $[Os_{10}C(CO)_{23}(NO)]^-$, which results from the closure of the Os_4 butterfly upon CO ejection and a change in coordination of the NO from bridging to terminal (132) (Scheme 12,H). The same mechanism involving coordination of NO^+ prior to CO ejection might also be involved in the reaction of $[Fe_6C(CO)_{16}]^{2-}$ with NO^+ to give $[Fe_6C(CO)_{15}(NO)]^-$. Further reaction of this monoanion with excess NO^+ affords $Fe_6C(CO)_{11}(NO)_4$ but the mechanism of the formation of this cluster is not clear, although it must involve an oxidative reduction step (17).

B. Protonation and Deprotonation Reactions

The protonation reactions of anionic HNCC are often complicated by further chemical transformations of unstable hydride derivatives, as discussed in Sections II,B,3 and III,A,1. There are, however, several examples of straightforward protonation reactions to yield hydrido clusters of unchanged nuclearity. In many cases, by choosing acids and solvents correctly, sequential protonation of the anions may be achieved.

Acidification of the dianions $[Os_5(CO)_{15}]^{2-}$ (58), $[Os_6(CO)_{18}]^{2-}$ (88), $[Os_7(CO)_{20}]^{2-}$ (410), $[Os_8(CO)_{22}]^{2-}$ (410), and $[Os_{10}C(CO)_{24}]^{2-}$ (130) with sulfuric acid in dichloromethane or tetrahydrofuran, for example, gives the respective monohydrides, while reaction in acetonitrile results in the precipitation of the dihydrides. The acidic nature of the dihydrides $[Os_m(CO)_nH_2]$ ($m = 6, 7, 8$; $n = 18, 20, 22$, respectively) and $[Os_{10}C(CO)_{24}H_2]$ is evidenced by their facile deprotonation by halides to give the corresponding monoanions; further deprotonation requires stronger bases such as KOH, DBU, or proton sponge (409, 410). The mechanism of this reaction is not understood but ^1H NMR studies on the reaction of $Os_7(CO)_{20}H_2$ with Cl^- suggest that formation of the monoanion $[Os_7(CO)_{20}H]^-$ occurs via the intermediate $[Os_7(CO)_{20}H_2Cl]^-$ (474). Interestingly, reaction of $Os_5(CO)_{15}H_2$ with I^- does not lead to deprotonation, but rather to formation of the adduct $[Os_5(CO)_{15}H_2I]^-$ (65), demonstrating that reduction in nuclearity and acidity occur concomitantly in these systems.

The isolation and full characterization of a number of large hydrido clusters have in many cases been hampered by their extremely high acidity. The clusters $[Ni_{38}Pt_6(CO)_{48}H_{(6-n)}]^{n-}$ ($n = 3-6$), for instance, exist as an equilibrium mixture of anions **a** ($n = 3$), **b** ($n = 4$), **c** ($n = 5$), and **d** ($n = 6$) in acetonitrile solution. This mixture has been found to be easily converted into one of its components by controlled addition of acid or base [Eq. (26)] (373).

$$\mathbf{a} \; \underset{H^+}{\overset{MeCN}{\rightleftarrows}} \; \mathbf{b} \; \underset{H^+}{\overset{CO_3^{2-}}{\rightleftarrows}} \; \mathbf{c} \; \underset{H_2O}{\overset{OH^-}{\rightleftarrows}} \; \mathbf{d} \qquad (26)$$

The hydrides **b** and **c** were isolated in a crystalline state. Isolation of **a**, however, was not possible due to the ready deprotonation of its salts on dissolution both in acetone and acetonitrile, the only solvents in which the mixture of **a–d** is soluble.

1. Sites of Protonation

It is not known whether proton transfer from the medium to the metal atoms through the carbonyl shell is direct or whether intermediates containing protonated ligands are formed. Protonation at the oxygen of a CO group in small clusters has been shown to occur [e.g., in $Fe_3(CO)_{10}H(COH)$ (440)] but not a single example has yet been reported for HNCC hydrides. Protonation of the exposed carbide atom in the pyramidal clusters $[M_5C(CO)_{14}]^{2-}$ [M = Fe (411, 414), Ru (412), and Os (78)] does not occur either. This is in contrast with the butterfly species $[Fe_4C(CO)_{14}]^{2-}$, which is attacked by acid to give first $[Fe_4C(CO)_{14}H]^-$, with the hydride attached to the metal framework, and then $Fe_4(CH)(CO)_{14}H$, in which the second hydride bridges an Fe—C vector (*413, 414*). Theoretical studies by Shriver *et al.* (*411*) have correlated the greater reactivity of the carbon atom in the butterfly cluster to the fact that the highest occupied molecular orbital (HOMO) of the butterfly carbide contains a significant negative charge on this atom. Hoffmann *et al.* (*415*) have also carried out calculations on this system and their results are in general agreement with the earlier calculations, except that the HOMO–LUMO (LUMO, lowest unoccupied molecular orbital) gap is less in the butterfly cluster than in the square-pyramidal species.

These and other MO calculations (*416*) indicate that the positions preferred by hydrogen atoms tend to reflect the electron distribution in clusters. In general, hydrides occupy edge-bridging sites, but a few examples of face-capping hydrides in HNCC are known [e.g., $Ru_6(CO)_{18}H_2$ (*35*) and $Os_8(CO)_{22}HI$ (*124*)]. The only reported species containing terminal hydrides are $Os_6(CO)_{19}H_2$ (*92*) and $[Rh_6(CO)_{15}H]^-$ (*397*). A number of clusters with semiinterstitial [e.g., $[Rh_{14}(CO)_{25}H]^{3-}$ (*223*)] and interstitial hydrides are also known (Table II). It is currently impossible to relate and predict the hydride position in a homologous series of compounds (Scheme 13). While $[Co_6(CO)_{15}H]^-$ and $[Ru_6(CO)_{18}H]^-$ contain interstitial hydrides, for example, the hydride in $[Rh_6(CO)_{15}H]^-$ occupies a terminal site in solution (*397*) and in $[Os_6(CO)_{18}H]^-$ the hydride caps an Os_3 face in the solid state (*88*).

2. Structural Characterization of Large Hydrido Clusters

X-Ray diffraction has often been successful in the determination of hydrogen atom position in HNCC hydrides. Direct location from Fourier

TABLE II

^{1}H NMR DATA OF INTERSTITIAL HYDRIDO HNCC

Cluster hydride	Environment	^{1}H NMR shift (δ)	Other studies[a]
$[Co_6(CO)_{15}H]^-$	Octahedral	23.2	N and X
$[Ru_6(CO)_{18}H]^-$	Octahedral	16.4	N and X
$[Os_{10}C(CO)_{24}H]^-$	Tetrahedral	-14.4	X
$[Os_{10}(CO)_{24}H_4]^{2-}$	?	-16.5 (broad) at RT, -14.7 and -19.1 (1:1) at 198 K	
$[Ni_{12}(CO)_{21}H]^{3-}$	Distorted, octahedral	-24	N and X
$[Ni_{12}(CO)_{21}H_2]^{2-}$	Distorted, octahedral	-18	N and X

[a] N, Neutron diffraction; X, X-ray diffraction.

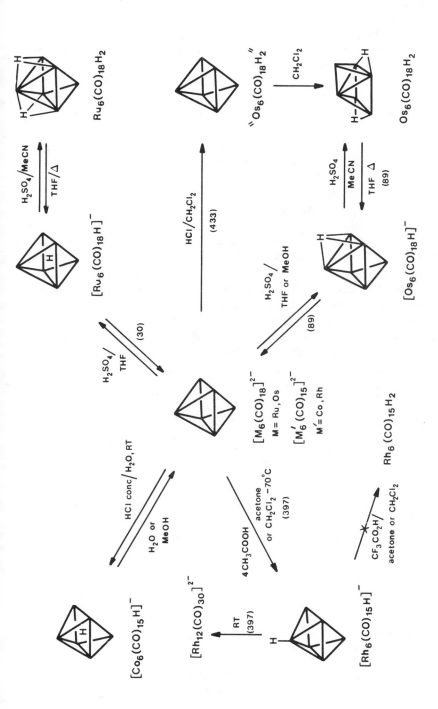

SCHEME 13. Protonation reactions of $[M_6(CO)_{18}H_{(2-n)}]^{n-}$ (M = Ru, Os; n = 2, 1) and $[M'_6(CO)_{15}H_{(2-n)}]^{n-}$ (M' = Co, Rh; n = 2, 1) and deprotonation reactions of the hydrido derivatives.

difference maps is rarely achieved, but instead indirect methods such as stereochemical considerations of the ligands' disposition about the metal atoms (446), M—M bond lengthening, metal cluster face or cavity expansion, and potential energy calculations (447) have been used with success. Precise representation of the hydrogen bonding in these systems, however, is possible only with the use of single-crystal neutron diffraction.

The use of this technique to study the hydrido clusters $[Ni_{12}(CO)_{21}H_{4-n}]^{n-}$ ($n = 2, 3$) (417) has provided the most detailed stereo-chemical study on the bonding of an interstitial hydrogen atom. The X-ray structures of $[Ni_{12}(CO)_{21}H_2]^{2-}$ and $[Ni_{12}(CO)_{21}H]^{3-}$ (256) consist of a 12-atom nickel fragment of a hexagonal-close-packed (hcp) metal lattice surrounded by 9 terminal and 12 bridging carbonyls (Fig. 6). Neutron diffraction studies have shown that the hydrogen atoms in both dianion and trianion are localized in octahedral sites as suggested by the X-ray analysis. However, the hydride in the trianion was found to be much closer to the central hexanickel plane (0.73 Å) than to the outer triangular plane (1.69 Å), effectively being coordinated to only three of the six nickel atoms. In the dihydride the two interstitial hydrogen atoms are closer to the centers of the octahedral sites. This has been ascribed both to competition for electron density on the central nickel triangle and mutual repulsion between the two hydrogen nuclei in neighboring octahedral holes.

It has been found (257) that once the two octahedral cavities have been occupied in $[Ni_{12}(CO)_{21}H_2]^{2-}$, this cluster cannot be protonated further. The fact that protonation of the monohydride $[Fe_6Pd_6(CO)_{24}H]^-$ also cannot be achieved has been used to support the tentative hydride location in the Pd_6 octahedron on the basis of the long average Pd—Pd distance (289).

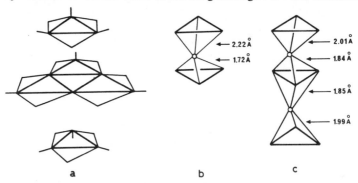

FIG. 6. (a) Schematic architecture of the hydrido clusters $[Ni_{12}(CO)_{21}H_{(4-n)}]^{n-}$ ($n = 2, 3$). Both anions are constructed from the planar $Ni_6(CO)_3(\mu_2\text{-}CO)_6$ capped by two $Ni_3(CO)_3(\mu_2\text{-}CO)_3$ fragments through Ni–Ni interactions. Fragments of the structures showing the hydrogen atoms in the octahedral interstices in (b) $[Ni_{12}(CO)_{21}H]^{3-}$ and in (c) $[Ni_{12}(CO)_{21}H_2]^{2-}$ (mean Ni-H distances given) (417).

Proton NMR cannot be used as an indication of the position occupied by hydrides in HNCC in solution as the range of chemical shifts observed is enormous (441). For example, all fully characterized carbonyl clusters that contain interstitial hydrides are listed in Table II, with chemical shifts from 23.2 δ in $[Co_6(CO)_{15}H]^-$ to $-24\ \delta$ in $[Ni_{12}(CO)_{21}H]^{3-}$. The octahedral $[Ru_6(CO)_{18}H]^-$ was the first reported cluster for which an interstitial hydride was assigned on the basis of X-ray (30, 31) and solid-state infrared spectroscopy studies (33). However, because of the extremely low field position of the NMR signal (16.4 δ), it was suspected to be of the formyl type (417). Its interstitial position was later unequivocally established by neutron diffraction studies (32). The observation of the ^{191}Os–^1H satellites in the proton NMR of $[Os_{10}C(CO)_{24}H]^-$ has allowed in this case conclusive location of the hydride in one of the tetrahedra of the tetracapped octahedral cluster. This supports the results of X-ray studies that evidenced slight expansion of the tetrahedral site (131).

Multinuclear NMR has been used extensively to investigate the structures of rhodium clusters in solution. Where X-ray quality crystals are not available, as is the case for $[Rh_6(CO)_{15}H]^-$, the study of $^1H,^1H$-$\{^{103}Rh\}$, ^{13}C, ^{13}C-$\{^{103}Rh\}$, and ^{103}Rh NMR has allowed determination of the cluster structure (Scheme 13) (397). A combination of this technique with X-ray diffraction studies in the hydrido species $[Rh_{13}(CO)_{24}H_{(5-n)}]^{n-}$ $(n = 2, 3)$ has helped to establish the nature of the hydrogen atoms in these clusters. Both clusters have a centered twinned cuboctahedral arrangement of rhodium atoms (Fig. 7). Their hydrides were located in the square-pyramidal holes for which a comparison of the mean Rh—Rh bonds of cavities showed that lengthening had occurred. A $^1H,^1H$-$\{^{103}Rh\}$ INDOR and ^{13}C NMR spectroscopic solution study was consistent with a rapid migration of the interstitial hydrogen atoms inside the hcp rhodium cluster. As the accepted radius of a hydrogen atom, 0.37 Å, is much greater than the hole in the triangular Rh_3 face (0.22 Å) in the inside of the cluster, these studies suggest

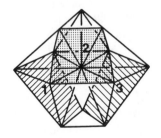

FIG. 7. The twinned cuboctahedral rhodium atoms arrangement in the hydrido clusters $[Rh_{13}(CO)_{25}H_{(5-n)}]^{n-}$ $(n = 2, 3, 4)$. The hydrogen atoms occupy the square-based pyramidal holes 1 $(n = 4)$, 1 and 2 $(n = 3)$, and 1, 2, and 3 $(n = 2)$.

that the hydrogen atom is protonic in character (*217*). A more attractive picture, however, is that of a "breathing cluster" with Rh—Rh bonds expanding and contracting as the CO ligands exchange and the hydrogens migrate.

3. Metal Polyhedral Rearrangements in Protonation Reactions

Addition of protons to anionic clusters to generate hydrides leaves the cluster electron count unaffected, yet the process is sometimes accompanied by structural changes. For example, both $[Os_6(CO)_{18}]^{2-}$ (*87*) and $[Os_6(CO)_{18}H]^-$ (*88*) have octahedral arrangements of metal atoms as predicted by Wade's rules. In the dihydride $Os_6(CO)_{18}H_2$, however, the metal atoms describe a capped square-based pyramid (*87*).

Recent molecular orbital (MO) calculations by Wade *et al.* (*408*) using the series $[B_6H_6]^{2-}$, $[B_6H_7]^-$, and B_8H_8 as models for the protonation of hexanuclear metal carbonyls have attempted to rationalize these findings. The charge distribution is symmetrical in an octahedral $[B_6H_6]^{2-}$ but asymmetrical in the capped square-based pyramidal isomer. It was found that upon protonation, significant charge redistribution occurs. This results in a substantial decrease in the symmetry of the octahedral cluster framework, which is disfavored in comparison with the capped square-based pyramidal structure much less affected by the protonation process.

Further evidence for the stabilization of this structure in the Os_6 system comes from the fact that the octahedral isomer of $Os_6(CO)_{18}H_2$, isolated by varying the conditions for protonation of $[Os_6(CO)_{18}]^{2-}$ (*433*), slowly rearranges in solution to the more thermodynamically stable capped square-based pyramidal structure. The dihydride $Ru_6(CO)_{18}H_2$, however, does not show preference for this structure and retains the octahedral metal atom geometry observed for $[Ru_6(CO)_{18}]^{2-}$ and $[Ru_6(CO)_{18}H]^-$ (*30, 31, 35*) even on warming.

The dihydride $Os_7(CO)_{20}H_2$ also exists in two isomeric forms, which, in this case, are formed by two very different routes (Scheme 14). Protonation of $[Os_7(CO)_{20}]^{2-}$ gives the brown isomer (*118*). Only the red isomer, synthesized via the addition of $Os(CO)_4H_2$ to $Os_6(CO)_{16}(MeCN)_2$, has been structurally characterized (*118*). Instead of the monocapped octahedron predicted by Wade's Rules and observed for the isoelectronic $Os_7(CO)_{21}$, it has a structure based on the $Os_6(CO)_{18}$ bicapped tetrahedral metal atoms arrangement, with an edge bridged by an $Os(CO)_4$ unit. These two species do not interconvert. Deprotonation of both isomers yields the same dianion and the red species, once deprotonated, even singly, cannot be reformed (*474*).

It has become evident that as cluster nuclearity increases, the simple relationship between their electron count and structure is lost, particularly

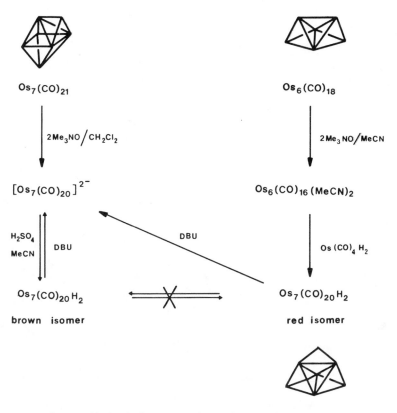

SCHEME 14. Synthetic routes to the two isomers of $Os_7(CO)_{20}H_2$.

when they contain ligands other than CO. In these cases the packing of the metal atoms and the ligand stereochemistries are optimized in a way that has not yet been rationalized. Comparison of the structures of $[Rh_{14}(CO)_{25}]^{4-}$ (220), $[Rh_{14}(CO)_{25}H]^{3-}$ (223), and $[Rh_{14}(CO)_{26}]^{2-}$ (219) (all with the same electron count) shows that isoelectronic replacement of two negative charges by a CO and protonation of the tetraanion lead to distortions from a body-centered cubic (bcc) arrangement of rhodium atoms to arrays intermediate between ccp and bcc (Scheme 15). In the similar isoelectronic series $[Os_8(CO)_{22}]^{2-}$, $[Os_8(CO)_{22}H]^{-}$, and $Os_8(CO)_{23}$, no significant distortion in the bicapped octahedral metal geometry is observed in going from the neutral carbonyl to the dianion. In contrast, the hydrido species $[Os_8(CO)_{22}H]^{-}$ has a fused tetrahedral structure (124). This emphasizes once again the ability of hydrides to stabilize alternative metal cluster geometries.

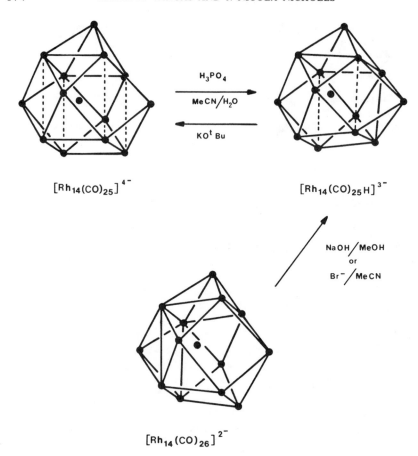

$$[Rh_{14}(CO)_{25}]^{4-}$$

$$[Rh_{14}(CO)_{25}H]^{3-}$$

$$[Rh_{14}(CO)_{26}]^{2-}$$

SCHEME 15. Structural transformations in the Rh_{14} system.

C. REDUCTION REACTIONS

Under this heading only reactions with hard nucleophiles that result in reduced species that do not incorporate the nucleophilic reagent are considered. The additions of soft nucleophiles to HNCC are discussed in Section III,E.

1. Two-Electron Reduction with Concomitant Loss of CO

As with oxidation, the reduction of HNCC can be complicated by redox condensation reactions and by cluster degradation induced by free carbon monoxide. The latter is most commonly observed when reduction is effected by alkali metals (see Section II,B,1) [e.g., Eq. (27)] (196).

$$10[Co_6(CO)_{15}]^{2-} + 22Na \longrightarrow$$
$$9[Co_6(CO)_{14}]^{4-} + 6[Co(CO)_4]^- + 22Na^+ \qquad (27)$$

The ejection of CO and subsequent cluster degradation may be avoided if nucleophiles such as OH^-, OR^-, H^-, and R^- are used instead. These species oxidize CO to CO_2 simultaneously with cluster reduction. It is believed that the mechanism for this reaction in HNCC is similar to that known for mononuclear complexes [Eqs. (28) and (29)] (418).

$$M_m(CO)_n + OH^- \longrightarrow [M_m(CO)_{(n-1)}(CO_2H)]^- \qquad (28)$$

$$[M_m(CO)_{(n-1)}(CO_2H)]^- \longrightarrow [M_m(CO)_{(n-1)}]^{2-} + CO_2 + H^+ \qquad (29)$$

Although no CO_2H-containing cluster has been isolated, there have been reports of related species in the reactions of OR^- and R^- with $Rh_6(CO)_{16}$ (168, 172, 173) (Scheme 16) and $Ru_6C(CO)_{17}$ [Eqs. (30) and (31)] (39).

$$Ru_6C(CO)_{17} \xrightarrow{Na_2CO_3/MeOH} [Ru_6C(CO)_{16}(COOMe)]^- \qquad (30)$$

$$[Ru_6C(CO)_{16}(COOMe)]^- \xrightarrow{OH^-/H_2O} [Ru_6C(CO)_{16}]^{2-} \qquad (31)$$

The reduction of neutral carbonyls with alkali metal hydroxides or carbonates is an important route to HNCC dianions. In a few instances reduction of rhodium clusters has been reported to proceed further with oxidation of another CO. Examples are the formation of $[Rh_6(CO)_{14}]^{4-}$ from $[Rh_6(CO)_{15}]^{2-}$ (170) (Scheme 16) and the reduction of the dicarbide $[Rh_{12}(C)_2(CO)_{24}]^{2-}$ to give $[Rh_{12}(C)_2CO_{23}]^{4-}$ (213). In the latter case, the unstable paramagnetic intermediate $[Rh_{12}(C)_2(CO)_{23}]^{3-}$ has also been isolated upon slight alteration of the reaction conditions [Eqs. (32) and (33)].

$$[Rh_{12}(C)_2(CO)_{24}]^{2-} \xrightarrow{OH^-/MeOH} [Rh_{12}(C)_2(CO)_{23}]^{3-} \qquad (32)$$

$$[Rh_{12}(C)_2(CO)_{23}]^{3-} \xrightarrow[i\text{-PrOH, }\Delta]{OH^-\text{ (excess)}} [Rh_{12}(C)_2(CO)_{23}]^{4-} \qquad (33)$$

The reduction of the trianion to $[Rh_{12}(C)_2(CO)_{23}]^{4-}$ is easily reversed by I_2 but the dianion $[Rh_{12}(C)_2(CO)_{24}]^{2-}$ cannot be regenerated under these reaction conditions.

2. Reduction Reactions of Osmium Binary Clusters $Os_m(CO)_n$

The reduction of $Os_5(CO)_{16}$ with excess potassium hydroxide proceeds to give $[Os_5(CO)_{15}]^{2-}$. In contrast the series $Os_m(CO)_n$ ($m = 6, 7, 8$; $n = 18$, 21, 23) undergoes fragmentation under similar conditions with the ejection

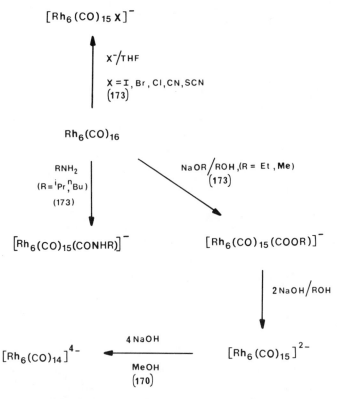

SCHEME 16. Reduction reactions of $Rh_6(CO)_{16}$.

of an "$Os(CO)_3$" unit, resulting in the formation of the dianions $[Os_{(m-1)}(CO)_{(n-3)}]^{2-}$ (Scheme 17) (386). It has been proposed that the difference in behavior on reduction may be related to the electron-precise nature of $Os_5(CO)_{16}$. Having six skeletal electron pairs ($s = 6$), this cluster adopts a fundamental polyhedron (trigonal-bipyramid). In contrast, the electron-deficient clusters $Os_6(CO)_{18}$ ($s = 6$), $Os_7(CO)_{21}$ ($s = 7$), and $Os_8(CO)_{23}$ ($s = 7$) possess capped polyhedral geometries, and, by loosing $Os(CO)_3$ groups, generate the electron-precise anions $[Os_5(CO)_{15}]^{2-}$ and $[Os_6(CO)_{18}]^{2-}$, and the less electron-deficient $[Os_7(CO)_{20}]^{2-}$, respectively.

Reduction of the binary carbonyls $Os_m(CO)_n$ ($m = 6, 7, 8; n = 18, 21, 23$) to their respective dianions $[Os_6(CO)_{18}]^{2-}$ and $[Os_m(CO)_{(n-1)}]^{2-}$ ($m = 7, 8$) is achieved with weaker bases such as nitriles (387) and halides (388, 335, 123). The importance in the choice of the reducing conditions for this cluster series is summarized in Scheme 17.

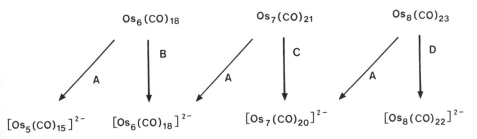

SCHEME 17. Reduction reactions of $Os_6(CO)_{18}$, $Os_7(CO)_{21}$, and $Os_8(CO)_{23}$. A, KOH/MeOH; B, I$^-$, Na–Hg, or Zn/CH$_2$Cl$_2$, nitriles; C, Me$_3$NO or [BH$_4$]$^-$/CH$_2$Cl$_2$; D, I$^-$ or [BH$_4$]$^-$/THF.

The reduction reactions of $Os_6(CO)_{18}$ have been investigated in great detail. Studies on the reduction of this cluster by donor ligands Y (Y = I$^-$, RCN, CN$^-$, pyridine, OMe$^-$, and OH$^-$) suggest that successive additions of Y probably occur at the same osmium atom to give $Os_6(CO)_{18}Y$ and $Os_6(CO)_{18}Y_2$ (Scheme 18). Kinetic studies of the nitrile reaction revealed a third-order rate dependence on the nitrile concentration, suggesting that the formation of the dianion $[Os_6(CO)_{18}]^{2-}$ is brought about by the attack of a third ligand (*387*). A similar process has been invoked to account for the formation of $[Os_5(CO)_{15}]^{2-}$ on reaction of $Os_6(CO)_{18}$ with strong bases such as OH$^-$, OMe$^-$, and concentrated CN$^-$ (*410*), discussed above. In this case, addition of a third molecule of Y serves to eliminate the electron deficient group. Evidence for these propositions comes from the reaction of $Os_6(CO)_{18}$ with pyridine, which leads to production either of $[Os_6(CO)_{18}]^{2-}$ or $[Os_5(CO)_{15}]^{2-}$, depending on the molar ratio of $Os_6(CO)_{18}$ over pyridine employed (*177*). The neutral $Os_6(CO)_{17}(C_5H_5N)_2$ was also isolated in low yield from this reaction (*112*). It adopts a spiked trigonal-bipyramidal structure with the two pyridine ligands on the terminal osmium. It may derive from $Os_6(CO)_{18}(C_5H_5N)_2$ through nucleophilic attack by the oxygen of a terminal CO at the terminal osmium of the trigonal bipyramid, and elimination of CO (*112*).

A different mechanism has been proposed for the reduction of $Os_6(CO)_{18}$ by R$_3$NO (two equivalents, R = Me, Et) to give $[Os_6(CO)_{17}H]^-$ (*419*). The reaction of $Os_6(CO)_{18}$ with trialkylamine oxide in the presence of reagents such as MeCN gives $Os_6(CO)_{(18-n)}(MeCN)_n$ ($n = 1, 2$) (see Section III,E,2,b). It is believed that, in the absence of acetonitrile, the initial product formed is the monosubstituted species $Os_6(CO)_{17}(R_3N)$, which, by transfer of an α hydrogen from the amine ligand to the metal cluster and accompanying elimination of the iminium ion $[R_2N=CHR]^+$, yields $[Os_6(CO)_{17}H]^-$ (Scheme 18).

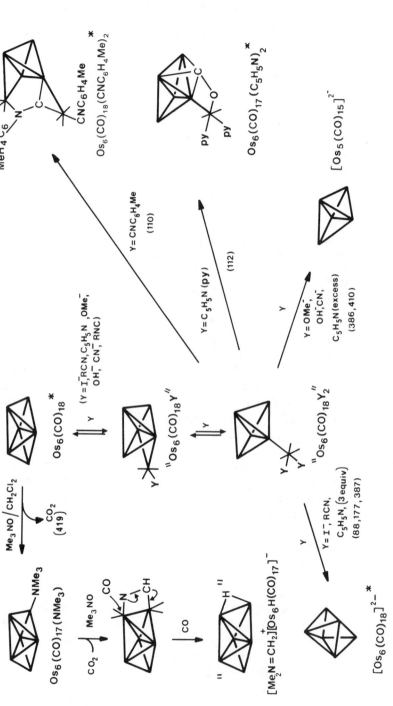

SCHEME 18. Reactions of $Os_6(CO)_{18}$ with nucleophiles Y (Y = I$^-$, RCN, C_5H_5N, OMe$^-$, OH$^-$, CN$^-$, RNC). Asterisks denote structures determined by X-ray studies.

D. ELECTRON-TRANSFER REACTIONS AND ELECTROCHEMICAL STUDIES

The electrochemistry of HNCC is a new and rapidly expanding field which has not been covered in previous reviews of electrochemical studies of transition metal clusters (420). In a similar way to smaller clusters, HNCC have been found to exhibit both reversible and irreversible redox behavior. These processes will be discussed separately in the following sections.

1. Characterization of Paramagnetic, Electrochemically Generated HNCC Species

It has been shown that clusters often undergo oxidation or reduction by single electron-transfer steps. A number of paramagnetic HNCC species have been obtained, particularly radical anion clusters. Characterization of these radicals is afforded by solution electron spin resonance (ESR) spectroscopy. Further characterization of these species has often been hampered by their instability toward oxidation or reduction and the difficulty of separating them from the base electrolyte. Only a few HNCC radicals have been found to be stable enough to allow isolation in the solid state. These species were in fact generated chemically by a variety of methods discussed in Sections III,A and III,C. In these cases, ESR studies were substantiated by X-ray structural data and a detailed picture of their bonding has been established, for example, for $[Co_6C(CO)_{14}]^-$, $[Rh_{12}(C)_2(CO)_{23}]^{3-}$, $[Co_{13}(C)_2(CO)_{24}]^{4-}$, (148) and $[Fe_3Pt_3(CO)_{15}]^-$ (284).

The g values for many radical HNCC species are close to the free electron values, e.g., $[Ru_6C(CO)_{17}]^-$ ($g = 2.001$ at 223 K) (385). This suggests that there is little mixing of excited or lower energy levels with the orbital containing the odd electron and that this orbital is delocalized and largely metallic in character. Deviations of g values from the free electron values denote the contribution from the spin–orbit coupling of metal atoms between various states, e.g., $[Rh_{12}(C)_2(CO)_{23}]^{3-}$ ($g_1 = 2.282$, $g_2 = 2.198$, $g_3 = 2.038$, from 120 to 4.2 K) (148).

2. Reversible Electrochemical Generation of Different Oxidation States

a. Redox Behavior of $Os_6(CO)_{18}$. Studies of the electrochemical reduction of $Os_6(CO)_{18}$ (421) have shown it to proceed in a chemically reversible two-electron wave to the cluster dianion $[Os_6(CO)_{18}]^{2-}$ at either Pt or Au electrodes in tetrahydrofuran solution [Eq. (34)].

$$Os_6(CO)_{18} + 2e^- \longrightarrow [Os_6(CO)_{18}]^{2-} \quad [E^\circ = 0.04 \text{ V (vs. SCE)}] \quad (34)$$

Cyclic voltammetry results were shown to be consistent with a redox model involving two stepwise one-electron transfers, in which the structural rearrangement of the bicapped tetrahedron of $Os_6(CO)_{18}$ occurs during the first electron-transfer step to give an octahedral structure. From the temperature dependence of the electron-transfer rate, the cluster reorganization energy was estimated to be $\Delta H^{\ddagger} = 8$ kcal/mol. This is lower than the activation energy for carbonyl scrambling in the cluster (85). In other words, the breaking and remaking of M—M bonds in the formation of a transient state in this cluster is energetically more favorable than similar processes involving M—CO bonds. This observation is consistent with both the $Os_6(CO)_{18}$ redox chemistry discussed in Sections III,A and III,B and with the fact that the reactivity of this cluster with nucleophiles under mild conditions is dominated by opening up of the cluster rather than CO substitution (see Section III,E).

Further two-electron reduction of the dianion at -2.15 V was also achieved electrochemically, and gave a highly unstable species formulated as $[Os_6(CO)_{18}]^{4-}$. This is in contrast with the chemical reduction of the related cluster $[Ru_6(CO)_{18}]^{2-}$ with stoichiometric amounts of $Na[(C_6H_5)_2CO]$ in tetrahydrofuran, which affords $[Ru_6(CO)_{17}]^{4-}$ and $[Ru_6(CO)_{16}]^{6-}$ sequentially with CO dissociation (423). The electrochemical oxidation of $[Ru_6(CO)_{18}]^{2-}$ is also different from that of $[Os_6(CO)_{18}]^{2-}$ as it is chemically irreversible (421).

b. Redox Behavior of "Raft" Os_6 Clusters. Predictions of the reduction potentials of the raft Os_6 clusters have been made by Evans and Mingos (422) on the basis of studies of the electronic requirements of the hypothetical planar $[Os_6(CO)_{24}]^{6+}$ cluster. These predictions may also be applied to $Os_6(CO)_{21}$, which can be derived from $[Os_6(CO)_{24}]^{6+}$ by isolobal replacement of three $[Os(CO)_4]^{2+}$ by three $Os(CO)_3$ groups. Their calculations have indicated the existence of a low-lying empty molecular orbital of a_2^1 symmetry in these clusters (Fig. 8), which is antibonding between the osmium atoms of the central triangle and bonding between the triangle and the bridging osmium groups. As a consequence, two-electron reduction of $Os_6(CO)_{21}$ should occur easily and the occupation of this orbital should lead to an increase in bond lengths in the central triangle and a decrease in the bond lengths to the bridging groups.

Electrochemical studies of the series of phosphite and phosphine substituted compounds $Os_6(CO)_{(21-n)}\{PR_3\}_n$ (R = OMe, $n = 1$-6; R = Ph, $n = 1$-3) (99) are in accord with the predictions of Evans and Mingos. These compounds undergo two well-defined reversible one-electron reductions and one irreversible oxidation (Table III). A correlation has been established between the values of $E_{1/2}$ for the first wave potential and the number of

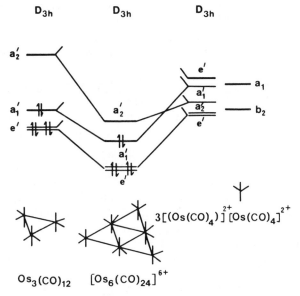

FIG. 8. Construction of the molecular orbitals for the "raft" cluster $[Os_6(CO)_{24}]^{6+}$ from $Os_3(CO)_{12}$ and three $[Os(CO)_4]^{2+}$ fragments.

TABLE III

REDUCTION POTENTIAL FOR SOME $P(OMe)_3$, $P(OPh)_3$, PPh_3, AND MeCN DERIVATIVES OF $Os_6(CO)_{21}$

Cluster	$E_{1/2}(1)$ $(V)^a$	E_p (mV)	$E_{1/2}(2)$ (V)	E_p (mV)
$Os_6(CO)_{20}P(OMe)_3$ (ax.)	−0.96	65	−1.12	90
$Os_6(CO)_{20}P(OPh)_3$ (ax.)	−1.01	62	−1.27	57
$Os_6(CO)_{20}(PPh_3)$ (eq.)	−1.02	67	−1.30	62
$Os_6(CO)_{20}P(OMe)_3$ (eq.)	−1.09	67	−1.28	85
$Os_6(CO)_{20}P(OPh)_3$ (eq.)	−1.15	64	−1.38	72
$Os_6(CO)_{19}\{P(OMe)_3\}_2$, isomer I	−1.24	80	−1.46	85
$Os_6(CO)_{19}\{P(OMe)_3\}_2$, isomer II	−1.24	60	−1.43	75
$Os_6(CO)_{18}\{P(OMe)_3\}_3$, isomer I	−1.42	68	−1.62	125
$Os_6(CO)_{18}\{P(OMe)_3\}_3$, isomer II	−1.28	75	−1.49	140
$Os_6(CO)_{17}\{P(OMe)_3\}_4$	−1.43	—	−1.60	—
$Os_6(CO)_{16}\{P(OMe)_3\}_5$	−1.58	64	−1.78	85
$Os_6(CO)_{15}\{P(OMe)_3\}_6$	−1.84	62	−2.04	75
$Os_6(CO)_{20}(MeCN)$	−0.98	68	−1.22	114
$Os_6(CO)_{19}(MeCN)_2$	−1.20	—	—	—

^{a} Conditions: MeCN solutions, 298 K, Pt electrode with Ag/Ag^+ reference electrode, $FeCp_2$ internal calibrant, and $[Bu_4N][BF_4]$ electrolyte.

phosphites and phosphines in the two series. As expected on the basis of electronic arguments, an increase in the degree of substitution of the cluster by PR_3 is accompanied by a decrease in the tendency to undergo further substitution; however, changing PPh_3 to $P(OMe)_3$ has no effect within experimental error.

The stability of the reduced products was found to be highly dependent on the nature of the ligands attached to these raft species. For example, the series $Os_6(CO)_{(21-n)}(MeCN)_n$ ($n = 1, 2$) does not undergo reversible reduction, that is, breakdown of the reduced species occurs. This is probably due to the fact that MeCN is a poorer π acceptor than PR_3 and therefore is not able to stabilize the reduced species produced during the cyclic voltammetry.

Reduced species of the type $[Os_6(CO)_{21}]^{2-}$ have not yet been characterized crystallographically, but evidence to support the existence of the orbital described above is provided by X-ray and ESR studies of the related species $[Fe_3Pt_3(CO)_{15}]^{n-}$ ($n = 1, 2$). The $a_2{}^1$ symmetry orbital is fully occupied in the dianion of this mixed-metal cluster, which is easily oxidized to the paramagnetic species $[Fe_3Pt_3(CO)_{15}]^-$ [Eq. (35)] (283).

$$[Fe_3Pt_3(CO)_{15}]^{2-} \underset{OH^-/MeOH}{\overset{\substack{H_3PO_4 \text{ or } H_2SO_4 \\ \text{or } Cu^+ \text{ or } Ag^+ \text{ salts}}}{\rightleftarrows}} [Fe_3Pt_3(CO)_{15}]^- \tag{35}$$

X-ray diffraction studies have indeed shown a significant shortening of the average Pt—Pt bond distance in going from $[Fe_3Pt_3(CO)_{15}]^{2-}$ (2.750 Å) to the monoanion (2.656 Å). Also, ESR and electronic diffuse-reflectance spectra (284) were found to agree with the location of the unpaired electron in $[Fe_3Pt_3(CO)_{15}]^-$ in an orbital corresponding to that described above.

c. *The Redox Reactions of $Os_4(CO)_{12}(AuPR_3)_2$.* The hexanuclear clusters $Os_4(CO)_{12}(AuPR_3)_2$ ($R_3 = Et_3$, Ph_2Me, Ph_3) may be considered as true electron reservoirs (Scheme 19) (425). They have been shown to undergo two fully reversible one-electron reduction steps corresponding to the formation of $[Os_4(CO)_{12}(AuPR_3)_2]^-$ ($g = 1.775$) and $[Os_4(CO)_{12}(AuPR_3)_2]^{2-}$. The reduced species $[Os_4(CO)_{12}(AuPR_3)_2]^{2-}$ were also generated chemically by reaction of $Os_4(CO)_{12}(AuPR_3)_2$ with Na/Hg, but these species were too unstable to be further characterized. The neutral compounds $Os_4(CO)_{12}(AuPR_3)_2$ also undergo two fully reversible one-electron oxidations (Scheme 19).

3. *Irreversible Electrochemical Redox Reactions*

a. *The reduction of $M_5C(CO)_{15}$ ($M = Fe, Ru, Os$).* Cyclic voltammetry of $M_5C(CO)_{15}$ ($M = Ru, Os$) at a Pt electrode in dichloromethane shows

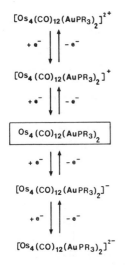

A-OXIDATION		
PR_3	$E_{1/2}(1)$ (V)[*]	$E_{1/2}(2)$ (V)
PEt_3	0.60	0.78
PPh_2Me	0.68	0.83
PPh_3	0.68	0.80

B-REDUCTION				
PR_3	$E_{1/2}(1)$ (V)[*]	Ep	$E_{1/2}(2)$(V)	Ep
PEt_3	−0.64	65mV	−0.89	60mV
PPh_2Me	−0.57	60mV	−0.79	60mV
PPh_3	−0.58	70mV	−0.81	65mV

SCHEME 19. Electrochemical reactions of $Os_4(CO)_{12}(AuPR_3)_2$ and potentials (relative to NHE) for the two-step two-electron oxidation (CH_2Cl_2 solution, 298 K) and reduction (258 K, $FeCp_2$ internal calibrant) processes.

two-electron irreversible reductions at -1.78 V (M = Ru) and -1.50 V (M = Os) versus Ag/Ag^+ at a scan rate of 1 V sec^{-1} at 298 K. No ESR-active intermediates were detected during the formation of the dianions $[M_5C(CO)_{14}]^{2-}$, which have also been generated chemically by the action of alkali metals, hydroxides, or carbonates on $M_5C(CO)_{15}$ (78). Cyclic voltammetry of $Fe_5C(CO)_{15}$ at a carbon electrode in the same solvent, however, suggests that the process proceeds through two steps at 253 K: one quasi-reversible, assigned to a two-electron transfer [step A, Eq. (36)], followed by irreversible loss of CO [step B, Eq. (36)] (13).

$$Fe_5C(CO)_{15} + 2e^- \xrightarrow{\text{A}}$$
$$[Fe_5C(CO)_{15}]^{2-} \xrightarrow{\text{B}} [Fe_5C(CO)_{14}]^{2-} + CO \quad (36)$$

If this is so, then the formation of $[Fe_5C(CO)_{15}]^{2-}$ could be compared with the generation of the bridged butterfly $M_5C(CO)_{15}L$ (M = Ru, Os) species from the square-based pyramid $M_5C(CO)_{15}$ upon addition of L as discussed in Section III,E,2,a.

The complexes $[M_5C(CO)_{14}]^{2-}$ show irreversible oxidation waves in dichloromethane at $+0.15$ V (M = Ru) and $+0.35$ V (M = Os) versus Ag/Ag^+, and the reformation of $M_5C(CO)_{15}$ has been attributed to scavenging of CO by the electron-deficient "$M_5C(CO)_{14}$" (78).

b. *Electrochemical Reactions of* $[Rh_{12}(CO)_{30}]^{2-}$ *(426).* Cyclic voltammetry of $[Bu_4N]_2[Rh_{12}(CO)_{30}]$ in tetrahydrofuran at 298 K shows two reductions at -1.00 and -1.45 V versus Ag/Ag^+. The first reduction is a two-electron process that gives $[Rh_6(CO)_{15}]^{2-}$. However, at 233 K and a scan rate of 2.0 V sec^{-1}, fragmentation is slower and the process is partly reversible, leading to the formation of $[Rh_{12}(CO)_{30}]^{4-}$. The second reduction is irreversible even at 223 K and is attributed to the reduction of $[Rh_6(CO)_{15}]^{2-}$ to unidentified products. The cyclic voltammogram of $[Rh_{12}(CO)_{30}]^{2-}$ also exhibits a two-electron oxidation at $+0.80$ V which is partly reversible at 233 K. These processes are summarized in Eqs. (37) and (38).

$$[Rh_{12}(CO)_{30}]^{2-} \underset{-1.00\ V}{\overset{+2e^-}{\rightleftharpoons}} [Rh_{12}(CO)_{30}]^{4-} \xrightarrow{k=14.5\ sec^{-1}}$$

$$2[Rh_6(CO)_{15}]^{2-} \xrightarrow[-1.45\ V]{} products \quad (37)$$

$$[Rh_{12}(CO)_{30}]^{2-} \overset{-2e^-}{\rightleftharpoons} Rh_{12}(CO)_{30} \xrightarrow{k=11.5\ sec^{-1}} products \quad (38)$$

c. *The Redox Chemistry of* $[Os_{10}C(CO)_{24}]^{2-}$ *and* $[Os_{10}(CO)_{24}H_4]^{2-}$. Cyclic voltammetry studies of $[Os_{10}X(CO)_{24}]^{2-}$ (X = C, H_4) in dichloromethane at a Pt electrode show that these species undergo two one-electron oxidation steps (427) (Scheme 20). The first of these is fully reversible for both compounds. As the potential required to oxidize $[Os_{10}C(CO)_{24}]^{2-}$ (0.78 V versus Ag/Ag^+) is much higher than that for $[Os_{10}(CO)_{24}H_4]^{2-}$ (0.39 V), it has been suggested that the carbide atom somehow stabilizes the paired electron with respect to oxidation. Both of the oxidized species $[Os_{10}X(CO)_{24}]^-$ (X = C, H_4) exhibit paramagnetic behavior up to 200 K. Their g values of 2.295 (X = C) and 2.277 (X = H_4) are indicative of spin–orbit coupling between the various states in these species. In the case of the hydride radical, two weak signals associated with the interstitial hydrogen atoms were also observed. The second oxidation step is irreversible in both species. The diamagnetic solid formed, presumably $Os_{10}X(CO)_{24}$, gives $[Os_{10}X(CO)_{24}]^{2-}$ when taken up in donor solvents.

The voltammetry of $[Os_{10}C(CO)_{24}]^{2-}$ at a hanging drop mercury electrode is a complex process whose nature is not fully understood. Controlled electrolysis at a mercury pool of 0.8 V leads to a product that upon standing reacts slowly to yield $[Os_{20}Hg(C)_2(CO)_{48}]^{2-}$. This compound has also been generated chemically by reaction of $[Os_{10}C(CO)_{24}]^{2-}$ with mercuric salts or $AgBF_4$ and mercury (338) (see Section II,B,2,a).

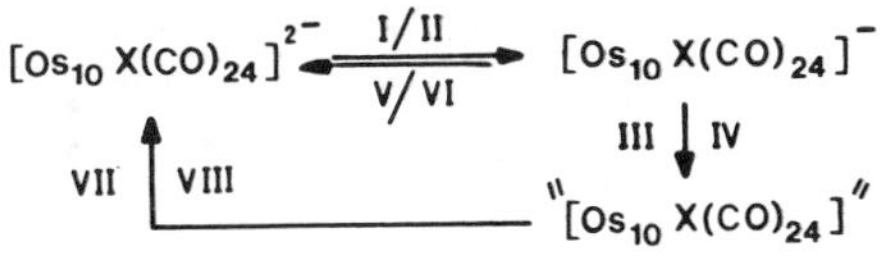

$$[Os_{10} X(CO)_{24}]^{2-} \xrightleftharpoons[V/VI]{I/II} [Os_{10} X(CO)_{24}]^{-}$$

VII $\uparrow$ VIII $\qquad$ III $\downarrow$ IV

$$\text{''}[Os_{10} X(CO)_{24}]\text{''}$$

Conditions for oxidation of the $[Os_{10}X(CO)_{24}]^{2-}$ species		
	CHEMICAL	ELECTROCHEMICAL
I	$X = C$, 1 equiv $AgBF_4$ or $FeCl_3$	+1.0V at Pt electrode
II	$X = H_4$, 1 equiv $AgBF_4$ or $FeCl_3$	+0.50V at Pt electrode
III	$X = C$, 2 equiv $AgBF_4$ or $FeCl_3$	+1.40V at Pt electrode
IV	$X = H_4$, 2 equiv $AgBF_4$ or $FeCl_3$	+0.80V at Pt electrode
V	$X = C$ air, 2–3 min	
VI	$X = H_4$ air, 12 hrs	
VII	$X = C$, acetone, CH_3CN or Δ	
VIII	$X = H_4$ acetone, CH_3CN or Δ	

SCHEME 20. Redox chemistry of $[Os_{10}X(CO)_{24}]^{2-}$ ($X = C$, H_4). Unless specified, chemical reactions are as in CH_2Cl_2. Electrochemical potentials versus Ag/Ag^+.

E. REACTIONS WITH SOFT NUCLEOPHILES

The reactions of HNCC with soft nucleophiles (CO, PR_3, RCCR, R_2CCR_2, pyridine, NO_2^-, and halides) result either in the formation of addition or substitution products, or in cluster breakdown. The nature of the species formed depends on the nucleophile (which may or may not undergo transformation on the cluster surface), the type of metal cluster, and the conditions employed in the reaction. Generally, clusters of the lighter elements tend to fragment even under mild conditions, while those of the heavier elements, which are more robust, often afford addition and substitution products.

1. Addition Reactions

Simple nucleophilic addition at the metal centers of a cluster results in an increase in the number of electrons available for skeletal bonding. As a consequence, breakage of M—M bonds resulting in an opening of the cluster framework, sometimes followed by cluster breakdown, may be observed.

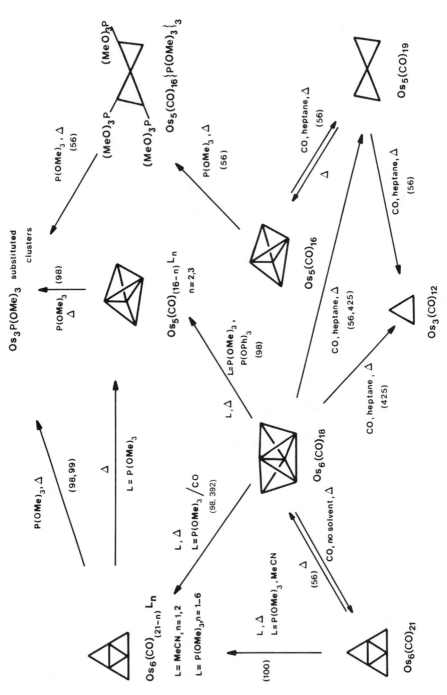

SCHEME 21. Reactions of $Os_6(CO)_{18}$ with CO and PR_3.

For example, the thermal reactions of $Os_6(CO)_{18}$ with CO (56, 425) and $P(OMe)_3$ (56, 98, 392) lead to the open "raft" species $Os_6(CO)_{21}$ and $[Os_6(CO)_{21-n}\{P(OMe)_3\}_n]$ ($n = 1-6$), respectively, probably by sequential bond fission in the bicapped tetrahedral cluster. As indicated in Scheme 21, under slightly different reaction conditions, elimination of an osmium vertex occurs, resulting in the formation of clusters containing a trigonal-bipyramidal metal framework. These react further with nucleophiles to give open Os_5 clusters and finally substituted triosmium species. Cluster unfolding has also beeen observed in the reactions of $Os_5(CO)_{15}H_2$ with a variety of nucleophiles (Scheme 22).

Certain ligands have been found to act as anchors by preventing cluster breakdown upon addition of nucleophiles. The acetylene–phosphido ligand in $Ru_5(CO)_{13}(\mu_5\text{-CCPPh}_2)(\mu\text{-PPh}_2)$, for example, allows extensive metal cluster rearrangement upon addition of two equivalents of CO (Scheme 23a) (22). Similarly, the square-based pyramidal clusters $Os_5C(CO)_{15}$ and $Ru_5(CO)_{13}(\mu_4\text{-}\eta^2\text{-CCPh})(\mu\text{-PPh}_2)$ (20) are safely opened up through breakage of axial and equatorial edges, respectively, in order to accommodate a variety of nucleophiles, under mild conditions (Scheme 23b). It is interesting to note that the related compounds $Os_5S(CO)_{15}$ and $Os_5(CO)_{15}PPh$ do not react with halides or phosphines under similar reaction conditions (428). This behavior must reflect a difference in the bonding of the capping groups in these clusters compared with $Os_5C(CO)_{15}$. As shown in Fig. 9, the S and P atoms in $Os_5S(CO)_{15}$ and $Os_5(CO)_{15}P(OMe)$ interact only with the four atoms of the basal plane. This seems to impose rigidity on the Os_5 skeleton, preventing the opening up of the polyhedron exhibited by the carbido species or by $Ru_5(CO)_{13}(\mu_4\text{-}\eta^2\text{-CCPh})(\mu\text{-PPh}_2)$ upon addition of nucleophiles.

2. Substitution Reactions

a. Thermal Activation. The addition of a nucleophile L is sometimes followed by thermally induced dissociation of a carbonyl ligand with reformation of an M—M bond, resulting in the generation of a substituted derivative. Substitution via an associative process has been widely observed in the chemistry of HNCC of osmium. For example, $Os_5(CO)_{15}H_2\{P(OMe)_3\}$ looses CO under thermolytic conditions to give $Os_5(CO)_{14}H_2\{P(OMe)_3\}$ (Scheme 22). Subsequent CO substitution by PR_3 to give $Os_5(CO)_{13}H_2\{P(OMe)_3\}(PR_3)$ (R = Et, OMe) probably follows the same mechanism, although a more drastic polyhedral rearrangement during the course of the reaction has been invoked to explain the structure of the products (66).

For most HNCC, however, it is difficult to establish whether substitution occurs via an associative or dissociative mechanism due to the absence of

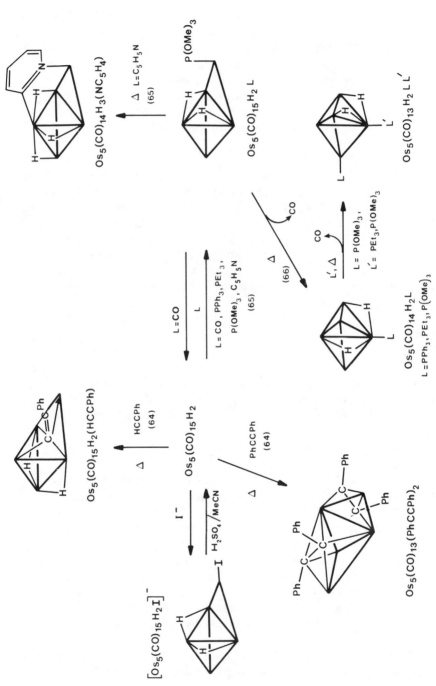

SCHEME 22. Reactions of $Os_5(CO)_{15}H_2$ with nucleophiles.

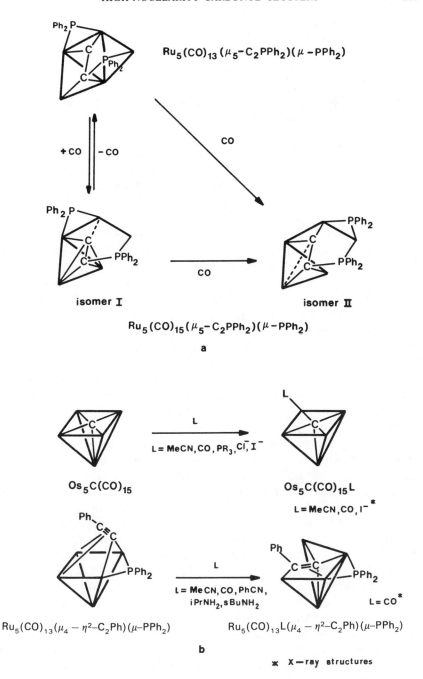

SCHEME 23. Structural transformations of some M_5 clusters upon addition of nucleophiles.

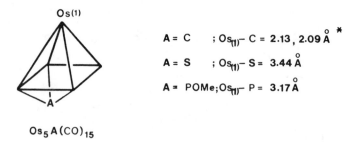

$$Os_5 A(CO)_{15}$$

FIG. 9. Distances between the apical Os atom and the capping atoms in $Os_5C(CO)_{15}$ (434), $Os_5S(CO)_{15}$ (73), and $Os_5(CO)_{15}(POMe)$ (71). Asterisk indicates that there are two independent $OS_5C(CO)_{15}$ molecules in the unit cell.

kinetic data and the infrequent isolation of the intermediates. The reactions of $Ru_5C(CO)_{15}$ with PPh_3 and diphosphines (diphos), for example, proceed directly to give the substituted species $Ru_5C(CO)_{(15-n)}(PPh_3)_n$ ($n = 1, 2$) (25) and $Ru_5C(CO)_{13}$(diphos) (28), respectively, no intermediate adducts having been detected. By analogy with $Os_5C(CO)_{15}$, the $Ru_5C(CO)_{15}$ cluster is nevertheless believed to react via an associative mechanism (Scheme 24).

Ligand reorganization on the cluster surface induced by thermolysis is a common process. The pyridine ligand in $Os_5(CO)_{15}H_2(C_5H_5N)$ (177) and $Os_5C(CO)_{15}(C_5H_5N)$ (429), for instance, undergoes ortho-metallation with hydrogen transfer to the metal core and ejection of a CO ligand from the clusters to give $Os_5(CO)_{14}H_3(C_5H_4N)$ and $Os_5C(CO)_{14}(H)(C_5H_4N)$, respectively. A similar mechanism has been proposed for the formation of $Os_6(CO)_{16}(H)(C_5H_3RN)$ (R = H, Me) (177) from the reactions of $Os_6(CO)_{18}$ with pyridine and picoline, respectively, in apolar solvents (Scheme 25). Under the extreme conditions used for CO substitution in $Os_6(CO)_{18}$ by acetylenes, even more drastic modifications have been observed in the ligands attached to the cluster. These changes may have involved splitting of a C—C bond in the organic fragment, and possibly of the C—O bond of a carbonyl leading to carbide formation (Scheme 25).

Among the cobalt subgroup, only the $M_6(CO)_{16}$ (M = Rh, Ir) clusters have been shown to react with phosphines and related nucleophiles (396) to afford substituted products (Scheme 26). Facile replacement of four carbonyl ligands by $P(OR)_3$ (R = Ph, Me) has been achieved in both systems (176, 239) but the substitution of a fifth ligand was observed only for the less bulky $P(OMe)_3$ ligand giving $Ir_6(CO)_{11}\{P(OMe)_3\}_5$ (240). Attempts at further substitution in the rhodium systems gave only breakdown products, unless the reaction was performed under vacuum and with stoichiometric amounts of $P(OMe)_3$ or 4-ethyl-2,6,7-trioxa-1-phosphabicyclo[2.2.2]octane (ETPO) in which case a sixth CO could be substituted (391). Similarly a maximum of six carbonyl ligands was substituted in the reactions of $Rh_6(CO)_{16}$ with

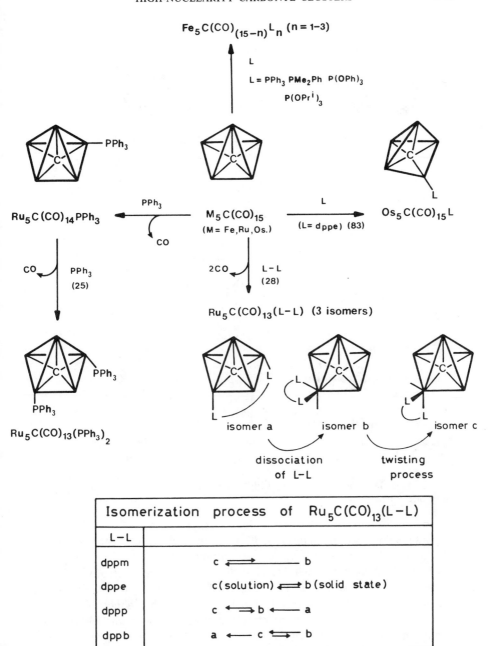

SCHEME 24. Reactions of $M_5C(CO)_{15}$ (M = Fe, Ru, Os) with phosphines and diphosphines and the mechanism of isomerization of the $Ru_5C(CO)_{13}$(L–L) species (L–L = dppm, dppe, dppp, and dppb).

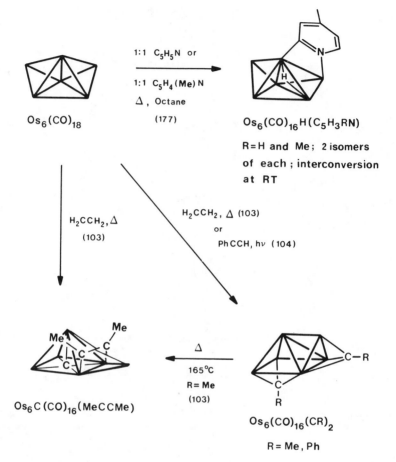

SCHEME 25. Reactions of $Os_6(CO)_{18}$ with C_5H_4RN (R = H, Me), PhCCH, and CH_2CH_2 under thermolytic conditions.

dppm (*179*) or norbonadiene (*175, 394*), but breakdown to a rhodium dimer also occurred if further substitution was attempted.

Generally, second-row metal clusters do not need to be thermally activated to afford CO substitution. The degree of substitution in these clusters therefore can often be controlled by kinetic factors, such as concentration and ratio of the reagents as well as reaction times. In this way high yields of each compound in the series $Ru_6C(CO)_{(17-n)}\{P(OMe)_3\}_n$ ($n = 1$–4) (*52*) and $Rh_6(CO)_{(16-2n)}(L-L)_n$ ($n = 1, 2, 3$, L–L = dppm) (*179*) have been obtained. For HNCC of the heavier metal atoms, however, thermal activation must be employed. Because the temperatures necessary to activate CO substitution

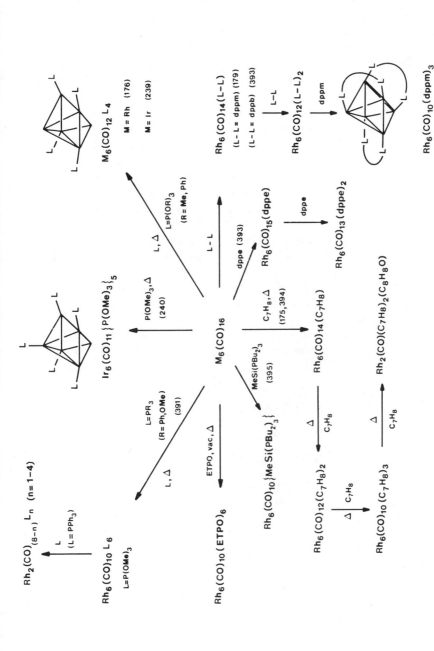

SCHEME 26. Reactions of $M_6(CO)_{16}$ (M = Rh, Ir) with phosphines, diphosphines, and norbornadiene. ETPO, 4-ethyl-2,6,7-trioxa-1-phosphabicyclo[2.2.2]octane.

often lead to degradation of the desired products, these are obtained in low yields. Photochemical activation has been applied with relative success only in isolated cases and is still an open field in HNCC chemistry. Chemical activation of osmium clusters has given spectacular results. The methods used for this purpose discussed below have great potential in the investigation of the chemistry of HNCC of rhenium and iridium.

b. Chemical Activation. Basically three strategies have been employed to activate HNCC of osmium in the reactions with nucleophiles.

i. Introduction of ligands, such as nitrosyl, halogens, and thiols, that can exhibit more than one oxidation state. A bridging halogen, for example, acts as a three-electron donor and can convert to a terminal one-electron donor; through this process of bridge opening, a vacant site may be created on one metal atom. Such a mechanism has been proposed in the reaction of $[Os_{10}C(CO)_{24}(\mu\text{-I})_2]$ with $P(OMe)_3$ (Scheme 27). (*133*). Activation can also occur through charge polarization in the molecule caused by the ligand, which may direct the nucleophilic attack as well as reduce the activation energy for the process. One example is the reaction of $Os_5C(CO)_{15}I_2$ with alcohols to give $Os_5C(CO)_{14}(COOR)I$. This reaction needs much milder conditions that the analogous reaction of $Os_5C(CO)_{15}$ to give $Os_5C(CO)_{14}(H)(COOR)$ (*79*) (Scheme 28).

ii. Synthesis of coordinatively unsaturated species. A series of unsaturated hexaosmium compounds has recently been reported (*313, 330*). The key

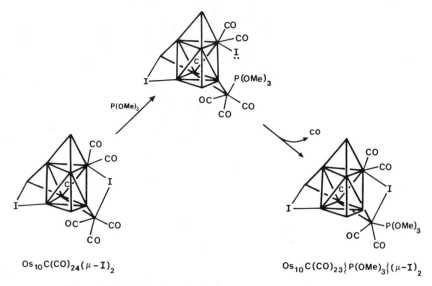

$Os_{10}C(CO)_{24}(\mu\text{-I})_2$ $Os_{10}C(CO)_{23}\{P(OMe)_3\}(\mu\text{-I})_2$

SCHEME 27. Substitution of CO by $P(OMe)_3$ in $Os_{10}C(CO)_{24}(\mu\text{-I})_2$.

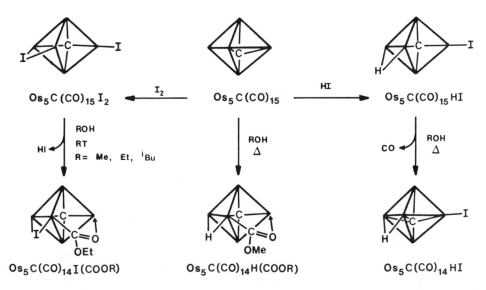

SCHEME 28. Chemical activation of $Os_5C(CO)_{15}$ by iodine ligand towards ROH addition.

species, $[Os_6(CO)_{17}H]^-$, was prepared from $Os_6(CO)_{18}$ by oxidation of a carbonyl ligand with trimethylamine oxide (see Section III,C,2); the neutral 84-electron species $Os_6(CO)_{17}H_2$ and $Os_6(CO)_{17}(AuPR_3)_2$ (R = Me, Ph) were obtained by addition of H^+ and $AuPR_3^+$ fragments, respectively, to the monoanion. These three bicapped tetrahedral compounds isoelectronic with $Os_6(CO)_{18}$ all add 2-electron donor ligands such as CO and PR_3 to afford 86-electron species (Scheme 29).

Other examples of formally unsaturated clusters are provided by the series $Os_4(CO)_{12}(AuYR_3)$ (Y = P, As; R_3 = Et_3, Ph_2Me, Ph_3) (328). Some of their reactions are depicted in Scheme 30. A parallel can be drawn between these unsaturated clusters and $Os_3(CO)_{10}H_2$, which has been used for the synthesis of a multitude of Os_3 derivatives.

iii. Synthesis of species containing ligands which can be displaced more easily than CO. This is also an extension of the technique employed to activate $Os_3(CO)_{12}$ and has been widely used in Os_6 chemistry. When oxidative displacement of carbon monoxide from $Os_6(CO)_{18}$ and $Os_5(CO)_{16}$ by trimethylamine oxide is carried out in the presence of acetonitrile, the substituted products $Os_6(CO)_{17}(MeCN)$, $Os_6(CO)_{16}(MeCN)_2$ (97, 118), and $Os_5(CO)_{15}(MeCN)$ (92) are isolated in good yields. The acetonitrile derivatives of the raft "$Os_6(CO)_{21}$" species $Os_6(CO)_{(21-n)}(MeCN)_n$ (n = 1, 2) have alternatively been synthesized by condensation of $Os_3(CO)_{10}(MeCN)_2$ using $PdCl_2$ (99).

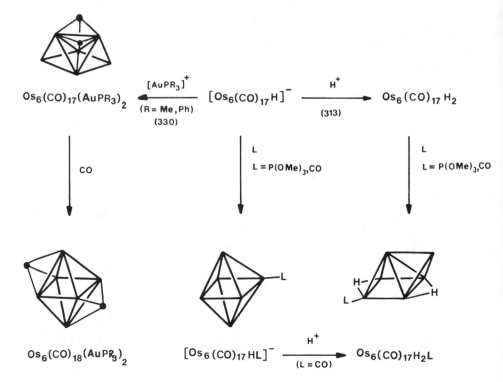

SCHEME 29. Reactions of $[Os_6(CO)_{17}H]^-$, $Os_6(CO)_{17}H_2$, and $Os_6(CO)_{17}(AuPR_3)_2$ (R = Me, Ph) with PR_3 and CO.

The reactions involving nucleophilic displacement of MeCN in the species $Os_5(CO)_{15}(MeCN)$, $Os_6(CO)_{(18-n)}(MeCN)_n$ ($n = 1, 2$), and $Os_6(CO)_{(21-n)}(MeCN)_n$ ($n = 1, 2$) have been studied extensively (Schemes 31, 32, and 33).

Facile substitution of the MeCN ligand by a variety of trisubstituted phosphines has been observed for the three systems with no structural changes in the metal geometry. The reactions of the raft $Os_6(CO)_{20}(MeCN)$ with primary and secondary phenylphosphines (93) show a pattern similar to those reported for $Os_3(CO)_{11}(MeCN)$ (442); initial substitution of MeCN by PPh_2H and $PPhH_2$ is followed by hydrogen transfer to the metal frame. Consequently a change in the coordination mode of the phosphines occurs from terminal to μ_2-PPh_2 and μ_3-PPh, respectively. This process is induced by thermal activation. As a result, unpredictable modifications in the phosphido-substituted Os_6 polyhedra, including cluster close-up with CO ejection or cluster breakdown, have been shown to occur (Scheme 33).

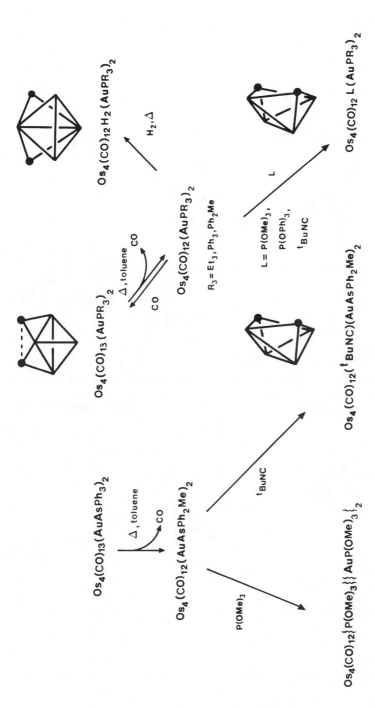

SCHEME 30. Reactions of the unsaturated clusters $Os_4(CO)_{12}(AuZR_3)_2$ (Z = P, As) *(328)*.

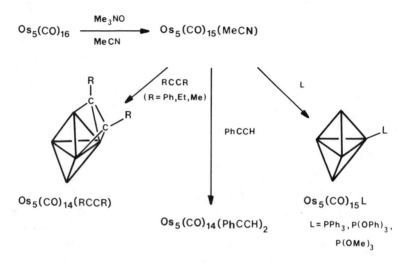

SCHEME 31. Reactions of $Os_5(CO)_{15}(MeCN)$ with nucleophiles (69).

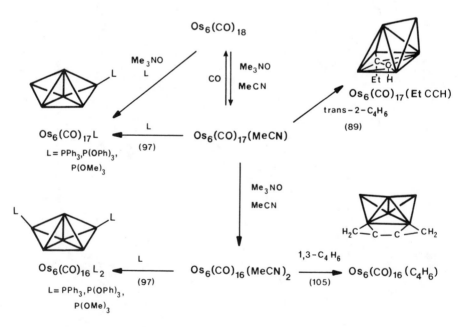

SCHEME 32. Reactions of $Os_6(CO)_{(18-n)}(MeCN)_n$ ($n = 1, 2$) with nucleophiles.

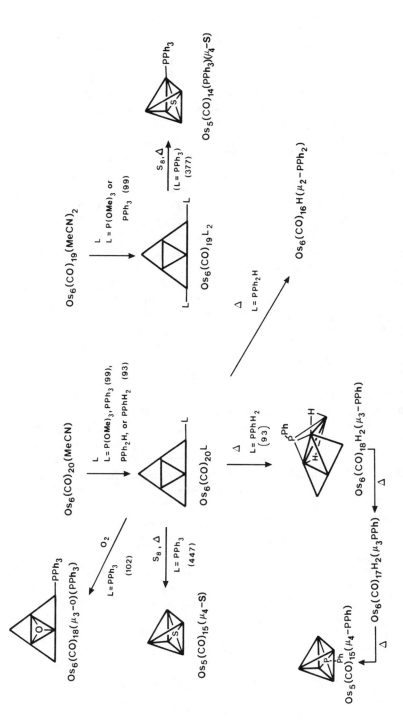

SCHEME 33. Reactions of $Os_6(CO)_{20}(MeCN)$ and $Os_6(CO)_{19}(MeCN)_2$ with phosphines.

Reactions of $Os_6(CO)_{(18-n)}(MeCN)_n$ ($n = 1, 2$) with a variety of acetylenes have also been investigated. Generally, analogous reactions of this species and $Os_6(CO)_{18}$ have been found to give similar final products. However, as a result of the much milder conditions employed in the reactions of the acetonitrile derivative, intermediates have been isolated, allowing a better understanding of these reactions. For example, the reactions of $Os_6(CO)_{(18-n)}(MeCN)_n$ ($n = 1, 2$) with RCCR' (R = R' = Me, Ph; R = Me, R' = Et) and RCCH (R = Me, Et, Ph) (89) have been shown to lead initially to substitution of MeCN and CO ligands to give the 84-electron species $Os_6(CO)_{16}(RCCR')$ and $Os_6(CO)_{16}(RCCH)$ with the same bicapped tetrahedron of osmium atoms as the starting material (Scheme 34). Upon heating $Os_6(CO)_{16}(RCCR')$, a two-electron reduction occurs as a consequence of the change in the donor properties of the acetylene upon splitting. The structural rearrangement that follows results in the formation of the capped square-based pyramidal cluster $Os_6(CO)_{16}(CR)_2$, which was previously isolated from the thermolytic reaction of $Os_6(CO)_{18}$ and ethylene (103) (Scheme 25).

Oxidative displacement of another CO ligand by trimethylamine oxide in $Os_6(CO)_{16}(MeCCMe)$ in the presence of acetonitrile has also been described (106). Subsequent substitution of the MeCN ligand in $Os_6(CO)_{15}(MeCN)$-(MeCCMe) with dimethylacetylene gave $Os_6(CO)_{15}(MeCCMe)_2$. This cluster was shown by X-ray crystallography to exist in two interconvertible isomeric forms, one of which exhibits a capped square based pyramidal arrangement of metal atoms, while the other has an edge-bridged trigonal-bipyramidal metal core (Scheme 34).

The terminal acetylene derivatives $Os_6(CO)_{16}(RCCH)$ (R = Me, Et, Ph) have been shown to react with CO to give $Os_6(CO)_{17}(RCCH)$, (89), in which the acetylene ligand is still intact and sits on the base of the capped pyramidal osmium polyhedron. This compound was found to convert by the action of heat to an isomer in which the proton from the acetylene has been transferred to the metal array. The process is accompanied by an opening up of the metal cluster (Scheme 34).

A different type of C—H bond activation has been observed upon substitution of MeCN in the raft $Os_6(CO)_{20}(MeCN)$ with phenylacetylene (70). X-ray analysis of $Os_6(CO)_{20}\{CC(H)Ph\}$ has revealed that a $1 \rightarrow 2$ hydrogen shift has occurred in this system to give a vinylidine ligand with a coordination mode similar to the one exhibited by the CCH_2 fragment on Pt (1:1:1) surfaces (431). This process was accompanied by reorganization of the raft Os_6 framework. Ejection of $Os(CO)_5$ from this cluster was found to occur upon heating to afford $Os_5(CO)_{15}\{CC(H)Ph\}$ (70) (Scheme 35).

The ligand 1,2-Cyclooctadiene (COD) has also been found to be as good a leaving group in HNCC as it is in $Os_3(CO)_{10}(COD)$. For example, substitution of one COD ligand in $Os_6(CO)_{16}(PtCOD)_2$ has been reported to occur

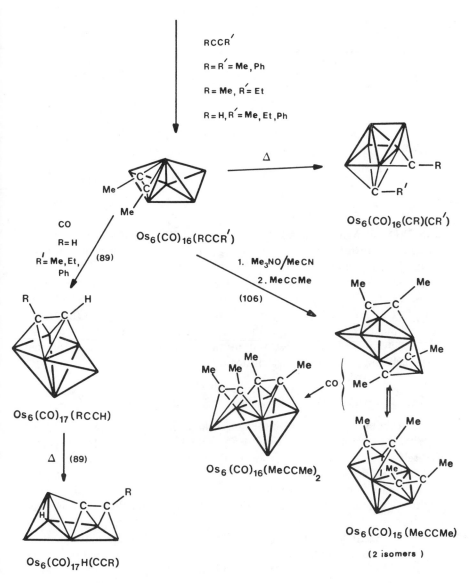

SCHEME 34. Reactions of $Os_6(CO)_{(18-n)}(MeCN)_n$ ($n = 1, 2$) with acetylenes RCCR'.

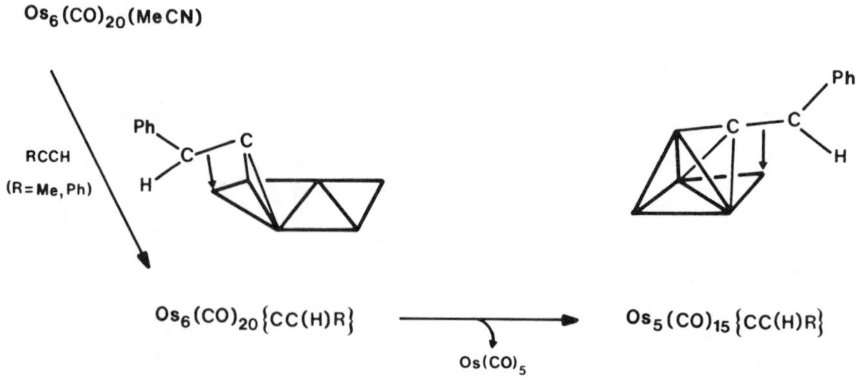

$Os_6(CO)_{20}(MeCN)$

SCHEME 35. Reactions of $Os_6(CO)_{20}(MeCN)$ with RCCH (R = Me, Ph) (70).

with $P(OMe)_3$, affording $Os_6(CO)_{16}(PtCOD)(Pt\{P(OMe)_3\}_2)$. (443). Although this process involves no change in the cluster electron count, it results in a rearrangement of the metal polyhedron (Scheme 36). The two Os_6Pt_2 species react readily with CO to yield $Os_6(CO)_{17}(PtCOD)_2$ and $Os_6(CO)_{17}(PtCOD)(Pt\{P(OMe)_3\}_2)$, respectively. Interestingly, substitution of a COD ligand by two $P(OMe)_3$ groups in $Os_6(CO)_{17}(PtCOD)_2$ is not accompanied by any drastic polyhedral change. As indicated in Scheme 36, the structural flexibility observed in this system is reminiscent of that previously observed in the Os_8 clusters (124).

3. Reactions with Halides and Nitrides

The reactions of HNCC with the nucleophiles X^- and $[NO_2]^-$ are treated separately because they may either lead to the formation of reduced species, as discussed in Section III,B, or follow the same trends as the reactions with CO and PR_3 discussed above. The outcome of the reaction seems to depend on the nature of the individual cluster considered. For example, simple addition of iodide with opening up of the metal polyhedron has been observed in the formation of $[Os_5C(CO)_{15}I]^-$ and $[Os_5(CO)_{15}H_2I]^-$ (65, 76) from $Os_5C(CO)_{15}$ and $Os_5(CO)_{15}H_2$, respectively. Alternatively, reaction of this nucleophile with $Rh_6(CO)_{16}$ results in the formation of the substituted products $[Rh_6(CO)_{15}I]^-$ and $[Rh_6(CO)_{14}I_2]^{2-}$ (Scheme 37). In contrast, the binary carbonyls $Os_m(CO)_n$ (m = 6, 7, 8; n = 18, 21, 23) are reduced to their respective dianions (Section II,C,2), while the hydrido clusters $Os_m(CO)_nH_2$ (m = 6, 7, 8; n = 18, 20, 22) are deprotonated by halides (Section II,B).

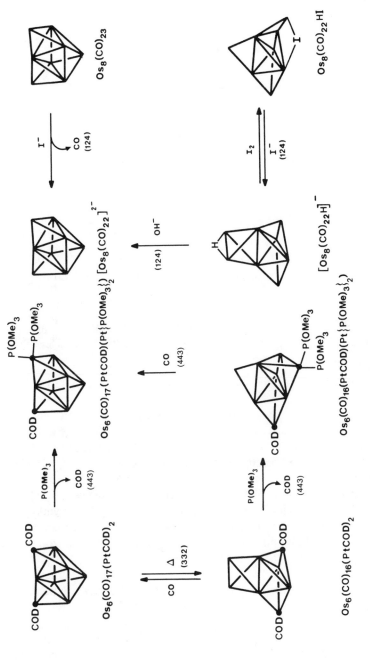

SCHEME 36. Structural transformations of the Os_6Pt_2 and Os_8 polyhedra.

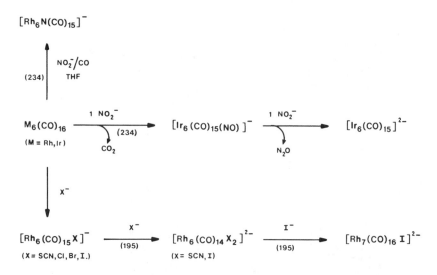

SCHEME 37. Reactions of $M_6(CO)_{16}$ (M = Rh, Ir) with NO_2^- and I^-.

Similarly, the reactions of neutral HNCC with $[NO_2]^-$ are complex (*460*). They have often been found to produce nitrosyl-substituted species according to Eq. (39) (*432*).

$$M_m(CO)_n + [NO_2]^- \longrightarrow [M_m(CO)_{(n-2)}(NO)]^- + CO_2 + CO \qquad (39)$$

Substitution of two carbonyls by a linear nitrosyl and a negative charge were observed in the reaction of $Ru_6C(CO)_{17}$ with $[NO_2]^-$, which yields $[Ru_6C(CO)_{15}(NO)]^-$ (*44*). In contrast, the $[NO_2]^-$ reactions of $M_6(CO)_{16}$ (M = Rh, Ir) follow different patterns. The monoanion $[Ir_6(CO)_{15}(NO)]^-$ is the product of the reaction of $Ir_6(CO)_{16}$ with one equivalent of $[NO_2]^-$; it reacts further with more $[NO_2]^-$ to give the reduced species $[Ir_6(CO)_{15}]^{2-}$ (*234*) (Scheme 37). The product of the reaction of $Rh_6(CO)_{16}$ with $[NO_2]^-$ under CO, however, is the nitrido species $[Rh_6N(CO)_{15}]^-$, formed via deoxygenation of NO and expulsion of CO_2 (*234*).

F. Oxidative Addition of the Small Molecules H_2, I_2, and HX

The reactions of HNCC with H_2, I_2, or HX (X = Cl, Br, I, SEt, SH, SeH) to give products of unchanged nuclearity may result in the oxidative addition of these molecules to M—M bonds. For example, the pentanuclear clusters

$M_5C(CO)_{15}$ have been shown to react with HX (M = Ru, X = Cl, Br; M = Os, X = I), H_2, and I_2 to yield $M_5C(CO)_{15}HX$, $M_5C(CO)_{15}H_2$, and $M_5C(CO)_{15}I_2$, respectively (Scheme 38). The structures of the $Ru_5C(CO)_{15}HX$ derivatives were elucidated by solid-state infrared spectroscopy (435). It has been shown that the metal–carbon stretching modes of carbido clusters give rise to absorptions of high intensity, which reflect the symmetry of the carbido environment (9, 149, 416, 424). In this case analysis of the vibrational modes associated with ruthenium–carbido and ruthenium–halogen stretching indicated that the five metal atoms in the compounds $Ru_5C(CO)_{15}HX$ (X = Cl, Br) adopt a bridged butterfly arrangement with a terminally bound halide, analogous to the anion $[Os_5C(CO)_{15}I]^-$, the structure of which has been determined by X-ray crystallography. Furthermore, consideration of the ruthenium–hydrido stretching frequencies in these complexes has led to the suggestion that the hydrogen ligand occupies a μ-bridging position across the shortest Ru—Ru bond in each cluster.

The hydrohalogenated compounds $M_5C(CO)_{15}HX$, which have also been obtained from the reactions of $M_5C(CO)_{15}$ with halides followed by protonation, have been shown to lose CO upon heating to give the species $M_5C(CO)_{14}HX$.

In the reactions of $M_5C(CO)_{15}$ with HX (M = Ru, X = SEt, SH, and SeH; M = Os, X = SH), the adduct $M_5C(CO)_{15}HX$ was isolated for M = Os. Upon heating $Os_5C(CO)_{15}H(SH)$, dissociation of a CO ligand was found to occur to give $Os_5C(CO)_{14}H(SH)$, a compound analogous to $Ru_5C(CO)_{14}HX$ (X = SEt, SH, SeH), the first-formed products in the reactions of $Ru_5C(CO)_{15}$. As shown in Scheme 38, ejection of CO is accompanied by the formation of a sulfur or selenium bridge (26).

In contrast, reaction of the sulfido cluster $Os_5S(CO)_{15}$ with H_2S gave the bisulfido compound $Os_5(CO)_{14}H_2S_2$ with a "bow tie" metal atom arrangement capped by sulfur atoms. Hydrogen transfer from the SH group to the metal atoms is probably a result of the forcing conditions employed for this reaction (67).

Reaction may also occur at a ligand. For example, hydrogenation of an acetylide ligand on a ruthenium cluster surface has also been reported (19). The reaction of $Ru_5(CO)_{13}(\mu_5\text{-CCPPh}_2)(\mu\text{-PPh}_2)$ with H_2 (1 atm) proceeds stepwise with absorption of three molecules of H_2 and successive formation of pentanuclear cluster complexes containing μ_5-vinylidene, -methylidene, and -carbide ligands (Scheme 39). Thus, the net reaction is the unusual conversion of the μ_5-CCPPh$_2$ ligand into carbon and MePPh$_2$. Under more forcing conditions (10 atm H_2), breakdown of $Ru_5(CO)_{13}(\mu_5\text{-CCPPh}_2)(\mu\text{-PPh}_2)$ to $Ru_4(CO)_{10}H_3(\mu_4\text{-CCPPh}_2)(\mu\text{-PPh}_2)$ has been observed (436).

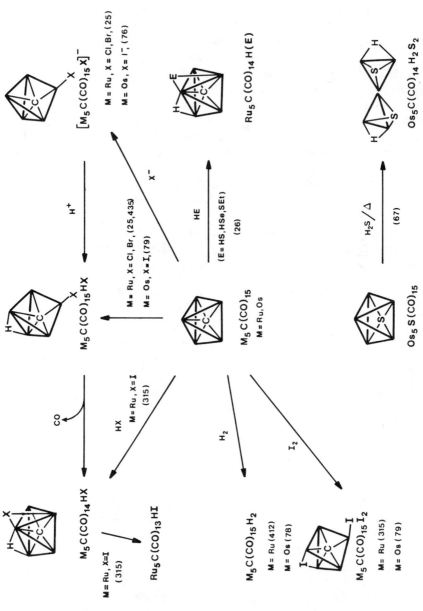

SCHEME 38. Reactions of $M_5C(CO)_{15}$ (M = Ru, Os) and $Os_5S(CO)_{15}$ with HX (X = Cl, Br, I, SH, SEt, SeH).

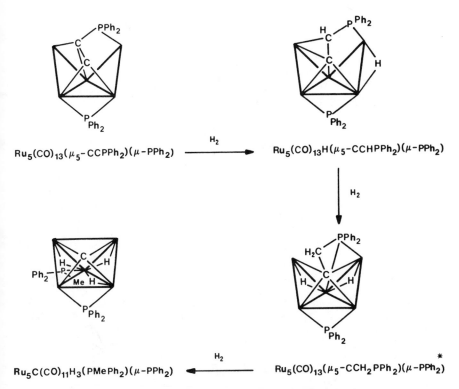

$Ru_5(CO)_{13}(\mu_5\text{-}CCPPh_2)(\mu\text{-}PPh_2)$ $\xrightarrow{\text{H}_2}$ $Ru_5(CO)_{13}H(\mu_5\text{-}CCHPPh_2)(\mu\text{-}PPh_2)$

$\downarrow \text{H}_2$

$Ru_5C(CO)_{11}H_3(PMePh_2)(\mu\text{-}PPh_2)$ $\xleftarrow{\text{H}_2}$ $Ru_5(CO)_{13}(\mu_5\text{-}CCH_2PPh_2)(\mu\text{-}PPh_2)$ *

Scheme 39. Stepwise hydrogenation of $Ru_5(CO)_{13}(\mu_5\text{-}CCPPh_2)(\mu\text{-}PPh_2)$ (19). Asterisk indicates structures proposed on the basis of NMR spectroscopy.

IV. Concluding Remarks

The potential that HNCC offer is beginning to be realized. It is clear that they have properties and exhibit reaction patterns that differ markedly from their mononuclear counterparts and even low-nuclearity clusters. It is also apparent that their reactivity patterns may be rationalized in simplistic terms and thereby extended to other systems. However, much work remains to be done, particularly in designed synthesis and studies of reaction mechanisms. The ability of these clusters to combine with important small substrates such as CO and H_2 need also to be explored in much more detail. The study of the reactivity of large mixed-metal systems, which, as expected, exhibit enhanced and modified reactivities, equally requires more detailed investigation. In fact it would be useful to have available HNCC, which contain early and late transition metal elements, in order to combine both Lewis basic and Lewis acidic properties in the same molecule. To date few examples of such systems have been reported.

In summary, despite the enormous effort that has been devoted to the synthesis and characterization of HNCC, there is a serious deficit in data related to their reactivity. Given the remarkable reactions that they are known to exhibit, efforts in this area are still needed and desirable.

ACKNOWLEDGMENTS

We wish to thank Professor Sir Jack Lewis and Dr. B. F. G. Johnson for all their help and encouragement over the last few years and for reading and making useful suggestions on the manuscript. We gratefully acknowledge the cooperation of J. A. Lunniss and P. A. Jackson in checking the manuscript. We also thank Sidney Sussex College, Cambridge (M.D.V.) and Gonville and Caius College, Cambridge (J.N.N.) for our fellowships.

REFERENCES

1. Taylor, N. J., *J. Chem. Soc., Chem. Commun.*, p. 478 (1985).
2. Ciani, G., D'Alfonso, G., Freni, M., Romiti, P., and Sironi, A., *J. Organomet. Chem.* **244**, C27 (1983).
3. Ciani, G., D'Alfonso, G., Freni, M., Romiti, P., and Sironi, A., *J. Chem. Soc., Chem. Commun.*, p. 339 (1982).
4. Ciani, G., D'Alfonso, G. Freni, M., Romiti, P., and Sironi, A., *J. Chem. Soc., Chem. Commun.*, p. 705 (1982).
5. Broadhurst, P. V., Johnson, B. F. G., Lewis, J., and Raithby, P. R., *J. Am. Chem. Soc.* **103**, 3198 (1981).
6. Braye, E. H., Dahl, L. F., Hubel, W., and Wampler, D. S., *J. Am. Chem. Soc.* **84**, 4633 (1962).
7. Bradley, J. S., Hill, E. W., Ansell, G. B., and Modrick, M. A., *Organometallics* **1**, 1634 (1982).
8. Bradley, J. S., *Philos. Trans. R. Soc. London Ser. A* **308**, 103 (1982).
9. Oxton, I. A., Powell, D. B., Goudsmit, R. J., Johnson, B. F. G., Lewis, J., Nelson, W. J. H., Nicholls, J. N., Rosales, M. J., Vargas, M. D., and Whitmire, K. H., *Inorg. Chim. Acta Lett.* **64**, L259 (1982).
10. Sosinski, B. A., Norem, N., Shelly, J., *Inorg. Chem.* **21**, 348 (1982).
11. Brint, R. P., O'Cuill, K., Spalding, T. R., and Deeney, F. A., *J. Organomet. Chem.* **247**, 61 (1983).
12. Brint, R. P., Collins, M. P., Spalding, T. R., and Deeney, F. A., *J. Organomet. Chem.* **258**, C57 (1983).
13. Gourdon, A., and Jeannin, Y., *C. R. Hebd. Seances Acad. Sci. Ser. B.* **295**, 1101 (1982).
14. Cooke, C. G., and Mays, M. J., *J. Organomet. Chem.* **88**, 231 (1975).
15. Tachikawa, M. Stein, J., Muetterties, E. L., Teller, R. G., Beno, M. A., Gebert, E., and Williams, J. M., *J. Am. Chem. Soc.* **102**, 6648 (1980).
16. Churchill, M. R., and Wormald, J., *J. Chem. Soc., Dalton Trans.*, p. 2410 (1974).
17. Gourdon, A., and Jeannin, Y., *J. Organomet. Chem.* **282**, C39 (1985).
18. Bruce, M. I., Matisons, J. G., Rodgers, J. R., and Wallis, R. C., *J. Chem. Soc., Chem. Commun.*, p. 1070 (1981).
19. Bruce, M. I., Skelton, B. W., White, A. H., and Williams, M. L., *J. Chem. Soc., Chem. Commun.*, p. 744 (1985).

20. MacLaughlin, S. A., Taylor, N. J., and Carty, A. J., *Organometallics* **2**, 1194 (1983).
21. Bruce, M. I., Williams, M. L., Patrick, J. M., and White, A. H., *J. Chem. Soc., Dalton Trans.*, p. 1229 (1985).
22. Bruce, M. I., and Williams, M. L., *J. Organomet. Chem.* **282**, C11 (1985).
23. Natarajan, L., Zsolnai, L., and Huttner, G., *J. Organomet. Chem.* **209**, 85 (1981).
24. Blohm, M. L., and Gladfelter, W. L., *Organometallics* **4**, 45 (1985).
25. Johnson, B. F. G., Lewis, J., Nicholls, J. N., Puga, J., Raithby, P. R., Rosales, M. J., McPartlin, M., and Clegg, W., *J. Chem. Soc., Dalton Trans.*, p. 277 (1983).
26. Cowie, A. G., Johnson, B. F. G., Lewis, J., Nicholls, J. N., Raithby, P. R., and Rosales, M. J., *J. Chem. Soc., Dalton Trans.*, p. 2311 (1983).
27. Johnson, B. F. G., Lewis, J., Nicholls, J. N., Oxton, I. A., Raithby, P. R., and Rosales, M. J., *J. Chem. Soc., Chem. Commun.*, p. 289 (1982).
28. Evans, J., Gracey, B. P., Gray, L. R., and Webster, M., *J. Organomet. Chem.* **240**, C61 (1982).
29. MacLaughlin, S. A., Taylor, N. J., and Carty, A. J., *Inorg. Chem.* **22**, 1409 (1983).
30. Jackson, P. F., Johnson, B. F. G., Lewis, J., McPartlin, M., and Nelson, W. J. H., *J. Chem. Soc., Chem. Commun.*, p. 735 (1979).
31. Eady, C. R., Jackson, P. F., Johnson, B. F. G., Lewis, J., Malatesta, M. C., McPartlin, M., and Nelson, W. J. H., *J. Chem. Soc., Dalton Trans.*, p. 383 (1980).
32. Jackson, P. F., Johnson, B. F. G., Lewis, J., Raithby, P. R., McPartlin, M., Nelson, W. J. H., Rouse, K. D., Allibon, J., and Mason, S. A., *J. Chem. Soc., Chem. Commun.*, p. 295 (1980).
33. Oxton, I. A., Kettle, S. F. A., Jackson, P. F., Johnson, B. F. G., and Lewis, J., *J. Chem. Soc., Chem. Commun.*, p. 687 (1979).
34. Boag, N. M., Knobler, C. B., and Kaesz, H. D., *Angew. Chem., Int. Ed. Engl.* **22**, 249 (1983).
35. Churchill, M. R., and Wormald, J., *J. Am. Chem. Soc.* **93**, 5670 (1971).
36. Churchill, M. R., Wormald, J., Knight, J., and Mays, M. J., *J. Chem. Soc., Chem. Commun.*, p. 458 (1970).
37. Andrews, J. A., Jayasooriya, V. A., Oxton, I. A., Powell, D. B., Sheppard, N., Jackson, P. F., Johnson, B. F. G., and Lewis, J., *Inorg. Chem.* **19**, 3033 (1980).
38. Sirugi, A., Bianchi, M., and Benedetti, E., *J. Chem. Soc., Chem. Commun.*, p. 596 (1969).
39. Johnson, B. F. G., Lewis, J., Sankey, S. W., Wong, K., McPartlin, M., and Nelson, W. J. H., *J. Organomet. Chem.* **191**, C3 (1980).
40. Johnson, B. F. G., Johnston, R. D., and Lewis, J., *J. Chem. Soc. A*, p. 2865 (1968).
41. Eady, C. R., Johnson, B. F. G., and Lewis, J., *J. Chem. Soc., Dalton Trans.*, p. 2606 (1975).
42. Ansell, G. B., and Bradley, J. S., *Acta Crystallogr., Sect. B* **36**, 726 (1980).
43. Bradley, J. S., Ansell, G. B., and Hill, E. W., *J. Organomet. Chem.* **184**, C33 (1980).
44. Johnson, B. F. G., Lewis, J., Nelson, W. J. H., Puga, P., McPartlin, M., and Sironi, A., *J. Organomet. Chem.* **253**, C5 (1983).
45. Johnson, B. F. G., Lewis, J., Wong, K., and McPartlin, M., *J. Organomet. Chem.* **185**, C17 (1980).
46. Adams, R. D., Mathur, P., and Segmüller, B. E., *Organometallics* **2**, 1258 (1983).
47. Jackson, P. F., Johnson, B. F. G., Lewis, J., Raithby, P. R., Will, G. J., McPartlin, M., and Nelson, W. J. H., *J. Chem. Soc., Chem. Commun.*, p. 1190 (1980).
48. Mason, R., and Robinson, W. R., *J. Chem. Soc., Chem. Commun.*, p. 468 (1968).
49. Gomez-Sal, M. P., Johnson, B. F. G., Lewis, J., Raithby, P. R., and Wright, A. H., *J. Chem. Soc., Chem. Commun.*, p. 1682 (1985).
50. Ansell, G. B., and Bradley, J. S., *Acta Crystallogr. Sect. B* **36**, 1930 (1980).
51. Johnson, B. F. G., Lewis, J., Nelson, W. J. H., Puga, P., Raithby, P. R., Braga, D., McPartlin, M., and Clegg, W., *J. Organomet. Chem.* **243**, C13 (1983).
52. Brown, S. C., Evans, J., and Webster, M., *J. Chem. Soc., Dalton Trans.*, p. 2263 (1981).
53. Hayward, C. M. T., Shapley, J. R., Churchill, M. R., Bueno, C., and Rheingold, M. R., *J. Am. Chem. Soc.* **104**, 7347 (1982).

54. Eady, C. R., Johnson, B. F. G., Lewis, J., Reichert, B. E., and Sheldrick, G. M., *J. Chem. Soc.,* *Chem. Commun.,* p. 271 (1976).
55. Housecroft, C. E., O'Neill, M., Wade, K., and Smith, B. C., *J. Organomet. Chem.* **213**. 35 (1981).
56. Farrar, D. H., Johnson, B. F. G., Lewis, J., Raithby, P. R., and Rosales, M. J., *J. Chem. Soc.,* *Dalton Trans.,* p. 2501 (1982).
57. Guy, J. J., and Sheldrick, G. M., *Acta Crystallogr., Sect. B* **34**, 1722 (1978).
58. Eady, C. R., Guy, J. J., Johnson, B. F. G., Lewis, J., Malatesta, M. C., and Sheldrick, G. M., *J. Chem. Soc., Chem. Commun.,* p. 807 (1976).
59. Dawoodi, Z., Mays, M. J., and Raithby P. R., *J. Chem. Soc., Chem. Commun.,* p. 712 (1980).
60. Dawoodi, Z., Mays, M. J., and Raithby, P. R., *J. Chem. Soc., Chem. Commun.,* p. 801 (1981).
61. Guy, J. J., and Sheldrick, G. M., *Acta Crystallogr., Sect. B* **34**, 1725 (1978).
62. Nicholls, J. N., Farrar, D. H., Johnson, B. F. G., and Lewis, J., *J. Chem. Soc., Dalton Trans.,* p. 1395 (1982).
63. Eady, C. R., Johnson, B. F. G., and Lewis, J., *J. Chem. Soc., Dalton Trans.,* p. 838 (1977).
64. Farrar, D. H., John, G. R., Johnson, B. F. G., Lewis, J., Raithby, P. R., and Rosales, M. J., *J. Chem. Soc., Chem. Commun.,* p. 886 (1981).
65. John, G. R., Johnson, B. F. G., Lewis, J., Nelson, W. J. H., and McPartlin, M., *J. Organomet. Chem.* **171**, C14 (1979).
66. Johnson, B. F. G., Lewis, J., Raithby, P. R., and Rosales, M. J., *J. Organomet. Chem.* **259**, C9 (1983).
67. Adams, R. D., Horvàth, I. T., and Yang, L. W., *Organometallics* **2**, 1257 (1983).
68. Johnson, B. F. G., Lewis, J., Raithby, P. R., and Rosales, M. J., *J. Chem. Soc., Dalton Trans.,* p. 2645 (1983).
69. Johnson, B. F. G., Khattar, R., Lewis, J., and Raithby, P. R., unpublished results.
70. Jeffrey, J. G., Johnson, B. F. G., Lewis, J., Raithby, P. R., and Welch, D. A., *J. Chem. Soc., Chem. Commun.* p. 318 (1986).
71. Fernandez, J. M., Johnson, B. F. G., Lewis, J., and Raithby, P. R., *Acta Crystallogr., Sect. B* **35**, 1711 (1979).
72. Fernandez, J. M., Johnson, B. F. G., Lewis, J., and Raithby, P. R., *J. Am. Soc., Dalton Trans.,* p. 2250 (1981).
73. Adams, R. D., Horvàth, I. T., Segmüller, B. E., and Yang, L. W., *Organometallics* **2**, 1301 (1983).
74. Rivera, A. V., Sheldrick, M., and Hursthouse, M. B., *Acta Crystallogr., Sec. B* **34**, 3376 (1978).
75. Jackson, P. F., Ph. D. thesis, Cambridge University, 1980.
76. Jackson, P. F., Johnson, B. F. G., Lewis, J., Nicholls, J. N., McPartlin, M., and Nelson, W. J. H., *J. Chem. Soc., Chem. Commun.,* p. 564 (1980).
77. Johnson, B. F. G., Lewis, J., Nelson, W. J. H., Nicholls, J. N., and Vargas, M. D., *J. Organomet. Chem.* **249**, 255 (1983).
78. Johnson, B. F. G., Lewis, J., Nelson, W. J. H., Nicholls, J. N., Puga, J., Raithby, P. R., Rosales, M. J., Schröder, M. J., and Vargas, M. D., *J. Chem. Soc., Dalton Trans.,* p. 2447 (1983).
79. Johnson, B. F. G., Lewis, J., Nelson, W. J. H., Nicholls, J. N., Vargas, M. D., Braga, D., Henrick, K., and McPartlin, M., *J. Chem. Soc., Dalton Trans.,* p. 1809 (1984).
80. Jackson, P. F., Johnson, B. F. G., Lewis, J., Nelson, W. J. H., and McPartlin, M., *J. Chem. Soc., Dalton Trans.,* p. 2099 (1982).
81. Fernandez, J. M., Johnson, B. F. G., Lewis, J., Raithby, P. R., and Sheldrick, G. M., *Acta Crystallogr., Sec. B* **34**, 1994 (1978).
82. Orpen, A. G., and Sheldrick, G. M., *Acta Crystallogr., Sect. B* **34**, 1992 (1978).

83. Johnson, B. F. G., Lewis, J., Raithby, P. R., Rosales, M. J., and Welch, D. A., *J. Chem. Soc.*, *Dalton Trans.*, p. 453 (1986).
84. Mason, R., Thomas, K. M., and Mingos, D. M. P., *J. Am. Chem. Soc.* **95**, 3802 (1973).
85. Eady, C. R., Jackson, W. G., Johnson, B. F. G., Lewis, J., and Matheson, T. W., *J. Chem. Soc.*, *Chem. Commun.*, p. 958 (1975).
86. Cook, S. L., Evans, J., Greaves, N., Johnson, B. F. G., Lewis, J., Raithby, P. R., Wells, P. B., and Worthington, P., *J. Chem. Soc.*, *Chem. Commun.*, p. 777 (1983).
87. McPartlin, M., Eady, C. R., Johnson, B. F. G., and Lewis, J., *J. Chem. Soc.*, *Chem. Commun.*, p. 883 (1976).
88. Eady, C. R., Johnson, B. F. G., and Lewis, J., *J. Chem. Soc.*, *Chem. Commun.*, p. 302 (1976).
89. Gomez-Sal, M. P., Johnson, B. F. G., Kamarudin, R. A., Lewis, J., and Raithby, P. R., *J. Chem. Soc.*, *Chem. Commun.* p. 1622 (1985).
90. Nicholls, J. N., Stewart, M., and Raithby, P. R., unpublished results.
91. Orpen, A. G., *J. Organomet. Chem.* **159**, C1 (1978).
92. Johnson, B. F. G., Khattar, R., Lewis, J., Powell, G. L., and McPartlin, M., *J. Chem. Soc.*, *Chem. Commun.* p. 50 (1986).
93. Hay, C., Jeffrey, J. G., Johnson, B. F. G., Lewis, J., and Raithby, P. R., unpublished results.
94. Colbran, S., Johnson, B. F. G., Lewis, J., and Raithby, P. R., unpublished results.
95. Adams, R. D., Horvàth, I. T., Mathur, P., and Segmüller, B. E., *Organometallics* **2**, 996 (1983).
96. Adams, R. D., Foust, D. F., and Segmüller, B. E., *Organometallics* **2**, 308 (1983).
97. Couture, C., Farrar, D. H., Gomez-Sal, M. P., Johnson, B. F. G., Kamarudin, R. A., Lewis, J., and Raithby, P. R., *Acta Crystallogr.*, Sect. C **42**, 163 (1986).
98. Goudsmit, R. J., Ph. D. thesis, Cambridge University, 1982.
99. Goudsmit, R. J., Jeffrey, J. G., Johnson, B. F. G., Lewis, J., McQueen, R. C. S., Saunders, A. J., and Liu, J. C., *J. Chem. Soc.*, *Chem. Commun.*, p. 24 (1986).
100. Goudsmit, R. J., Johnson, B. F. G., Lewis, J., Raithby, P. R., and Whitmire, K. H., *J. Chem. Soc.*, *Chem. Commun.*, p. 640 (1982).
101. Johnson, B. F. G., Kamarudin, R. A., Lewis, J., and Raithby, P. R., unpublished results.
102. Hay, C. A., Jeffrey, J. G., Johnson, B. F. G., Lewis, J., and Raithby, P. R., unpublished results.
103. Eady, C. R., Fernandez, J. M., Johnson, B. F. G., Lewis, J., Raithby, P. R., and Sheldrick, G. M., *J. Chem. Soc.*, *Chem. Commun.*, p. 421 (1978).
104. Fernandez, J. M., Johnson, B. F. G., Lewis, J., and Raithby, P. R., *Acta Crystallogr.*, Sect. B **34**, 3086 (1978).
105. Johnson, B. F. G., Kamarudin, R. A., Lewis, J., and Raithby, P. R., unpublished results.
106. Johnson, B. F. G., Kamarudin, R. A., Lewis, J., and Raithby, P. R., *J. Chem. Soc.*, *Chem. Commun.*, p. 1622 (1985).
107. Orpen, A. G., and Sheldrick, G. M., *Acta Crystallogr.*, Sect. B **34**, 1989 (1978).
108. Mays, M. J., and Gavens, P. D., *J. Organomet. Chem.* **124**, C37 (1977).
109. Rivera, A. V., Sheldrick, G. M., and Hursthouse, M. B., *Acta Crystallogr.*, Sect. B **34**, 1985 (1978).
110. Eady, C. R., Gavens, P. D., Johnson, B. F. G., Lewis, J., Malatesta, M. C., Mays, M. J., Orpen, A. G., Rivera, A. V., Sheldrick, G. M., and Hursthouse, M. B., *J. Organomet. Chem.* **149**, C43 (1978).
111. Adams, R. D., Dawoodi, Z., Foust, D. F., and Segmüller, B. E., *J. Am. Chem. Soc.* **105**, 831 (1983).
112. Johnson, B. F. G., Lewis, J., McPartlin, M., Pearsall, M. A., and Sironi, A., *J. Chem. Soc.*, *Chem. Commun.*, p. 1089 (1984).

113. Goudsmit, R. J., Johnson, B. F. G., Lewis, J., Raithby, P. R., and Whitmire, K. H., *J. Chem. Soc., Chem. Commun.*, p. 246 (1983).

114. Adams, R. D., Horvàth, I. T., and Mathur, P., *Organometallics* **3**, 623 (1984).

115. Adams, R. D., Horvàth, I. T., and Yang, L. W., *J. Am. Chem. Soc.* **105**, 1533 (1983).

116. Adams, R. D., and Horvàth, I. T., *J. Am. Chem. Soc.* **106**, 1869 (1984).

117. Eady, C. R., Johnson, B. F G., Lewis, J., Mason, R., Hitchcock, P. B., and Thomas, K. M., *J. Chem. Soc., Chem. Commun.*, p. 385 (1977).

118. Ditzel, E. J., Holden, H. D., Johnson, B. F. G., Lewis, J., Saunders, A., and Taylor, M. J., *J. Chem. Soc., Chem. Commun.*, p. 1373 (1982).

119. Johnson, B. F G., Lewis, J., Powell, G. L., Raithby, P. R., Vargas, M. D., Morris, J., and McPartlin, M., *J. Chem. Soc., Chem. Commun.*, p. 429 (1986).

120. Adams, R. D., Foust, D. F., and Mathur, P., *Organometallics* **2**, 990 (1983).

121. Adams, R. D., Horvàth, I. T., Mathur, P., Segmüller, B. E., and Yang, L. W., *Organometallics* **2**, 1078 (1983).

122. Jackson, P. F., Johnson, B. F. G., Lewis, J., Nelson, W. J. H., Pearsall, M. A., Raithby, P. R., and Vargas, M. D., *J. Chem. Soc., Dalton Trans.*, (1987), in press.

123. Jackson, P. F., Johnson, B. F. G., Lewis, J., and Raithby, P. R., *J. Chem. Soc., Chem. Commun.*, p. 60 (1980).

124. Johnson, B. F. G., Lewis, J., Nelson, W. J. H., Vargas, M. D., Braga, D., Henrick, K., and McPartlin, M., *J. Chem. Soc., Dalton Trans.*, p. 2151 (1984).

125. Johnson, B. F. G., Lewis, J., Nelson, W. J. H., Raithby, P. R., Vargas, M. D., McPartlin, M., and Sironi, A., *J. Chem. Soc., Chem. Commun.*, p. 1476 (1983).

126. Guy, J. J., and Sheldrick, G. M., *Acta Crystallogr., Sect. B* **34**, 1718 (1978).

127. Eady, C. R., Guy, J. J., Johnson, B. F. G., Lewis, J., Malatesta, M. C., and Sheldrick, G. M., *J. Chem. Soc., Dalton Trans.*, p. 1358 (1978).

128. Johnson, B. F. G., Lewis, J., Nelson, W. J. H., Vargas, M. D., Braga, D., and McPartlin, M., *J. Chem. Soc., Chem. Commun.*, p. 241 (1983).

129. Attard, J. P., Johnson, B. F. G., Lewis, J., Mace, J. M., McPartlin, M., and Sironi, A., *J. Chem. Soc., Chem. Commun.*, p. 595 (1984).

130. Jackson, P. F., Johnson, B. F. G., Lewis, J., McPartlin, M., and Nelson, W. J. H., *J. Chem. Soc., Chem. Commun.*, p. 48 (1982).

131. Constable, E. C., Johnson, B. F. G., Lewis, J., Pain, G. N., and Taylor, M. J., *J Chem. Soc., Chem. Commun.*, p. 754 (1982).

132. Braga, D., Henrick, K., Johnson, B. F. G., Lewis, J., McPartlin, M., Nelson, W. J. H., and Puga, J., *J. Organomet. Chem.* **266**, 173 (1984).

133. Goudsmit, R. J., Johnson, B. F. G., Lewis, J., Nelson, W. J. H., Vargas, M. D., Braga, D., McPartlin, M., and Sironi, A., *J. Chem. Soc., Dalton Trans.*, p. 1795 (1985).

134. Braga, D., Henrick, K., Johnson, B. F. G., Lewis, J., McPartlin, M., Nelson, W. J. H., Sironi, A., Vargas, M. D., *J. Chem. Soc., Chem. Commun.*, p. 1132 (1983).

135. Keller, E., and Vahrenkamp, H., *Chem. Ber.* **112**, 2347 (1979).

136. Keller, E., and Vahrenkamp, H., *Angew. Chem., Int. Ed. Engl.* **16**, 731 (1977).

137. Albano, V. G., Chini, P., and Scatturin, V., *J. Chem. Soc., Chem. Commun.*, p. 163 (1968).

138. Chini, P., *Inorg. Chem.* **8**, 1206 (1969).

139. Johnson, B. F. G., Johnston, R. D., and Lewis, J., *J. Chem. Soc., Chem. Commun.*, p. 1057 (1967).

140. Chini, P., *J. Chem. Soc., Chem. Commun.*, p. 29 (1967).

141. Albano, V. G., Bellon, P. L., Chini, P., and Scatturin, V., *J. Organomet. Chem.* **16**, 461 (1969).

142. Chini, P., Albano, V. G., and Martinengo, S., *J. Organomet. Chem.* **16**, 471 (1969).

143. Mingos, D. M. P., *J. Chem. Soc., Dalton Trans.*, p. 133 (1974).

144. Hart, D. W., Teller, R. G., Wei, C. Y., Bau, R., Longoni, G., Campanella, S., Chini, P., and Koetzle, T. F., *J. Am. Chem. Soc.* **103**, 1458 (1981).

145. Martinengo, S., Strumolo, D., Chini, P., Albano, V. G., and Braga, D., *J. Chem. Soc., Dalton Trans.*, p. 35 (1985).

146. Creighton, J. A., Della Pergola, R., Heaton, B. T., Martinengo, S., Strona, L., and Willis, D. A., *J. Chem. Soc., Chem. Commun.*, p. 864 (1982).

147. Albano, V. G., Chini, P., Ciani, G., Sansoni, M., and Martinengo, S., *J. Chem. Soc., Dalton Trans.*, p. 163 (1980).

148. Beringhelli, T., Morazzoni, F., and Strumolo, D., *J. Organomet. Chem.* **236**, 108 (1982).

149. Bor, G., Dietler, U. K., Stanghellini, P. L., Gervasio, G., Rossetti, R., Sbrignadello, G., and Battiston, G. A., *J. Organomet. Chem.* **213**, 277 (1981).

150. Bor, G., Gervasio, G., Rossetti, R., and Stanghellini, P. L., *J. Chem. Soc., Chem. Commun.*, p. 841 (1978).

151. Gervasio, G., Rosetti, R., Stanghellini, P. L., and Bor, G., *Inorg. Chem.* **23**, 2073 (1984).

152. Martinengo, S., Ciani, G., Sironi, A., Heaton, B. T., and Mason, J., *J. Am. Chem. Soc.* **101**, 7095 (1979).

153. Ciani, G., and Sironi, A., *J. Organomet. Chem.* **241**, 385 (1983).

154. Chini, P., Ciani, G., Martinengo, S., Sironi, A., Longhetti, L., and Heaton, B. T., *J. Chem. Soc., Chem. Commun.*, p. 188 (1979).

155. Albano, V. G., Chini, P., Martinengo, S., Sansoni, M., Strumolo, D., and Ciani, G., *J. Chem. Soc., Dalton Trans.*, p. 463 (1978).

156. Albano, V. G., Chini, P., Ciani, G., Sansoni, M., Strumolo, D., Heaton, B. T., and Martinengo, S., *J. Am. Chem. Soc.* **98**, 5027 (1976).

157. Mackay, K. M., Nicholson, B. K., Robinson, W. T., and Sims, A. W., *J. Chem. Soc., Chem. Commun.*, p. 1276 (1984).

158. Albano, V. G., Braga, D., Ciani, G., and Martinengo, S., *J. Organomet. Chem.* **213**, 293 (1981).

159. Albano, V. G., Braga, D., Chini, P., Ciani, G., and Martinengo, S., *J. Chem. Soc., Dalton Trans.*, p. 645 (1982).

160. Albano, V. G., Braga, D. Fumagalli, A., and Martinengo, S., *J. Chem. Soc., Dalton Trans.*, p. 1137 (1985).

161. Reinghold, A. L., and Sullivan, P. J., *J. Chem. Soc., Chem. Commun.*, p. 39 (1983).

162. Fumagalli, A., Koetzle, T. F., Takusgawa, F., Chini, P., Martinengo, S., and Heaton, B. T., *J. Am. Chem. Soc.* **102**, 1740 (1980).

163. Martinengo, S., Ciano, G., and Sironi, A., *J. Chem. Soc., Chem. Commun.*, p. 1059 (1979).

164. Corey, E. R., Dahl, L. F., and Beck, W., *J. Am. Chem. Soc.* **85**, 1202 (1963).

165. Chaston, S. H. H., and Stone, F. G. A., *J. Chem. Soc. A*, p. 500 (1969).

166. James, B. R., Rempel, G. L., and Teo, W. K., *Inorg. Synth.* **16**, 49 (1976).

167. Kettle, S. F. A., *J. Chem. Soc. A*, p. 314 (1967).

168. Ciani, G., Sironi, A., Chini, P., and Martinengo, S., *J. Organomet. Chem.* **213**, C37 (1981).

169. Chini, P., and Martinengo, S., *J. Chem. Soc., Chem. Commun.*, p. 1092 (1969).

170. Martinengo, S., Fumagalli, A., and Chini, P., *J. Organomet. Chem.* **284**, 275 (1985).

171. Heaton, B. T., Towl, A. D. C., Chini, P., Fumagalli, A., McCaffrey, D. J. A., and Martinengo, S., *J. Chem. Soc., Chem. Commun.*, p. 523 (1975).

172. Chini, P., Martinengo, S., and Garlaschelli, L., *J. Chem. Soc., Chem. Commun.*, p. 709 (1972).

173. Chini, P., Martinengo, S., and Giordano, G., *Gazz. Chim. Ital.* **102**, 330 (1972).

174. Ciani, G., Sironi, A., Chini, P., Ceriotti, A., and Martinengo, S., *J. Organomet. Chem.* **192**, C39 (1980).

175. Antony, J., Jarvis, J., and Whyman, R., *J. Chem. Soc., Chem. Commun.*, p. 563 (1975).

176. Ciani, G., Manassero, M., and Albano, V. G., *J. Chem. Soc., Dalton Trans.*, p. 515 (1981).

177. Johnson, B. F. G., Nelson, W. J. H., Pearsall, M. A., Raithby, P. R., and McPartlin, M., unpublished results.

178. Ceriotti, A., Ciani, G., Garlaschelli, L., Sartorelli, V., and Sironi, A., *J. Organomet. Chem.* **229**, C9 (1982).
179. Forster, D. F., Nicholls, B. S., and Smith, A. K., *J. Organomet. Chem.* **236**, 395 (1982).
180. Jones, R. A., and Whittlesey, B. R., *J. Am. Chem. Soc.* **107**, 1078 (1985).
181. Albano, V. G., Bellon, P. L., and Sansoni, M., *J. Chem. Soc. A*, p. 678 (1971).
182. Evans, J., Johnson, B. F. G., Lewis, J., Matheson, T. W., and Norton, J. R., *J. Chem. Soc., Dalton Trans.*, p. 626 (1978).
183. Albano, V. G., Sansoni, M., Chini, P., and Martinengo, S., *J. Chem. Soc., Dalton Trans.*, p. 651 (1973).
184. Albano, V. G., Chini, P., Martinengo, S., McCaffrey, D. J. A., Strumolo, D., and Heaton, B. T., *J. Am. Chem. Soc.* **96**, 8106 (1974).
185. Heaton, B. T., Strona, L., and Martinengo, S., *J. Organomet. Chem.* **215**, 415 (1981).
186. Gansow, O. A., Gill, D. S., Bennis, F. J., Hutchinson, J. R., Vidal, J. L., and Schoening, R. C., *J. Am. Chem. Soc.* **102**, 2449 (1980).
187. Martinengo, S., Strumolo, D., and Chini, P., *Inorg. Synth.* **20**, 212 (1980).
188. Albano, V. G., Braga, D., and Martinengo, S., *J. Chem. Soc., Dalton Trans.*, p. 717 (1981).
189. Bonfichi, R., Ciani, G., Sironi, A., and Martinengo, S., *J. Chem. Soc., Dalton Trans.*, p. 253 (1983).
190. Martinengo, S., Ciani, G., Sironi, A., Heaton, B. T., and Mason, J., *J. Am. Chem. Soc.* **101**, 7095 (1979).
191. Albano, V. G., Bellon, P. L., and Ciani, G. F., *J. Chem. Soc., Chem. Commun.*, p. 1024 (1969).
192. Martinengo, S., and Chini, P., *Gazz. Chim. Ital.* **102**, 344 (1972).
193. Brown, C., Heaton, B. T., Longhetti, L., Smith, D. O., Chini, P., and Martinengo, S., *J. Organomet. Chem.* **169**, 309 (1979).
194. Albano, V. G., Ciani, G., Martinengo, S., Chino, P., and Giordano, G., *J. Organomet. Chem.* **88**, 381 (1975).
195. Albano, V. G., Ceriotti, A., Ciani, G., Martinengo, S., Chini, P., and Giordano, G., *J. Organomet. Chem.* **88**, 375 (1975).
196. Chini, P., Longoni, G., and Albano, V. G., *Adv. Organomet. Chem.* **14**, 285 (1976).
197. Martinengo, S., Ciani, G., and Sironi, A., *J. Chem. Soc., Chem. Commun.*, p. 1577 (1984).
198. Albano, V. G., Sansoni, M., Chini, P., Martinengo, S., and Strumolo, D., *J. Chem. Soc., Dalton Trans.*, p. 305 (1975).
199. Albano, V. G., Chini, P., Martinengo, S., Sansoni, M., and Strumolo, D., *J. Chem. Soc., Chem. Commun.*, p. 299 (1974).
200. Martinengo, S., Fumagalli, A., Bonfichi, R., Ciani, G., and Sironi, A., *J. Chem. Soc., Chem. Commun.*, p. 825 (1982).
201. Vidal, J. L., Walker, W. E., Pruett, R. L., and Schoening, R. C., *Inorg. Chem.* **18**, 129 (1979).
202. Ciani, G., Garlaschelli, L., Sironi, A., and Martinengo, S., *J. Chem. Soc., Chem. Commun.*, p. 536 (1981).
203. Garlaschelli, L., Fumagalli, A., Martinengo, S., Heaton, B. T., Smith, D. O., and Strona, L., *J. Chem. Soc., Dalton Trans.*, p. 2265 (1982).
204. Vidal, J. L., Walker, W. E., and Schoening, R. C., *Inorg. Chem.* **20**, 238 (1981).
205. Vidal, J. L., *Inorg. Chem.* **20**, 243 (1981).
206. Fumagalli, A., Martinengo, S., Ciani, G., and Sironi, A., *J. Chem. Soc., Chem. Commun.*, p. 453 (1983).
207. Albano, V. G., and Bellon, P. S., *J. Organomet. Chem.* **19**, 405 (1969).
208. Martinengo, S., and Chino, P., *Inorg. Synth.* **20**, 215 (1980).
209. Chini, P., Martinengo, S., McCaffrey, D. J. A., and Heaton, B. T., *J. Chem. Soc., Chem. Commun.*, p. 310 (1974).
210. Albano, V. G., Chini, P., Martinengo, S., Sansoni, M., and Strumolo, D., *J. Chem. Soc., Dalton Trans.*, p. 459 (1978).

211. Albano, V. G., Braga, D., Chini, P., Strumolo, D., and Martinengo, S., *J. Chem. Soc., Dalton Trans.*, p. 249 (1983).

212. Albano, V. G., Braga, D., Martinengo, S., Seregni, C., and Strumolo, D., *J. Organomet. Chem.* **252**, C93 (1983).

213. Albano, V. G., Braga, D., Strumolo, D., Seregni, C., and Martinengo, S., *J. Chem. Soc., Dalton Trans.*, p. 1309 (1985).

214. Ciani, G., Sironi, A., and Martinengo, S., *J. Chem. Soc., Dalton Trans.*, p. 519 (1981).

215. Albano, V. G., Ciani, G., Martinengo, S., and Sironi, A., *J. Chem. Soc., Dalton Trans.*, p. 978 (1979).

216. Vidal, J. L., and Schoening, R. C., *J. Organo. Chem.* **218**, 217 (1981).

217. Martinengo, S., Heaton, B. T., Goodfellow, R. J., and Chini, P., *J. Chem. Soc., Chem. Commun.*, p. 39 (1977).

218. Albano, V. G., Anker, W. M., Ceriotti, A., Chini, P., Ciani, G., and Martinengo, S., *J. Chem. Soc., Chem. Commun.*, p. 859 (1975).

219. Martinengo, S., Ciani, G., and Sironi, A., *J. Chem. Soc., Chem. Commun.*, p. 1140 (1980).

220. Ciani, G., Sironi, A., and Martinengo, S., *J. Chem. Soc., Dalton Trans.*, p. 1099 (1982).

221. Vidal, J. L., and Schoening, R. C., *Inorg. Chem.* **20**, 265 (1981).

222. Heaton, B. T., Brown, C., Smith, D. O., Strona, L., Goodfellow, R. J., Chini, P., and Martinengo, S., *J. Am. Chem. Soc.* **102**, 6175 (1980).

223. Ciani, G., Sironi, A., and Martinengo, S., *J. Organomet. Chem.* **192**, C42 (1980).

224. Martinengo, S., Strumolo, D., Chini, P., Albano, V. G., and Braga, D., *J. Chem. Soc., Dalton Trans.*, p. 1837 (1984).

225. Martinengo, S., Ciani, G., Sironi, A., and Chini, P., *J. Am. Chem. Soc.* **100**, 7096 (1978).

226. Vidal, J. L., and Schoening, R. C., *Inorg. Chem.* **21**, 438 (1982).

227. Vidal, J. L., Kapicak, L. A., and Troup, J. M., *J. Organomet. Chem.* **215**, C11 (1981).

228. Albano, V. G., Sansoni, M., Chini, P., Martinengo, S., and Strumolo, D., *J. Chem. Soc., Dalton Trans.*, p. 970 (1976).

229. Ciani, G., Magni, A., Sironi, A., and Martinengo, S., *J. Chem. Soc., Chem. Commun.*, p. 1280 (1981).

230. Vidal, J. L., Fiato, R. A., Cosby, L. A., and Pruett, R. L., *Inorg. Chem.* **17**, 2574 (1978).

231. Vidal, J. L., Schoening, R. C., Pruett, R. L., and Fiato, R. A., *Inorg. Chem.* **18**, 1821 (1979).

232. Martinengo, S., Ciani, G., and Sironi, A., *J. Am. Chem. Soc.* **102**, 7564 (1980).

233. Vidal, J. L., Schoening, R. C., and Troup, J. M., *Inorg. Chem.* **20**, 227 (1981).

234. Stevens, R. E., Liu, P. C. C., and Gladfelter, W. L., *J. Organomet. Chem.* **287**, 133 (1985).

235. Garlaschelli, L., Martinengo, S., Bellon, P. L., Demartin, F., Manassero, M., Chang, M. Y., Wei, C.-Y., and Bau, R., *J. Am. Chem. Soc.* **106**, 6664 (1984).

236. Angoletta, M., Caglio, G., and Malatesta, L., *J. Organomet. Chem.* **94**, 99 (1975).

237. Demartin, F., Manassero, M., Sansoni, M., Garlaschelli, L., and Martinengo, S., *J. Chem. Soc., Chem. Commun.*, p. 903 (1980).

238. Demartin, F., Manassero, M., Sansoni, M., Garlaschelli, L., Raimondi, C., and Martinengo, S., *J. Organomet. Chem.* **243**, C10 (1983).

239. Demartin, F., Manassero, M., Sansoni, M., Garlaschelli, L., Sartorelli, U., and Tagliabue, F., *J. Organomet. Chem.* **234**, C39 (1982).

240. Demartin, F., Manassero, M., Sansoni, M., Garlaschelli, L., Malatesta, M. C., and Sartorelli, U., *J. Organomet. Chem.* **248**, C17 (1983).

241. Pierpont, C. G., *Inorg. Chem.* **18**, 2972 (1979).

242. Pierpont, C. G. Stuntz, G. F., and Shapley, J. R., *J. Am. Chem. Soc.* **100**, 616 (1978).

243. Demartin, F., Manassero, M., Sansoni, M., Garlaschelli, L., Raimondi, C., Martinengo, S., and Canziani, F., *J. Chem. Soc., Chem. Commun.*, p. 528 (1981).

244. Angoletta, M., Malatesta, L., and Caglio, G., *J. Organomet. Chem.* **94**, 99 (1975).

245. Longoni, G., Chini, P., Lower, L. D., and Dahl, L. F., *J. Am. Chem. Soc.* **97**, 5034 (1975).

246. Longoni, G., Chini, P., and Cavalieri, A., *Inorg. Chem.* **15**, 3025 (1976).
247. Longoni, G., Heaton, B. T., and Chini, P., *J. Chem. Soc., Dalton Trans.*, p. 1537 (1980).
248. Olmstead, M. M., and Power, P. P., *J. Am. Chem. Soc.* **106**, 1495 (1984).
249. Calabrese, J. C., Dahl, L. F., Cavalieri, A., Chini, P., Longoni, G., and Martinengo, S., *J. Am. Chem. Soc.*, **96**, 2616 (1974).
250. Hall, T. L., and Ruff, J. K., *Inorg. Chem.* **20**, 4444 (1981).
251. Lower, L. D., and Dahl, L. F., *J. Am. Chem. Soc.*, **98**, 5046 (1976).
252. Ceriotti, A., Longoni, G., Manassero, M., Perego, M., and Sansoni, M., *Inorg. Chem.* **24**, 117 (1985).
253. Lower, L. D., Ph.D. Dissertation, University of Wisconsin, Madison, 1976.
254. Ceriotti, A., Longoni, G., Manassero, M., Masciocchi, N., Resconi, L., and Sansoni, M., *J. Chem. Soc., Chem. Commun.*, p. 181 (1985).
255. Chini, P., Longoni, G., Manassero, M., and Sansoni, M., *Abstr. Eighth Meet. Ital. Assoc. Crystallogr., Ferrara, Commun.* **34** (1977).
256. Broach, R. W., Dahl, L. F., Longoni, G., Chini, P., Schultz, A. J., and Wiliams, J. M., *Adv. Chem. Ser.* **167**, 93 (1978).
257. Ceriotti, A., Chini, P., Della Pergola, R., and Longoni, G., *Inorg. Chem.* **22**, 1595 (1983).
258. Goddard, R., Jolly, P. W., Kruger, C., Schick, K. P., and Wilke, G., *Organometallics* **1**, 1709 (1982).
259. Mednikov, E. G., Eremenko, N. K., Slovokhotov, Y. L., Shruchkov, Y. T., and Gubin, S. P., *J. Organomet. Chem.* **258**, 247 (1983).
260. Mednikov, E. G., Eremenko, N. K., Mikhalov, V. A., Gubin, S. P., Slovokhotov, Y. L., and Shruchkov, Y. T., *J. Chem. Soc., Chem. Commun.*, p. 989 (1981).
261. Mednikov, E. G., Eremenko, N. K., Gubin, S. P., Slovokhotov, Y. L., and Struchkov, Y. T., *J. Organomet. Chem.* **239**, 401 (1982).
262. Barbier, J. P., Bender, R., Braunstein, P., Fischer, J., and Ricard, L., *J. Chem. Res.* **230**, 2910 (1978).
263. Bender, R., Braunstein, P., Fischer, J., Ricard, L., and Mitschler, A., *Nouv. J. Chim.* **5**, 81 (1981).
264. Briant, C. E., Evans, D. G., and Mingos, D. M. P., *J. Chem. Soc., Chem. Commun.*, p. 1144 (1982).
265. Calabrese, J. C., Dahl, L. F., Chini, P., Longoni, G., and Martinengo, S., *J. Am. Chem. Soc.*, **96**, 2614 (1974).
266. Brown, C., Heaton, B. T., Chini, P., Fumagalli, A., and Longoni, G., *J. Chem. Soc., Chem. Commun.*, p. 309 (1977).
267. Washecheck, D. M., Wucherer, E. J., Dahl, L. F., Ceriotti, A., Longoni, G., Manassero, M., Sansoni, M., and Chini, P., *J. Am. Chem. Soc.*, **101**, 6110 (1979).
268. Tachikawa, M., Sievert, A. C., Thompson, M. R., Day, C. S., Day, V. W., and Muetterties, E. L., *J. Am. Chem. Soc.*, **102**, 1725 (1980).
269. Tachikawa, M., Geerts, R. L., and Muetterties, E. L., *J. Organomet. Chem.* **213**, 11 (1981).
270. Ermer, S., King, K., Hardcastle, K. I., Rosenberg, E., Lanfredi, A. M. M., Tiripicchio, A., and Camellini, M. T., *Inorg. Chem.* **22**, 1339 (1983).
271. Ruff, J. K., White, R. P., and Dahl, L. F., *J. Am. Chem. Soc.* **93**, 2159 (1971).
272. Bunkhall, S. R., Holden, H. D., Johnson, B. F. G., Lewis, J., Pain, G. N., Raithby, P. R., and Taylor, M. J., *J. Chem. Soc., Chem. Commun.*, p. 25 (1984).
273. Adams, R. D., Horvath, I. T., and Mathur, P., *J. Am. Chem. Soc.* **106**, 6296 (1984).
274. Awang, M. R., Carriedo, G. A., Howard, J. A. K., Mead, K. A., Moore, I., Nunn, C. M., and Stone, F. G. A., *J. Am. Chem. Soc., Chem. Commun.*, p. 964 (1983).
275. Gade, W., and Weiss, E., *Angew. Chem., Int. Ed. Engl.* **20**, 803 (1981).
276. Churchill, M. R., and Hollander, F. J., *Inorg. Chem.* **17**, 3546 (1978).

277. Shapley, J. R., Pearson, G. A., Tachikawa, M., Schmidt, G. E., Churchill, M. R., and Hollander, F. J., *J. Am. Chem. Soc.* **99**, 8064 (1977).
278. Haupt, H. J., and Preut, H., *Acta Crystallogr., Sect. B* **35**, 1205 (1979).
279. Haupt, H. J., Neumann, F., and Preut, H., *J. Organomet. Chem.* **99**, 439 (1975).
280. Lauher, J. W., and Wald, K., *J. Am. Chem. Soc.* **103**, 7645 (1981).
281. Ceriotti, A., Longoni, G., Manassero, M., Sansoni, M., Della Pergola, R., Heaton, B. T., and Smith, D. O., *J. Chem. Soc., Chem. Commun.*, p. 886 (1982).
282. Slovokhotov, Y. L., Struchkov, Y. T., Lopatin, V. E., and Gubin, S. P., *J. Organomet. Chem.* **266**, 139 (1984).
283. Longoni, G., Manassero, M., and Sansoni, M., *J. Am. Chem. Soc.* **102**, 7973 (1980).
284. Longoni, G., and Morazzioni, F., *J. Chem. Soc., Dalton Trans.*, p. 1735 (1981).
285. Briant, C. E., Smith, R. G., and Mingos, D. M. P., *J. Chem. Soc., Chem. Commun.*, p. 586 (1984).
286. Roland, E., Fisher, K., and Vahrenkamp, H., *Angew. Chem., Int. Ed. Engl.* **22**, 326 (1983).
287. Churchill, M. R., Fettinger, J. C., and Whitmire, K. H., *J. Organomet Chem.* **284**, 13 (1985).
288. Hriljac, J. A., Swepston, P. N., and Shriver, D. F., *Organometallics* **4**, 158 (1985).
289. Longoni, G., Manassero, M., and Sansoni, M., *J. Am. Chem. Soc.* **102**, 3242 (1980).
290. Johnson, B. F. G., Kaner, D. A., Lewis, J., Raithby, P. R., and Rosales, M. J., *J. Organomet. Chem.* **231**, C59 (1982).
291. Ernst, R. D., Marks, T. J., and Ibers, J. A., *J. Am. Chem. Soc.* **99**, 2090 (1977).
292. Whitmire, K. H., Churchill, M. R., and Fettinger, J. C., *J. Am. Chem. Soc.* **107**, 1057 (1985).
293. Johnson, B. F. G., Kaner, D. A., Lewis, J., and Rosales, M. J., *J. Organomet. Chem.* **238**, C73 (1982).
294. McNeese, T. J., Wreford, S. S., Tipton, D. L., and Bau, R., *J. Chem. Soc., Chem. Commun.*, p. 390 (1977).
295. Braunstein, P., Rosé, J., Dedieu, A., Dusausoy, Y., Mangeot, J. P., Tiripicchio, A., Tiripicchio-Camellini, M., *J. Chem. Soc., Dalton Trans.*, p. 225 (1986).
296. Braunstein, P., Rosé, J., Tiripicchio, A., and Tiripicchio-Camellini, M., *J. Chem. Soc., Chem. Commun.*, p. 391 (1984).
297. Fumagalli, A., and Ciani, G., *J. Organomet. Chem.* **272**, 91 (1984).
298. Fumagalli, A., Koetzle, T. F., and Takusagawa, F., *J. Organomet. Chem.* **213**, 365 (1981).
299. Bruce, M. I., and Nicholson, B. K., *J. Organometallics* **3**, 101 (1984).
300. Mays, M. J., Raithby, P. R., Taylor, P. L., and Henrick, K., *J. Chem. Soc., Dalton Trans.*, p. 959 (1984).
301. Tiripicchio, A., Tiripicchio-Camellini, M., and Sappa, E., *J. Chem. Soc., Dalton Trans.*, p. 627 (1984).
302. Farrugia, L. J., Freeman, M. J., Green, M., Orpen, A. G., Stone, F. G. A., and Salter, I. D., *J. Organomet. Chem.* **249**, 273 (1983).
303. Blumenthal, T., Bruce, M. I., Shawkataly, O. B., Green, B. N., and Lewis, I., *J. Organomet. Chem.* **269**, C10 (1984).
304. Bruce, M. I., Shawkataly, O. B., and Nicholson, B. K., *J. Organomet. Chem.* **286**, 427 (1985).
305. Bruce, M. I., Horn, E., Shawkataly, O. B., and Snow, M. R., *J. Organomet. Chem.* **280**, 289 (1985).
306. Bateman, L. W., Green, M., Mead, K. A., Mills, R. M., Salter, I. D., Stone, F. G. A., and Woodward, P., *J. Chem. Soc., Dalton Trans.*, p. 2599 (1983).
307. Bruce, M. I., Shawkataly, O. B., and Nicholson, B. K., *J. Organomet. Chem.* **275**, 223 (1984).
308. Lanfranchi, M., Tiripicchio, A. Sappa, E., MacLaughlin, S. A., and Carty, A. J., *J. Chem. Soc., Chem. Commun.*, p. 538 (1982).
309. Freeman, M. J., Green, M., Orpen, A. G., Salter, I. D., and Stone, F. G. A., *J. Chem. Soc., Chem. Commun.*, p. 1332 (1983).

310. Curtis, H., Johnson, B. F. G., Lewis, J., McPartlin, M., and Raithby, P. R., unpublished results.

311. Cowie, A. G., Johnson, B. F. G., Lewis, J., and Raithby, P. R., *J. Chem. Soc., Chem. Commun.,* p. 1790 (1984).

312. Howard, J. A. K., Salter, I. D., and Stone, F. G. A., *Polyhedron* 3, 567 (1984).

313. Johnson, B. F. G., Lewis, J., McPartlin, M., and Pearsall, M. A., unpublished results.

314. Cowie, A. G., Johnson, B. F. G., Lewis, J., Nicholls, J. N., Raithby, P. R., and Swanson, A. G., *J. Chem. Soc., Chem. Commun.,* p. 637 (1984).

315. Johnson, B. F. G., Lewis, J., Nicholls, J. N., Puga, J., and Whitmire, K. H., *J. Chem. Soc., Dalton Trans.,* p. 787 (1983).

316. Bradley, J. S., Pruett, R. L., Hill, E., Ansell, G. B., Leonowicz, M. E., and Modrick, M. A., *J. Organometallics* 1, 748 (1982).

317. Bradley, J. S., Ansell, G. B., and Modrick, M. A., *Acta Crystallogr., Sect. C* 40, 365 (1984).

318. Gomez-Sal, M. P., Johnson, B. F. G., Lewis, J., Raithby, P. R., and Mustaffa, S. N. A. B. S., *J. Organomet. Chem.* 272, C21 (1984).

319. Ansell, G. B., Modrick, M. A., and Bradley, J. S., *Acta Crystallogr., Sect. C* 40, 1315 (1984).

320. Bradley, J. S., and Hill, E. W., U. S. Patent 4,301,086 (1981).

321. Braunstein, P., Rosé, J., Manotti-Lanfredi, A. M., Tiripicchio, A., and Sappa, E., *J. Chem. Soc., Dalton Trans.,* p. 1843 (1984).

322. Sappa, E., Lanfranchi, M., Tiripicchio, A., and Camellini, M. T., *J. Chem. Soc., Chem. Commun.,* p. 995 (1981).

323. Castiglioni, M., Sappa, E., Valle, M., Lanfranchi, M., and Tiripicchio, A., *J. Organomet. Chem.* 241, 99 (1983).

324. Howard, J. A. K., Farrugia, L., Foster, C., Stone, F. G. A., and Woodward, P., *Eur. Crystallogr. Meet.* 6, 73 (1980).

325. Burgess, K., Johnson, B. F. G., Kaner, D. A., Lewis, J., Raithby, P. R., and Mustaffa, S. N. A. B., *J. Chem. Soc., Chem. Commun.,* p. 455 (1983).

326. Johnson, B. F. G., Lewis, J., Kaner, D. A., Raithby, P. R., and Taylor, M. J., *J. Chem. Soc., Chem. Commun.,* p. 314 (1982).

327. Johnson, B. F. G., Kaner, D. A., Lewis, J., Raithby, P. R., and Taylor, M. J., *Polyhedron* 1, 105 (1982).

328. Hay, C. M., Johnson, B. F. G., Lewis, J., McQueen, R. C. S., Raithby, P. R., Sorrell, R. M., and Taylor, M. J., *Organometallics* 4, 202 (1985).

329. Ang, H. G., Johnson, B. F. G., Lewis, J., and Raithby, P. R., unpublished results.

330. Johnson, B. F. G., Lewis, J., Raithby, P. R., and Vargas, M. D., *J. Chem. Soc., Chem. Commun.,* (1986), in press.

331. Couture, C., Farrar, D. H., and Goudsmit, R. J., *Inorg. Chim. Acta,* 89, L29 (1984).

332. Couture, C., and Farrar, D. H., *J. Chem. Soc., Chem. Commun.,* p. 197 (1985).

333. Fajardo, M., Gomez-Sal, M. P., Holden, H. D., Johnson, B. F. G., Lewis, J., McQueen, R. C. S., and Raithby, P. R., *J. Organomet. Chem.* 267, C25 (1984).

334. Johnson, B. F. G., Kaner, D. A., Lewis, J., and Raithby, P. R., *J. Chem. Soc., Chem. Commun.,* p. 753 (1981).

335. Johnson, B. F. G., Lewis, J., Raithby, P. R., and Vargas, M. D., unpublished results.

336. Fajardo, M., Holden, H. D., Johnson, B. F. G., Lewis, J., and Raithby, P. R., *J. Chem. Soc., Chem. Commun.,* p. 24 (1984).

337. Braga, D., Henrick, K., Johnson, B. F. G., Lewis, J., McPartlin, M., Nelson, W. J. H., Vargas, M. D., *J. Chem. Soc., Dalton Trans.,* p. 975 (1986).

338. Drake, S. R., Johnson, B. F. G., Lewis, J. McPartlin, M., Henrick, K., and Morris, J., *J. Chem. Soc., Chem. Commun.,* p. 928 (1986).

339. Johnson, B. F. G., Lewis, J., Nelson, W. J. H., Raithby, P. R., and Vargas, M. D., *J. Chem. Soc., Chem. Commun.*, p. 608 (1983).

340. Martinengo, S., Chini, P., Albano, V. G., Cariati, F., and Salvatori, T., *J. Organomet. Chem.* **59**, 379 (1973).

341. Barbier, J. P., Braunstein, P., Fischer, J., and Ricard, L., *Inorg. Chim. Acta* **31**, L361 (1978).

342. Van Der Velden, J. W. A., Bour, J. J., Bosman, W. P., and Noordick, J. H., *Inorg. Chem.* **22**, 1913 (1983).

343. Longoni, G., Ceriotti, A., Della Pergola, R., Manassero, M., Perego, M., Piro, G., and Sansoni, M. , *Philos. Trans. R. Soc. London, Ser. A* **308**, 47 (1982).

344. Ceriotti, A., Della Pergola, R., Longoni, G., Manassero, M., and Sansoni, M., *J. Chem. Soc., Dalton Trans.*, p. 1181 (1984).

345. Albano, V. G., Ciani, G., and Chini, P., *J. Chem. Soc., Dalton Trans.*, p. 432 (1974).

346. Chini, P., Cavalieri, A., and Martinengo, S., *Coord. Chem. Rev.* **8**, 3 (1972).

347. Klufers, P., *Angew. Chem., Int. Ed. Engl.* **23**, 307 (1984).

348. Klufers, P., *Z. Kristallogr.*, in press.

349. Burlitch, J. M., Hayes, S. E., and Lemley, J. T., *J. Organometallics* **4**, 167 (1985).

350. Gerlach, R. F., D. Philos. thesis, University of Waikato, 1979.

351. Gerlach, R. F., Mackay, K. M., and Nicholson, B. K., *J. Chem. Soc., Dalton Trans.*, p. 80 (1981).

352. Schmid, G., and Etzrodt, G., *J. Organomet. Chem.* **137**, 367 (1977).

353. Foster, S. P., Mackay, K. M., and Nicholson, B. K., *J. Chem. Soc., Chem. Commun.*, p. 1156 (1982).

354. Duffy, D. N., Mackay, K. M., Nicholson, B. K., and Robinson, W. T., *J. Chem. Soc., Dalton Trans.*, p. 381 (1981).

355. Croft, R. A., Duffy, D. N., and Nicholson, B. K., *J. Chem. Soc., Dalton Trans.*, p. 1023 (1982).

356. Curnow, O. J., and Nicholson, B. K., *J. Organomet. Chem.* **267**, 257 (1984).

357. Arrigoni, A., Ceriotti, A., Della Pergola, R., Longoni, G., Manassero, M., Masciocchi, N., and Sansoni, M., *Angew. Chem., Int. Ed. Engl.* **23**, 322 (1984).

358. Foster, S. P., Mackay, K. M., and Nicholson, B. K., *Inorg. Chem.* **24**, 909 (1985).

359. Green, M., Howard, J. A. K., Mills, R. M., Pain, G. N., Stone, F. G. A., and Woodward, P., *J. Chem. Soc., Chem. Commun.*, p. 869 (1981).

360. Fumagalli, A., Martinengo, S., Chini, P., Albinati, A., and Bruckner, S., *Proc. A11 XIII Congr. Naz. Chim. Inorg., Camerino, Italy, 1980*, Soc. Chim. Ital., 1980.

361. Fumagalli, A., Martinengo, S., Chini, P., Galli, D., Heaton, B. T., and Della Pergola, R., *Inorg. Chem.* **23**, 2947 (1984).

362. Fumagalli, A., Martinengo, S., Chini, P., Albinati, A., Bruckner, S., and Heaton, B. T., *J. Chem. Soc., Chem. Commun.*, p. 195 (1978).

363. Fumagalli, A., Longoni, G., Chini, P., Albinati, A., and Bruckner, S., *J. Organomet. Chem.* **202**, 329 (1980)

364. Heaton, B. T., Della Pergola, R., Strona, L., Smith, D. O., and Fumagalli, A., *J. Chem. Soc., Dalton Trans.*, p. 2553 (1982).

365. Albano, V. G., Braga, D., Martinengo, S., Chini, P., Sansoni, M., and Strumolo, D., *J. Chem. Soc., Dalton Trans.*, p. 52 (1980).

366. Fumagalli, A., Martinengo, S., and Ciani, G., *J. Organomet. Chem.* **273**, C46 (1984).

367. Martinengo, S., Ciani, G., and Sironi, A., *J. Am. Chem. Soc.*, **104**, 328 (1982).

368. Fumagalli, A., Martinengo, S., and Ciani, G., *J. Chem. Soc., Chem. Commun.*, p. 1381 (1983).

369. Heaton, B. T., Strona, L., Martinengo, S., Strumolo, D., Albano, V. G., and Braga, D., *J. Chem. Soc., Dalton Trans.*, p. 2175 (1983).

370. Vidal, J. L., and Troup, J. M., *J. Organomet. Chem.* **213**, 351 (1981).

371. Freeman, M. J., Miles, A. D., Murray, M., Orpen, A. G., and Stone, F. G. A., *Polyhedron* **3**, 1093 (1984).

372. Nicholls, J. N., Raithby, P. R., and Vargas, M. D., *J. Chem. Soc., Chem. Commun.*, (1986), in press.
373. Ceriotti, A., Demartin, F., Longoni, G., Manassero, M., Marchionna, M., Piva, G., and Sansoni, M., *Angew. Chem., Int. Ed. Engl.* **24**, 697 (1985).
374. Heaton, B. T., Ochiello, E., and Strona, L., *Angl.-Jpn. Adv. Res. Meet., Osaka Univ., Toyonaka, Osaka*, **49** (1985).
375. Mednikov, E. G., Bashilov, V. V., Sokolov, V. I., Slovokhotov, Y. L., and Struchkov, Y. T., *Polyhedron* **2**, 141 (1983).
376. Beringhelli, T., Ciani, G., D'Alfonso, G., Sironi, A., and Freni, M., *J. Chem. Soc., Chem. Commun.*, p. 978 (1985).
377. Jeffrey, J. G., Johnson, B. F. G., Lewis, J., and Raithby, P. R., unpublished results.
378. Kaner, D. A., Ph. D. thesis, University of Cambridge, 1982.
379. Doyle, G., Eriksen, K. A., and Van Engen, D., *J. Chem. Soc., Chem. Commun.* **108**, 445 (1986).
380. Ceriotti, A., Longoni, G., Manassero, M., Masciocchi, N., Piro, G., Resconi, L., and Sansoni, M., *J. Chem. Soc., Chem. Commun.*, p. 1402 (1985).
381. Leopatin, V. E., Gubin, S. P., Mikova, N. M., Tsybenov, M. T. S., Slovokhotov, Y. L., and Struchkov, Y. T., *J. Organomet. Chem.* **292**, 275 (1985).
382. Hayward, C. M. T., and Shapley, J. R., *Inorg. Chem.* **21**, 3816 (1982).
383. Johnson, B. F. G., Kattar, R., Lewis, J., and Rosales, M. J., unpublished results.
384. Johnson, B. F. G., *J. Chem. Soc., Chem. Commun.*, p. 211 (1976).
385. Peake, B. M., Robinson, B. H., Simpson, J., and Watson, D. J., *J. Chem. Soc., Chem. Commun.*, p. 945 (1974).
386. John, G. R., Johnson, B. F. G., and Lewis, J., *J. Organomet. Chem.* **169**, C9 (1979).
387. Cox, D. J., John, G. R., Johnson, B. F. G., and Lewis, J., *J. Organomet. Chem.* **186**, C69 (1979).
388. John, G. R., Johnson, B. F. G., Lewis, J., and Mann, A. L., *J. Organomet. Chem.* **171**, C9 (1979).
389. Chini, P., *J. Chem. Soc., Chem. Commun.*, p. 440 (1967).
390. Ceriotti, A., Chini, P., Longoni, G., and Piro, G., *Gazz. Chim. Ital.* **112**, 353 (1982).
391. Booth, R. L., Else, M. J., Fields, R., and Haszeldine, R. N., *J. Organomet. Chem.* **27**, 119 (1971).
392. Jeffrey, J. G., Johnson, B. F. G., Lewis, J., and Raithby, P. R., unpublished results.
393. Nomiya, K., and Suzuki, H., *J. Organomet. Chem.* **168**, 115 (1979).
394. Kitamura, T., and Joh, T., *J. Organomet. Chem.* **65**, 235 (1974).
395. Foster, D. F., Nicholls, B. S., and Smith, A. K., *J. Organomet. Chem.* **244**, 159 (1983).
396. Nomiya, K., and Suzuki, H., *Bull. Chem. Soc. Jpn.* **52**, 623 (1979).
397. Heaton, B. T., Strona, L., Martinengo, S., Strumolo, D., Goodfellow, R. J., and Sadler, I. H., *J. Chem. Soc., Dalton Trans.*, p. 1499 (1982).
398. Vidal, J. L., and Schoening, R. C., *J. Organomet. Chem.* **241**, 395 (1983).
399. Heaton, B. T., Strona, L., Jonas, J., Eguchi, T., and Hoffman, G. A., *J. Chem. Soc., Dalton Trans.*, p. 1159 (1982).
400. Hall, T. L., and Ruff, J. K., *Inorg. Chem.* **20**, 4444 (1981).
401. Longoni, G., and Chini, P., *Inorg. Chem.* **15**, 3029 (1976).
402. Churchill, M. R., Hollander, F. J., Lashewycz, R. A., Pearson, G. A., and Shapley, J. R., *J. Am. Chem. Soc.* **103**, 2430 (1981).
403. Cowie, A. G., Johnson, B. F. G., Lewis, J., and Raithby, P. R., *J. Chem. Soc., Chem. Commun.*, (1986), in press.
404. Christie, J. A., Duffy, D. N., Mackay, K. M., and Nicholson, B. K., *J. Organomet. Chem.* **226**, 165 (1982).
405. Chini, P., Longoni, G., Martinengo, S., and Ceriotti, A., *Adv. Chem. Ser.* **167**, 1 (1978).
406. Holt, E. M., Whitmire, K. H., and Shriver, D. F., *J. Organomet. Chem.* **213**, 125 (1981).

407. Farrar, D. H., Johnson, B. F. G., Lewis, J., Nicholls, J. N., Raithby, P. R., and Rosales, M. J., *J. Chem. Soc., Chem. Commun.*, p. 273 (1981).

408. Cavanaugh, M. A., Fehlner, T. F., Stramel, R., O'Neill, M. E., and Wade, K., *Polyhedron* **4**, 687 (1985).

409. Vargas, M. D., Ph. D. thesis, University of Cambridge, 1983.

410. Jackson, P. F., John, G. R., Johnson, B. F. G., and Lewis, J., unpublished results.

411. Kolis, J. W., Basolo, F., and Shriver, D. F., *J. Am. Chem. Soc.* **104**, 5626 (1982).

412. Nicholls, J. N., Ph. D. thesis, University of Cambridge, 1982.

413. Beno, M. A., Williams, J. M., Tachikawa, M., and Muetterties, E. L., *J. Am. Chem. Soc.* **103**, 1485 (1981); *ibid.* **102**, 4542 (1980).

414. Tachikawa, M., and Muetterties, E. L., *J. Am. Chem. Soc.* **102**, 4541 (1980).

415. Wijeyesekera, S. D., Hoffmann, R., and Wilker, C. N., *Organometallics* **3**, 362 (1984).

416. D'Alfonso, G., Longoni, G., Rossetti, R., and Stanghellini, P. L., *Abstr. Int. Conf. 12th Organomet. Chem.*, Vienna, No. 528 (1985).

417. Broach, R. W., Dahl, L. F., Longoni, G., Chini, P., Schultz, A. J., and Williams, J. M., *Adv. Chem. Ser.* **167**, 93 (1978).

418. Kruck, T., Hofler, M., and Noack, M., *Chem. Ber.* **99**, 1153 (1966).

419. Johnson, B. F. G., Lewis, J., Pearsall, M. A., and Scott, L. G., unpublished results.

420. Lemoine, P., *Coord. Chem. Rev.* **47**, 55 (1982).

421. Geiger, W. E., and Tulyathan, B., *J. Am. Chem. Soc.* **107**, 5960 (1985).

422. Evans, D. G., Mingos, D. M. P., *Organometallics* **2**, 435 (1983).

423. Battacharya, A. A., and Shore, S. G., *Organometallics* **2**, 1251 (1983).

424. Oxton, I. A., Kettle, S. F., Jackson, P. F., Johnson, B. F. G., and Lewis, J., *J. Mol. Struct.* **71**, 117 (1981).

425. Nicholls, J. N., Farrar, D. H., Jackson, P. F., Johnson, B. F. G., and Lewis, J., *J. Chem. Soc., Dalton Trans.*, p. 1395 (1982).

426. Bonny, A., Crane, T. J., and Kane-Maguire, N. A. P., *Inorg. Chim. Acta* **65**, L83 (1982).

427. Drake, S. R., Johnson, B. F. G., Lewis, J., and McQueen, R. C. S., *J. Chem. Soc., Chem. Commun.*, (1986), in press.

428. Colbran, S. B., Jeffrey, J. G. J., Johnson, B. F. G., and Lewis, J., unpublished results.

429. Johnson, B. F. G., Lewis, J., and Nelson, W. J. H., unpublished results.

430. Brown, S. S. D., Salter, I. D., and Smith, B. M., *J. Chem. Soc., Chem. Commun.*, p. 1439 (1985).

431. Sinfelt, J. H., Slichter, C. P., and Wang, P. K., *Phys. Rev. Lett.* **53**, 82 (1984).

432. Stevens, R. E., Yanta, T. J., and Gladfelter, W. L., *J. Am. Chem. Soc.* **103**, 4981 (1981).

433. Farrar, D. H., Johnson, B. F. G., and Lewis, J., unpublished results.

434. Nelson, W. J. H., Ph. D. thesis, The Polytechnic of North London, 1980.

435. Oxton, I. A., Powell, D. B., Farrar, D. H., Johnson, B. F. G., Lewis, J., and Nicholls, J. N., *Inorg. Chem.* **20**, 4302 (1981).

436. Bruce, M. I., Williams, M. L., Skelton, B. W., and White, A. H., *J. Organomet. Chem.* **282**, C53 (1985).

437. Amer, S., Kramer, G., and Poe, A., *J. Organomet. Chem.* **220**, 75 (1981).

438. Hodali, H. A., and Shriver, D. F., *Inorg. Chem.* **18**, 1236 (1979).

439. Manning, A. R., *Coord. Chem. Rev.* **51**, 41 (1983).

440. Hodali, H. A., Shriver, D. F., and Ammlung, C. A., *J. Am. Chem. Soc.* **100**, 5239 (1978).

441. Kaesz, H. D., and Saillant, R. B., *Chem. Rev.* **72**, 231 (1972); Jesson, J. P., *in* "Transition Metal Hydrides" (E. L. Muetterties, ed.), p. 75. Dekker, New York, 1971.

442. Natarajan, K., Zolnai, L., and Huttner, G., *J. Organomet. Chem.* **220**, 365 (1981); Iwasaki, F., Mays, M. J., Raithby, P. R., Taylor, P. L., and Wheatley, P. L., *J. Organomet. Chem.* **213**, 185 (1981); Colbran, S. B., Johnson, B. F. G., Lewis, J., and Sorrell, R. M., *J. Organomet. Chem.*, **296**, C1 (1985).

443. Couture, C., and Farrar, D. H., personal communication.

444. Johnson, B. F. G., Lewis, J., Nelson, W. J. H., and Vargas, M. D., unpublished results.
445. Adams, R. D., and Yang, L. W., *J. Am. Chem. Soc.* **104**, 4115 (1982).
446. Churchill, M. R., De Boer, B. G., and Rotella, F. J., *Inorg. Chem.* **15**, 1843 (1976).
447. Orpen, A. G., *J. Chem. Soc., Dalton Trans.*, p. 2509 (1980).
448. Eady, C. R., Johnson, B. F. G., and Lewis, J., *J. Chem. Soc., Dalton Trans.*, p. 838 (1977).
449. Albano, V. G., Chini, P., Ciani, G., Fumagalli, A., and Martinengo, S., *J. Organomet. Chem.* **116**, 333 (1976).
450. Richter, F., and Vahrenkamp, H., *Angew. Chem., Int. Ed. Engl.* **18**, 351 (1979); Vahrenkamp, H., and Wucherer, E. J., *Angew. Chem., Int. Ed. Engl.* **20**, 680 (1981).
451. Evans, D. G., and Mingos, D. M. P., *J. Organomet. Chem.* **232**, 171 (1982).
452. Bruce, M. I., and Nicholson, B. K., *J. Organomet. Chem.* **252**, 243 (1983).
453. Braunstein, P., Rosé, J., Tiripicchio, A., and Tiripicchio Camellini, M., *Angew. Chem., Int. Ed. Engl.* **24**, 767 (1985).
454. Connor, J. A., *in* "Transition Metal Clusters" (B. F. G. Johnson, ed.), p. 345. Wiley, New York, 1980.
455. Hoffmann, R., *Angew. Chem., Int. Ed. Engl.* **21**, 711 (1982).
456. Johnson, B. F. G., and Lewis, J., *Adv. Inorg. Chem. Radiochem.* **24**, 225 (1981).
457. Deeming, A. J., *in* "Transition Metal Clusters" (B. F. G. Johnson, ed.), p. 391. Wiley, New York, 1980.
458. Vahrenkamp, H., *Adv. Organomet. Chem.* **22**, 169 (1983).
459. Geiger, W. E., and Connelly, N. G., *Adv. Organomet. Chem.* **24**, 87 (1985).
460. Gladfelter, W. L., *Adv. Organomet. Chem.* **24**, 41 (1985).
461. Bradley, J. S., *Adv. Organomet. Chem.* **22**, 1 (1983).
462. Tachikawa, M., and Muetterties, E. L., *Prog. Inorg. Chem.* **28**, 203 (1981).
463. Nicholls, J. N., *Polyhedron* **3**, 1307 (1984).
464. Hall, K. P., and Mingos, D. M. P., *Prog. Inorg. Chem.* **32**, 237 (1984).
465. Muetterties, E. L., *J. Organomet. Chem.* **200**, 177 (1980).
466. Chini, P., *J. Organomet. Chem.* **200**, 37 (1980).
467. Gladfelter, W. L., and Geoffroy, G. L., *Adv. Organomet. Chem.* **18**, 207 (1980).
468. Braunstein, P., *Gold Bull.* **18**, 17 (1985).
469. Raithby, P. R., and Rosales, M. J., *Adv. Inorg. Chem. Radiochem.* **29**, 170 (1985).
470. Doyle, G., Heaton, B. T., and Occhiello, E., *Organometallics* **4**, 1224 (1985).
471. Stone, F. G. A., *ACS Symp. Ser.* **211**, 383 (1983); *Philos. Trans. R. Soc. London Ser. A* **308**, 87 (1982).
472. Mlekuz, M., Bougeard, P., Sayer, B. G., Faggiani, R., Lock, C. J. L., McGlinchey, M. J., and Jaouen, G., *Organometallics* **4**, 2046 (1985), and references therein.
473. Cowie, A. G., Ph. D. thesis, Univ. Cambridge (1985).
474. Johnson, B. F. G., Lewis, J., Powell, G. L., Vargas, M. D., and McPartlin, M., *J. Chem. Soc., Dalton Trans.*, (1987), in press.

INORGANIC CHEMISTRY OF HEXAFLUOROACETONE

M. WITT, K. S. DHATHATHREYAN, and H. W. ROESKY

Institut für Anorganische Chemie der Universität Göttingen,
D-3400 Göttingen, Federal Republic of Germany

I. Introduction

Hexafluoroacetone (HFA) was discovered by Fukuhara and Bigelow in 1941 (109), but more than two decades passed before HFA gained wide interest in inorganic, organic, and technical chemistry. The chemical and physical properties of HFA were reviewed by Krespan and Middleton (168) in 1967 and more recently by Middleton (185).

The difference in reactivity compared to organic ketones is caused by the strong electron-withdrawing effect of the fluorine atoms in HFA, which leads to an electron-deficient carbonyl group. This is manifested in the inability to protonate the oxygen atom in super acidic media (203).

Depending on the nature of the reaction partners, there are various pathways leading to different products:

i. Insertion into activated single bonds:

$$A-B + HFA \longrightarrow A-O-\underset{\underset{CF_3}{|}}{\overset{\overset{CF_3}{|}}{C}}-B \tag{1}$$

ii. Oxidative addition to low-valent atoms (with ligand displacement in organometallic compounds) and multiple bonds to yield heterocycles of different size and geometry:

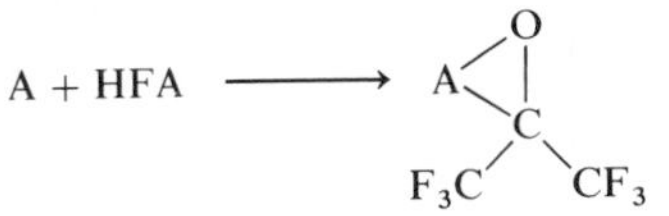

$$A + HFA \longrightarrow A\overset{O}{\underset{F_3C}{\diagdown}}\underset{CF_3}{\diagup}C \tag{2}$$

$$A{=}B + HFA \longrightarrow \overset{A-O}{\underset{B-C}{|}} \overset{}{\underset{CF_3}{\diagup}}\overset{CF_3}{\diagdown} \tag{3}$$

223

Copyright © 1986 by Academic Press, Inc.
All rights of reproduction in any form reserved.

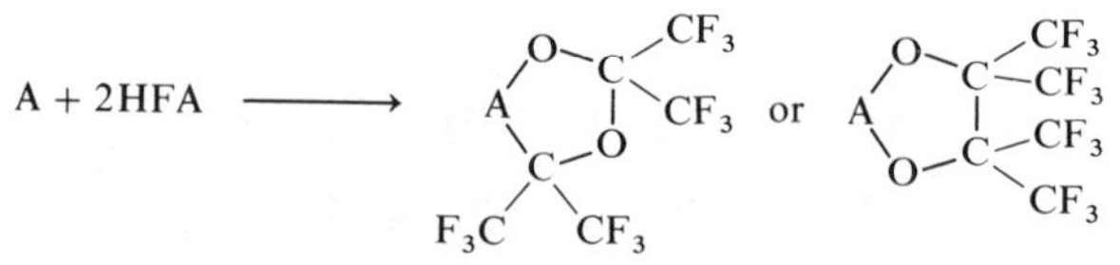

$$A + 2HFA \longrightarrow$$

(4)

$$A{=}B + 2HFA \longrightarrow$$

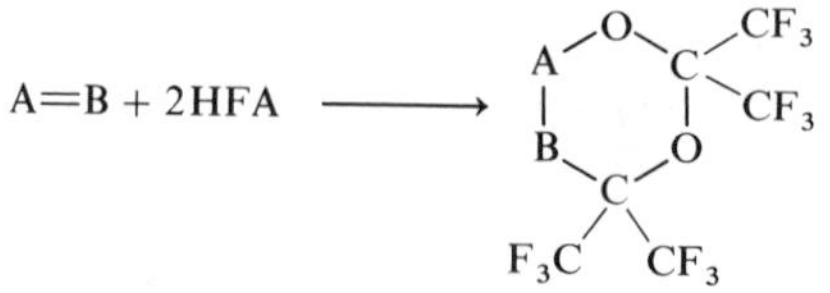

(5)

In addition to these common types of reactions, several syntheses are known in which HFA causes changes in the substrate (e.g., isomerization of nitriles to isonitriles) or rearrangements of initially unstable products. However, only few exceptions to the above-mentioned examples have been found so far in HFA chemistry.

This article deals mainly with synthetic "inorganic" aspects of HFA chemistry. Reaction conditions and spectroscopic data are mentioned if necessary for structural and mechanistic considerations. Particular attention has been drawn to literature coverage since 1966; earlier works reviewed by Krespan and Middleton (*168*) have been included only for completion of some sections.

II. Reactions of HFA with Compounds of Group IV Elements

The chemistry of group IV elements and HFA is dominated by insertion reactions according to Eq. (1). Cycloadditions involving the elements of group IV are rather seldom observed.

A. SILICON

1. Reactions with Si—H Bonds

Under free radical conditions HFA adds to a variety of silanes containing Si—H bonds to form hexafluoroisopropoxysilanes (*143a, 156*). Ionic conditions (dark, liquid phase, low temperatures) lead in the case of trimethylsilane to adducts **1** and **2** (*73, 156*).[1]

[1] Abbreviations: Me, CH_3; Et, C_2H_5; Pr, C_3H_7; *i*-Pr, $CH(CH_3)_2$; Bu, C_4H_9; *t*-Bu, $C(CH_3)_3$; Ph, C_6H_5; Cp, η^5-cyclopentadienyl; COD, 1,5-cyclooctadiene; Hfp, hexafluoroisopropyl; Pfp, perfluoropinacolyl; Py, pyridine; Ar, aryl; Al, alkyl; acac, acetyl acetonate.

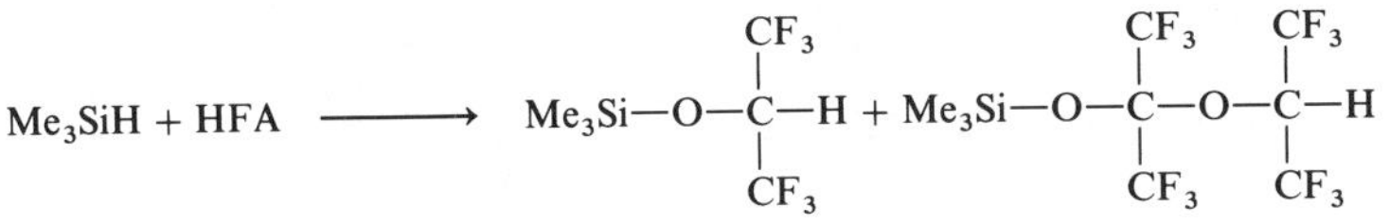

$$\text{Me}_3\text{SiH} + \text{HFA} \longrightarrow \underset{(1)}{\text{Me}_3\text{Si}-\text{O}-\underset{\underset{\text{CF}_3}{|}}{\overset{\overset{\text{CF}_3}{|}}{\text{C}}}-\text{H}} + \underset{(2)}{\text{Me}_3\text{Si}-\text{O}-\underset{\underset{\text{CF}_3}{|}}{\overset{\overset{\text{CF}_3}{|}}{\text{C}}}-\text{O}-\underset{\underset{\text{CF}_3}{|}}{\overset{\overset{\text{CF}_3}{|}}{\text{C}}}-\text{H}} \quad (6)$$

A mechanism proceeding via abstraction of a hydride ion by HFA with formation of isopropoxide ion as the initial step, followed by reversible addition of another molecule of HFA, has been proposed. The resulting anions combine with the Me$_3$Si cation to yield **1** and **2** (*158*). The "1:1 adduct" of **1** with HFA reported by Cullen and Styan (*73*) has been proved by Janzen and Willis (*158*) to be identical with **2**.

2. Reactions with Si—O Bonds

Similarly, hydroxy- and alkoxysilanes are attacked by HFA to form hemiketals and ketals (*35, 157*). A 1,3,4-dioxasilepane (**3**) is accessible by insertion of HFA into the Si—O bond of a 1,2-oxasilolane (*67*) [Eq. (7)] via nucleophilic attack of oxygen at the carbonyl carbon atom (*35*). Attempts to synthesize perfluoroisopropoxysilanes from alkylhalogenosilanes with HFA in the presence of KF via intermediate perfluoroisopropoxide ion proceed with elimination of HFA and formation of the corresponding fluorosilanes (*208*).

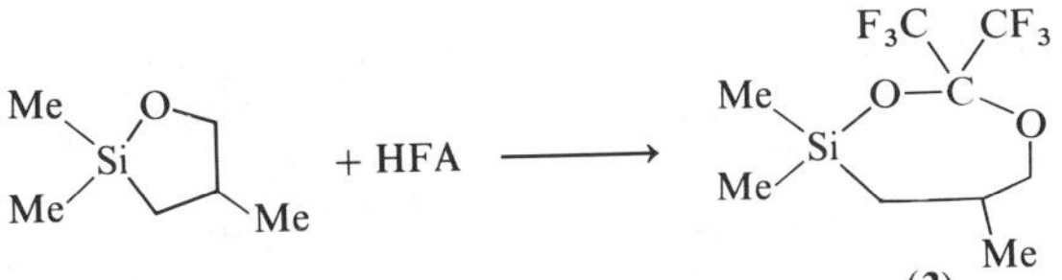

$$\underset{\text{Me}}{\overset{\text{Me}}{\diagdown}}\text{Si}\overset{\text{O}}{\diagup}\diagdown_{\text{Me}} + \text{HFA} \longrightarrow \underset{(3)}{\overset{\text{Me}}{\underset{\text{Me}}{\diagdown}}\text{Si}} \quad (7)$$

A silene was assumed to be involved in the reaction of dichlorodimethylsilane with two equivalents of lithium and HFA to yield a 1,3,4-dioxasilolane (*34*). Further investigations showed the mechanism, as well as the structure assigned, to be incorrect.

Frye *et al.* (*108*) obtained bis(trimethylsilyl) oxyperfluoropinacolate (**4**) and 2,2-dimethyl-4,4,5,5-tetrakistrifluoromethyl-1,3,2-dioxasilolane (**5**) by reacting HFA, lithium, and the corresponding chlorosilanes.

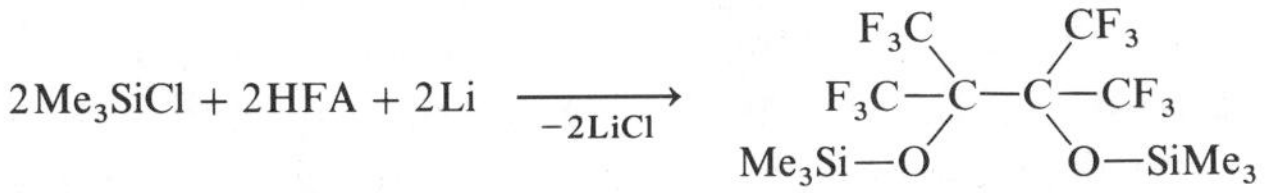

$$2\text{Me}_3\text{SiCl} + 2\text{HFA} + 2\text{Li} \xrightarrow[-2\text{LiCl}]{} \underset{(4)}{\underset{\text{Me}_3\text{Si}-\text{O}}{\overset{\text{F}_3\text{C}}{\text{F}_3\text{C}-\text{C}-\text{C}-\text{CF}_3}}} \quad (8)$$

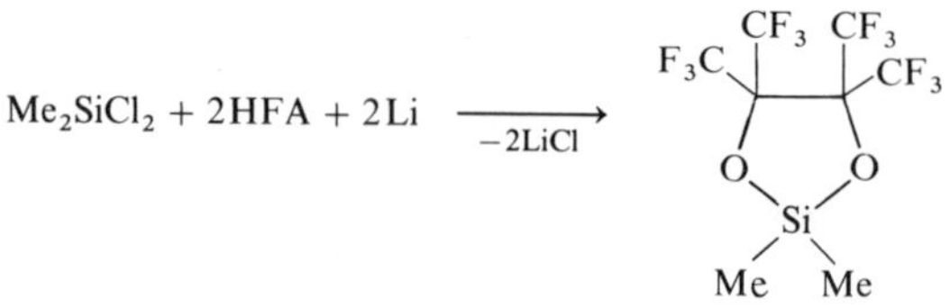

$$Me_2SiCl_2 + 2\,HFA + 2\,Li \xrightarrow[-2LiCl]{} \underset{(5)}{\begin{array}{c} \text{structure} \end{array}} \tag{9}$$

(5)

The structure of **5** was confirmed by metathesis of dimethyldiacetoxysilane with perfluoropinacol (*108*). The reactions according to Eqs. (8) and (9) proceed via the alkali salts of perfluoropinacolate dianion, which in the case of sodium can be obtained as a pure white powder (*153*). Compound **4** is also accessible by reaction of HFA and bis(trimethylsilyl)mercury under mild conditions (*153*). As the reaction rate is increased by UV radiation, formation of trimethylsilyl radicals is assumed to be the first step.

3. Reactions with Si—N Bonds

HFA inserts into one Si—N bond of hexamethyldisilazane at 50°C in a sealed tube with formation of **6** (*254*). However, no products have been obtained in the reactions of HFA with heptamethyldisilazane and a cyclic trisilazane (*2*).

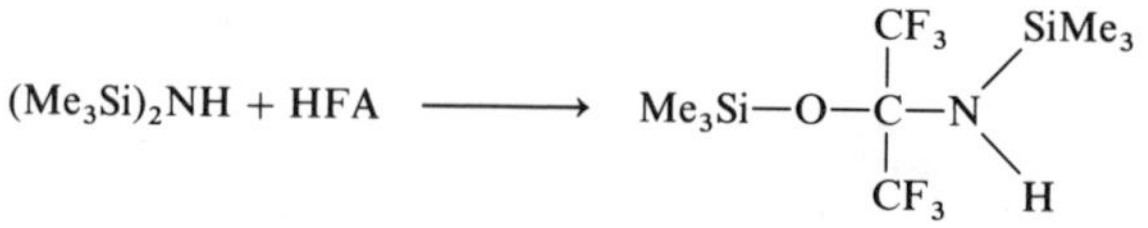

$$(Me_3Si)_2NH + HFA \longrightarrow Me_3Si\!-\!O\!-\!\underset{\underset{CF_3}{|}}{\overset{\overset{CF_3}{|}}{C}}\!-\!N\!\underset{H}{\overset{SiMe_3}{\big\langle}} \tag{10}$$

(6)

A polymeric 2:1 addition product is formed with a 1,3-diaza-2-silolidine (*2*).

Further examples of insertion into Si—N bonds have been found in the reactions of HFA with dimethylaminotrimethylsilane, phenylaminotrimethylsilane (*2*), azidotrimethylsilane (*1, 270*), and bis(trimethylsilyl)carbodiimide (*102*). All products show the insertion of HFA only in one of the Si—N bonds.

HFA also cleaves the Si—N bond in trimethylsilylaminotriphenyliminophosphorane to form **7** in high yield. A minor side reaction results in the formation of **8**. The nitrogen–oxygen exchange prevails with the corresponding tin compound (*vide infra*). Unlike its tin analogue, **8** does not add another molecule of HFA (*1*). In contrast to organic fluoroimines, ^{19}F NMR of **8** shows only one signal, suggesting a very low nitrogen inversion barrier.

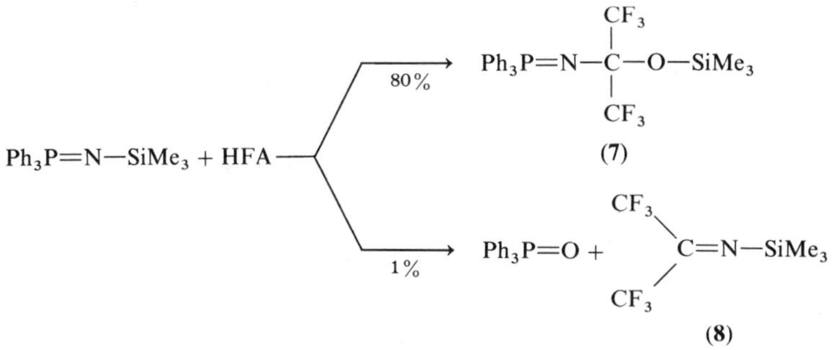

$$Ph_3P=N-SiMe_3 + HFA \xrightarrow{\quad} \begin{cases} \xrightarrow{80\%} Ph_3P=N-\underset{\underset{CF_3}{|}}{\overset{\overset{CF_3}{|}}{C}}-O-SiMe_3 \\ \qquad\qquad\qquad (7) \\ \\ \xrightarrow{1\%} Ph_3P=O + \underset{CF_3}{\overset{CF_3}{>}}C=N-SiMe_3 \\ \qquad\qquad\qquad (8) \end{cases} \quad (11)$$

4. Reactions with Si—C Bonds

HFA and cyanotrimethylsilane react stoichiometrically with formation of the substituted cyanhydrin **9** (*175*). Increasing the molar ratio of the reactants to 4:1 yields, in addition to **9**, compound **10**, with nitrile–isonitrile equilibrium competing with direct attack of HFA (*242*). The five-membered ring is also formed in the reaction of organic iso-nitriles with HFA (*188*). The same structural feature in addition to insertion has been found when triethylamine is present as a catalyst, as well as minor amounts of **9** (*83, 242*).

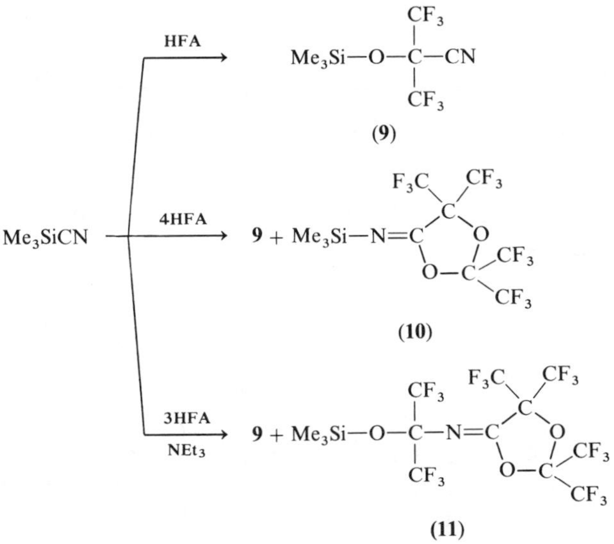

$$Me_3SiCN \xrightarrow{\quad} \begin{cases} \xrightarrow{HFA} Me_3Si-O-\underset{\underset{CF_3}{|}}{\overset{\overset{CF_3}{|}}{C}}-CN \\ \qquad\qquad (9) \\ \\ \xrightarrow{4HFA} 9 + Me_3Si-N=C\underset{O-C}{\overset{C}{<}} \\ \qquad\qquad (10) \\ \\ \xrightarrow[NEt_3]{3HFA} 9 + Me_3Si-O-\underset{\underset{CF_3}{|}}{\overset{\overset{CF_3}{|}}{C}}-N=C \\ \qquad\qquad (11) \end{cases} \quad (12)$$

Abel and Rowley (*4*) have done extensive work on the interaction of HFA and silanes with allylic substituents. At 100°C the reaction occurs according to Eq. (13). Decreasing the temperature leads to the formation of oxetane **13**.

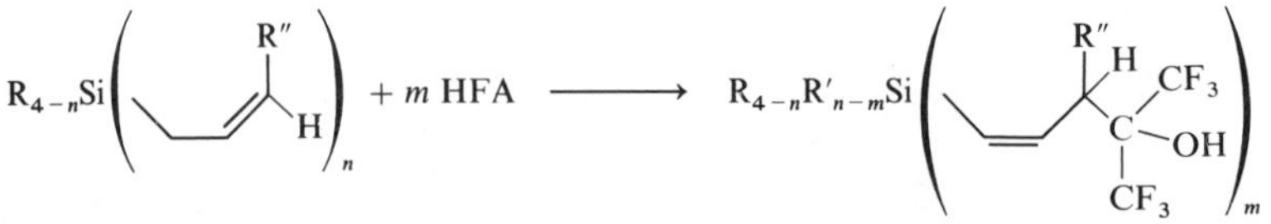

$$R_{4-n}Si\left(\underset{H}{\diagdown}\overset{R''}{\diagup}\right)_n + m\ HFA \longrightarrow R_{4-n}R'_{n-m}Si\left(\underset{}{\diagdown}\overset{R''}{\underset{CF_3}{\overset{H}{\diagup}\underset{C}{\diagup}}}\overset{CF_3}{\underset{OH}{}}\right)_m \qquad (13)$$

$$(12)$$

12	R	R'	R''	n	m
a	Ph		H	1	1
b	Me		Me	1	1
c	Me	CH_2CHCH_2	H	1-4	$\leq n$

Catalytic amounts of $AlCl_3$ yield, in addition to **12c** ($n = 1$), an insertion product into the Si—C bond (**14**) and an alcohol without isomerization of the double bond (**15**).

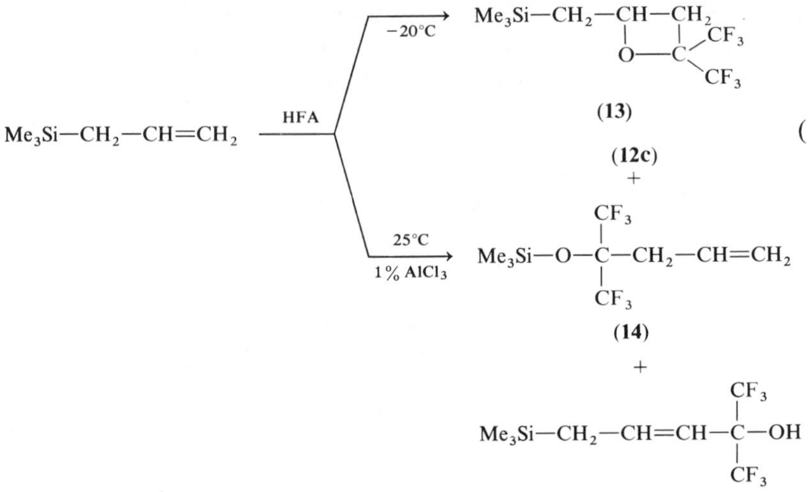

$$Me_3Si-CH_2-CH=CH_2 \xrightarrow{\ HFA\ } \quad (14)$$

$$\xrightarrow[]{-20°C} Me_3Si-CH_2-\underset{\underset{\displaystyle O-C}{|}}{CH}-\underset{\underset{\displaystyle CF_3}{\diagup}\underset{CF_3}{\diagdown}}{CH_2}$$

$$\textbf{(13)}$$

$$\textbf{(12c)}$$

$$+$$

$$\xrightarrow[1\%\ AlCl_3]{25°C} Me_3Si-O-\underset{\underset{\displaystyle CF_3}{|}}{\overset{\overset{\displaystyle CF_3}{|}}{C}}-CH_2-CH=CH_2$$

$$\textbf{(14)}$$

$$+$$

$$Me_3Si-CH_2-CH=CH-\underset{\underset{\displaystyle CF_3}{|}}{\overset{\overset{\displaystyle CF_3}{|}}{C}}-OH$$

$$\textbf{(15)}$$

Mechanisms of these reactions have been discussed in detail (4). Interestingly, the phenyl·substituted 2-butenylsilane failed to react with HFA even at 140°C. These observations have been explained by a mechanism involving a six-center intermediate with significant polar contribution to the transition state.

The reaction of HFA with substituted vinyltrimethylsilyl ethers in the presence of Lewis acids with subsequent hydrolysis provides a good route to alcohols containing the hexafluoroisopropyl group (148).

Cyclopentadienyltrimethylsilane yields the two isomeric alcohols **16a** and **16b**.

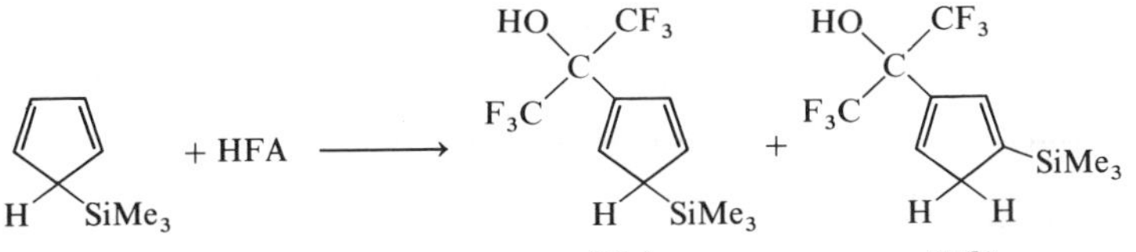

$$\text{(15)}$$

A series of carbonyl compounds including HFA has been found to react with a silirene with ring expansion to yield 1,2-oxasilolenes (*248a*).

5. Reactions with Si—S Bonds

HFA causes fission of Si—S bonds in acyclic (*109*) and cyclic (*6*) silthians.

$$\text{Me}_3\text{SiSR} + \text{HFA} \longrightarrow \text{Me}_3\text{Si}-\text{O}-\underset{\underset{\text{CF}_3}{|}}{\overset{\overset{\text{CF}_3}{|}}{\text{C}}}-\text{S}-\text{R} \qquad \text{(16)}$$

$$(\text{R} = t\text{-Bu}, \text{C}_6\text{F}_5)$$

$$\text{(17)}$$

$$[\text{Me}_2\text{SiS}]_n + \qquad \longrightarrow \quad n \, \text{Me}_2\text{Si} \qquad \qquad \text{(17)}$$

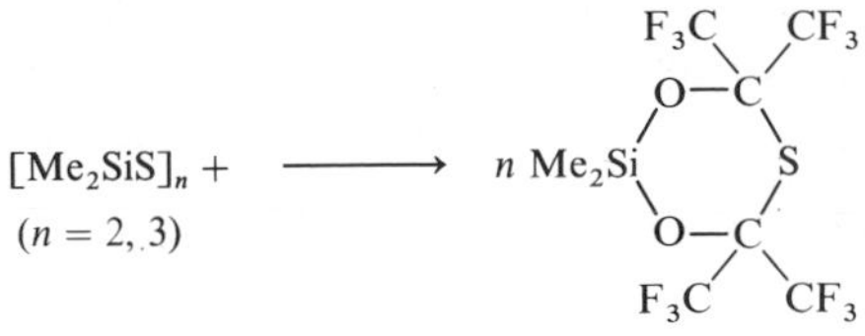

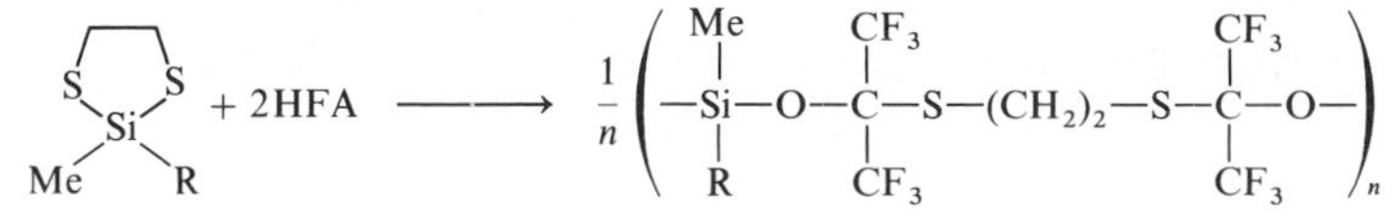

$$\text{(18)}$$

$$\text{(R = Me, CHCH}_2) \qquad\qquad\qquad \text{(19)}$$

$$\text{(18)}$$

In contrast to Eq. (18), the homologue dithiasilacyclohexane forms a monomeric heterocycle (**20**) (*6*).

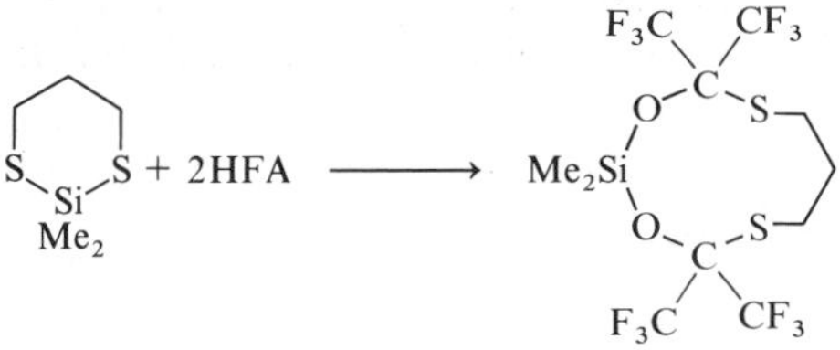

$$\text{(19)}$$

$$\text{(20)}$$

6. Reactions with Si—P and Si—As Bonds

The action of HFA on a silylphosphane has been reported (5). Both possible insertion modes have been found. Compound **21b** undergoes an intramolecular Arbuzov rearrangement, which is evident from the large $^{19}F-^{31}P$ and $^1H-^{31}P$ coupling constants (5).

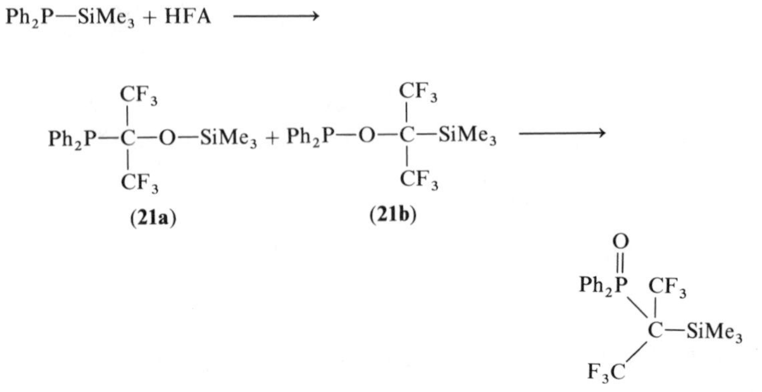

$$Ph_2P\text{—}SiMe_3 + HFA \longrightarrow$$

(20)

Similarly, permethylated silaarsanes with one to three arsenic atoms bound to silicon suffer bond cleavage, but only the silyl ethers analogous to **21a** are formed (3).

7. Miscellaneous

Bell and co-workers (24, 25) have investigated the generation of trifluoromethyl radicals from photolysis of HFA in the presence of silanes. Abstraction of the proton is observed in the case of trichlorosilane (24), while methyl(fluoro)silanes lead to the formation of CF_3H, C_2F_6, and $CF_2CH_2(25)$.

B. GERMANIUM AND TIN

The reactions of germanium and tin compounds with HFA are very similar to those of silicon compounds. But because of differences in polarity and bond strengths some reactions yield products different from those of their silicon analogues.

1. Insertion into E—H Bonds

The action of HFA on tin and germanium hydrides has been reported (73). In the case of tin, double insertion has been observed, the products of which can be cleaved by excess hydride to form the monoaddition product (73).

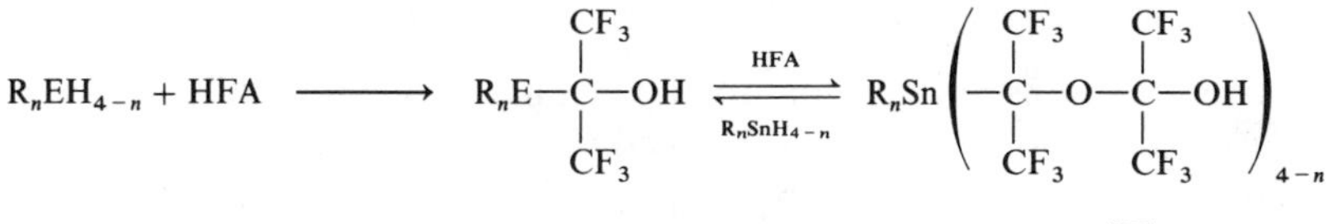

$$R_nEH_{4-n} + HFA \longrightarrow R_nE\underset{\underset{CF_3}{|}}{\overset{\overset{CF_3}{|}}{\underset{|}{C}}}\!-OH \underset{R_nSnH_{4-n}}{\overset{HFA}{\rightleftharpoons}} R_nSn\left(-\underset{\underset{CF_3}{|}}{\overset{\overset{CF_3}{|}}{C}}\!-O\!-\underset{\underset{CF_3}{|}}{\overset{\overset{CF_3}{|}}{C}}\!-OH\right)_{4-n}$$

(22)   (23)

(R = Me;   n = 2, 3)

(21)

E	R	n
Ge	Me	3
Sn	Me	2, 3
	Bu	3

The reversibility of the last step in Eq. (21) in contrast to Eq. (6) is further evidence for the ionic mechanism proposed by Janzen and Willis (158), since trialkylsilicon compounds form cations more easily.

2. Reactions with Ge—O Bonds

Reactions of open-chain and cyclic germoxanes and cyclic germadioxanes with HFA have been investigated.

Insertion of one molecule of HFA in the Ge—O bond of methoxytriethylgermane (87), hexaethyldigermoxane (87), some 1,2-oxagermetanes (22), and 1,2-oxagermolanes (22, 182) has been reported. 2,2-Diethyl-1,3,2-dioxagermolane can add either one or two molecules of HFA. The bis adduct 25a releases one molecule of HFA on heating (85).

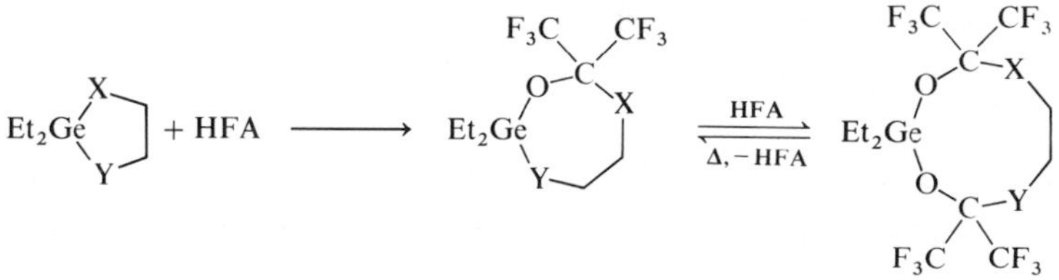

(24)   (25a)

(22)

24, 25	a	b	c
X	O	NMe	NMe
Y	O	O	NMe

3. Reactions with E—N Bonds

The same reactions have been found for N-methyl-substituted 2,2-diethyl-1,3,2-oxaazagermolidine (86) and 2,2-diethyl-1,3,2-diazagermolidine (172); dimethylamino(tri-n-butyl)germane and dimethylamino(trimethyl)stannane also undergo insertion into the E—N bond (2).

In analogy to the corresponding silicon compound, bis(tri-n-butyl)stannyl-carbodiimide adds only one molecule of HFA (102).

Whereas in Eq. (11) (vide supra) P=N bond breaking plays only a minor role, in the homologous trialkylstannyltriphenylphosphorane/HFA reaction, the ketimines 26 are the only detectable products which insert another HFA molecule into the Sn—N bonds (1).

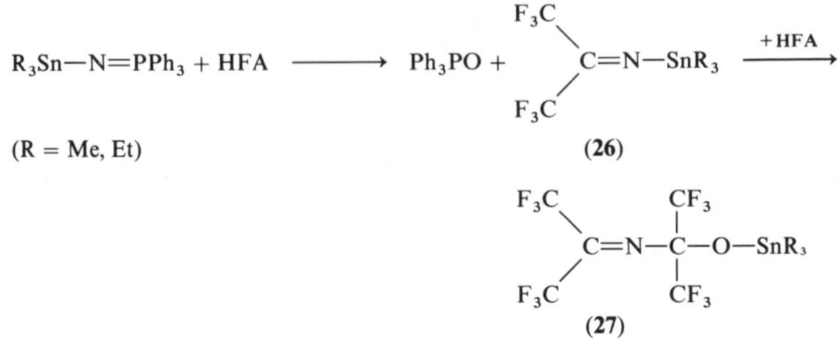

$$R_3Sn—N{=}PPh_3 + HFA \longrightarrow Ph_3PO + \underset{F_3C}{\overset{F_3C}{>}}C{=}N—SnR_3 \xrightarrow{\ +HFA\ }$$

(R = Me, Et)   (26)

$$\underset{F_3C}{\overset{F_3C}{>}}C{=}N—\overset{\overset{CF_3}{|}}{\underset{\underset{CF_3}{|}}{C}}—O—SnR_3 \qquad (23)$$

(27)

Although ketimines (e.g., 8, 26) generally have a low inversion barrier, the three ^{19}F NMR signals of 27 collapse only at 100°C to form two signals (1).

4. Reactions with Sn—C Bonds

In sharp contrast to the reactions of allyl- and cyclopentadienylsilanes with HFA [Eqs. (13)–(15)], where alcohol formation and double bond isomerization prevail, the corresponding tin systems exhibit only Sn—C bond rupture without shifting the double bond (4).

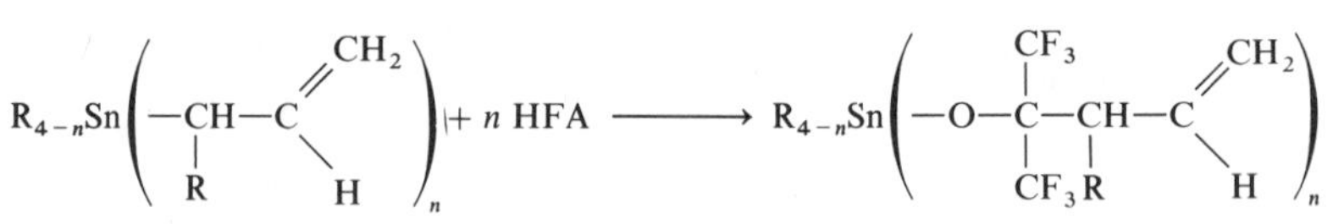

$$R_{4-n}Sn\left(\underset{R}{\overset{}{\underset{|}{-}CH}}-C\overset{\overset{\displaystyle CH_2}{\diagup\!\!/}}{\diagdown_H}\right)_n + n\,HFA \longrightarrow R_{4-n}Sn\left(-O-\underset{\underset{CF_3R}{|}}{\overset{\overset{CF_3}{|}}{C}}-CH-C\overset{\overset{\displaystyle CH_2}{\diagup\!\!/}}{\diagdown_H}\right)_n \qquad (24)$$

(28)

R	R'	n
Ph	H	1
Me	H	1, 2
Me	Me	1

An interesting skeletal rearrangement has been found in the triphenyl-2-butenylstannane/HFA system (4).

$$R_3Sn-CH_2-CH=CH-CH_3 + HFA \longrightarrow R_3Sn-O-\underset{\underset{CF_3}{|}}{\overset{\overset{F_3C}{|}}{C}}-\underset{\underset{CH_3}{|}}{CH} \overset{CH=CH_2}{\diagup}$$

(R = Me, Ph)   (29)

$$(25)$$

While cyclopentadienyltrimethylstannane forms two valence isomers under the influence of HFA, the double bond in 1-indenyltrimethylstannane is not shifted (4).

III. Reactions of HFA with Compounds of Group V Elements

A. PHOSPHORUS

HFA has been found to be a versatile reagent in phosphorus chemistry. Insertion reactions into P—E bonds (Section III,A,1, 2, 6, and 7) as well as oxidative ring formation and reactions in the coordination sphere of phosphorus (Section III,A,9) are described. Reactions involving an Si—P bond are discussed in Section II,A,6.

1. Insertion into P—H and P—O Bonds

The reaction of HFA with phosphanes has been reported by Bruker et al. (43) and reinvestigated by Röschenthaler (227).

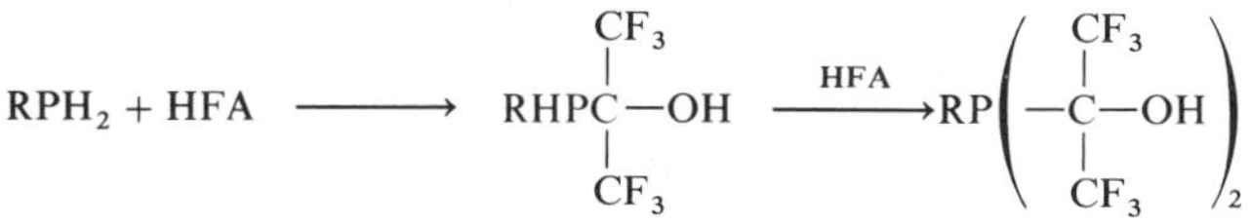

$$RPH_2 + HFA \longrightarrow RHP\underset{\underset{CF_3}{|}}{\overset{\overset{CF_3}{|}}{C}}-OH \xrightarrow{HFA} RP\left(\underset{\underset{CF_3}{|}}{\overset{\overset{CF_3}{|}}{-C}-OH}\right)_2 \qquad (26)$$

(R = H, Me)   (30)   (31a, R = H; 31b, R = Me)

Both observed the monoaddition products **30**, but the phosphanediols **31** have been found only by Röschenthaler (227). While the methylated compound **31b** decomposes above 45°C, the phosphane derivative **31a** has been found to be air-stable. Dimethylphosphane was claimed by Bruker and co-workers (43) to yield the alcohol **32a**, which could not be confirmed (227). Instead, the oxidation product **33a** has been found together with a diphosphane **34** and fluorophosphorane **35** (227).

$$\text{Me}_2\text{PH} + \text{HFA} \quad \left\langle \begin{array}{l} \longrightarrow \left[\begin{array}{c} \text{CF}_3 \\ | \\ \text{Me}_2\text{P}-\text{C}-\text{OH} \\ | \\ \text{CF}_3 \end{array} \right] \xrightarrow{\text{O}_2} \begin{array}{c} \text{O} \quad \text{CF}_3 \\ \| \quad | \\ \text{Me}_2\text{P}-\text{O}-\text{CH} \\ | \\ \text{CF}_3 \end{array} \\ \qquad\qquad (\mathbf{32a}) \qquad\qquad\qquad (\mathbf{33a}) \\[2em] \longrightarrow \mathbf{33a} + \text{Me}_2\text{PPMe}_2 + \text{Me}_2\text{P}\left(\begin{array}{c} \text{CF}_3 \\ | \\ \text{OCH} \\ | \\ \text{F} \quad \text{CF}_3 \end{array} \right)_{\!/2} \\ \qquad\qquad (\mathbf{34}) \qquad\qquad (\mathbf{35}) \end{array} \right. \qquad (27)$$

The primary addition product from HFA and diphenylphosphane (**32b**) has been unambiguously characterized by several groups (*97, 155*). It is easily oxidized by atmospheric oxygen and dinitrogen tetroxide (*97, 155*) to form phosphonous acid ester **33b** via phosphane oxide **36** (*155*), which can also be synthesized from HFA and diphenylphosphane oxide. Kinetics of the base-catalyzed rearrangement have been studied (*154*). Further action of HFA on **32b** yields difluorophosphorane **37** (*155*). A dipolar intermediate with tetra-coordinated phosphorus has been postulated by Stockel (*260*) in the formation of **33b** and **33c** (R = c-C$_6$H$_{11}$).

$$\text{Ph}_2\text{PH} + \text{HFA} \longrightarrow$$

$$\begin{array}{ccccc} \begin{array}{c} \text{CF}_3 \\ | \\ \text{Ph}_2\text{P}-\text{C}-\text{OH} \\ | \\ \text{CF}_3 \end{array} & \xrightarrow[\text{N}_2\text{O}_4]{\text{O}_2} & \begin{array}{c} \text{O} \quad \text{CF}_3 \\ \| \quad | \\ \text{Ph}_2\text{P}-\text{C}-\text{OH} \\ | \\ \text{CF}_3 \end{array} & \xrightarrow{\text{B}^-} & \begin{array}{c} \text{O} \quad \text{CF}_3 \\ \| \quad | \\ \text{Ph}_2\text{P}-\text{OCH} \\ | \\ \text{CF}_3 \end{array} \\ (\mathbf{32b}) & & (\mathbf{36}) & & (\mathbf{33b}) \end{array} \qquad (28)$$

$$\begin{array}{ccc} \Big\downarrow \text{HFA} & \Big\uparrow \text{HFA} & \Big\uparrow \\[1em] \begin{array}{c} \text{F} \quad \text{CF}_3 \\ | \quad | \\ \text{Ph}_2\text{P}-\text{C}-\text{OH} \\ | \quad | \\ \text{F} \quad \text{CF}_3 \end{array} & \begin{array}{c} \text{O} \\ \| \\ \text{Ph}_2\text{PH} \end{array} & \begin{array}{c} \text{CF}_3 \qquad \text{O} \\ | \qquad\quad \| \\ \text{HC}-\text{OH} + \text{Ph}_2\text{PCl} \\ | \\ \text{CF}_3 \end{array} \\ (\mathbf{37}) & & \end{array}$$

Compound **33b** is also accessible from hexafluoropropanol and chlorodi-phenyloxophosphorane (*97*). Thermal decomposition of **33b** yields tetra-phenyldiphosphane and hexafluoropropanol.

Depending on the substituents, different product distributions of phos-phonic **36d–f** and phosphoric acid esters **33d–f** have been found in the reaction of HFA and phosphinic acid esters (*152*).

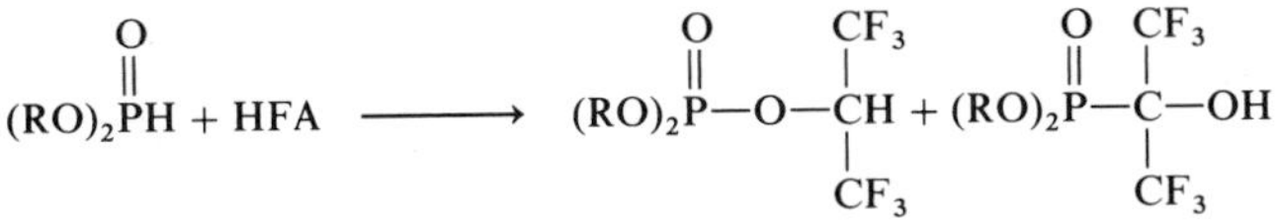

$$(RO)_2\overset{\overset{O}{\|}}{P}H + HFA \longrightarrow (RO)_2\overset{\overset{O}{\|}}{P}-O-\overset{\overset{CF_3}{|}}{C}H + (RO)_2\overset{\overset{O}{\|}}{P}-\overset{\overset{CF_3}{|}}{C}-OH \qquad (29)$$

(33c–e)　　　　　(36c–e)

YIELD
(%)

	R	33	36
d	Me	94	6
e	Et	88	12
f	Bu	5	95

Only O–addition products **33d, e, g,** and **h** have been found by Ivin *et al.* (*149*) in a series of reactions of HFA with phosphinates and phosphinites.

$$\overset{R'}{\underset{RO}{>}}\overset{}{P}-H + HFA \longrightarrow \overset{R'}{\underset{RO}{>}}P-O-\overset{\overset{CF_3}{|}}{C}H \qquad (30)$$

(33)

33	d	e	g	h
R	Me	Et	*i*-Pr	Et
R′	OMe	OEt	*i*-PrO	Me

Formation of **33i** (*37*) and **33j** (*104*) from oxidative rearrangement has been reported. Triethylamine has been employed as a catalyst in the formation of **33j.** No reaction has been found with $P(OCH(CF_3)_2)_3$ (*226*).

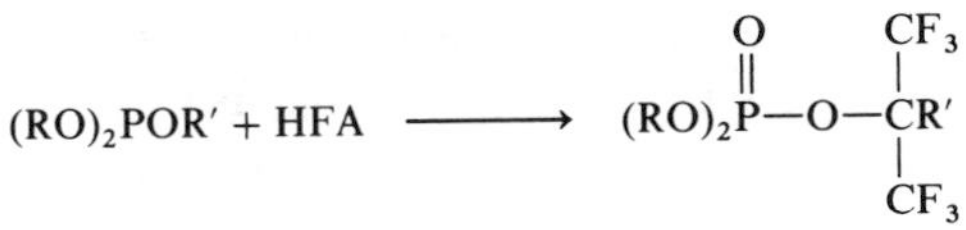

$$(RO)_2POR' + HFA \longrightarrow (RO)_2\overset{\overset{O}{\|}}{P}-O-\overset{\overset{CF_3}{|}}{C}R' \qquad (31)$$

(33)

	33i	33j
R	$-OCMe_2CMe_2O-/2$	$-OCH_2CF_2CF_2H$
R′	$-OC(O)-C_6H_2Me_3$	H

The reaction of a spirobicyclic phosphorane with HFA leads to the formation of the alcohol **38** (*113*).

$$
\begin{array}{c}
\text{Me}_2\text{C}\!-\!\text{O} \quad\ \text{O}\!-\!\text{CMe}_2 \\
\text{Me}_2\text{C}\!-\!\text{O}\!\diagdown\!\overset{|}{\underset{\text{H}}{\text{P}}}\!\diagdown\!\text{O}\!-\!\text{CMe}_2
\end{array}
+ \text{HFA} \longrightarrow
\begin{array}{c}
\text{Me}_2\text{C}\!-\!\text{O} \quad\ \text{O}\!-\!\text{CMe}_2 \\
\text{Me}_2\text{C}\!-\!\text{O}\!\diagdown\!\overset{|}{\text{P}}\!\diagdown\!\text{O}\!-\!\text{CMe}_2 \\
\text{F}_3\text{C}\!-\!\overset{|}{\underset{\text{OH}}{\text{C}}}\!-\!\text{CF}_3 \\
\textbf{(38)}
\end{array}
\tag{32}
$$

Insertion of HFA into P—O—P (*106*) and P—O—Si bonds (*161, 211*) of phosphites results in the formation of **39** and **40**. A mechanism involving a dipolar intermediate **40a** [Eq. (34)] has been discussed for the silicon compound **40** (*161, 211*).

$$
(\text{EtO})_2\text{P}\!-\!\text{O}\!-\!\text{P(OEt)}_2 + \text{HFA} \longrightarrow
(\text{EtO})_2\overset{\overset{\text{O}}{\|}}{\text{P}}\!-\!\text{O}\!-\!\overset{\overset{\text{CF}_3}{|}}{\underset{\underset{\text{CF}_3}{|}}{\text{C}}}\!-\!\text{P(OEt)}_2
\tag{33}
$$

$$
\textbf{(39)}
$$

$$
(\text{EtO})_2\text{P}\!-\!\text{O}\!-\!\text{SiMe}_3 + \text{HFA} \longrightarrow
\begin{array}{c}
\text{O}\!-\!\text{SiMe}_3 \\
(\text{EtO})_2\text{P}^{+}\quad\ \text{O}^{-} \\
\diagdown\ \underset{\text{C}}{\diagup} \\
\text{F}_3\text{C}\diagup\ \diagdown\text{CF}_3
\end{array}
\longrightarrow
$$

$$
\textbf{(40a)}
$$

$$
(\text{EtO})_2\overset{\overset{\text{O}}{\|}}{\text{P}}\!-\!\overset{\overset{\text{CF}_3}{|}}{\underset{\underset{\text{CF}_3}{|}}{\text{C}}}\!-\!\text{O}\!-\!\text{SiMe}_3
\tag{34}
$$

$$
\textbf{(40)}
$$

No spectroscopic evidence for the structure of **40** (e.g., P–F coupling constants) has been reported; the mechanism has been verified by Evans *et al.* (*98*) in the reaction of silyl phosphites with a wide variety of different ketones. At elevated temperatures C—O inversion has been postulated, and a vinyl phosphate **41** is formed with elimination of trimethylfluorosilane (*161*).

$$
\textbf{40} \;\overset{\Delta}{\rightleftharpoons}\; \textbf{40a} \;\overset{\ominus}{\longrightarrow}\;
\begin{array}{c}
\text{O}\!-\!\text{SiMe}_3 \\
\text{EtO})_2\text{P}^{+}\quad\overset{\uparrow}{\underset{|}{\text{F}}} \\
\overset{|}{\text{O}}\!\diagdown\!\!\overset{\frown}{\underset{\text{C}}{}}\!\diagup\!\text{CF}_2 \\
\overset{|}{\text{CF}_3}
\end{array}
\longrightarrow
(\text{EtO})_2\overset{\overset{\text{O}}{\|}}{\text{P}}\!-\!\text{O}\!-\!\text{C}\overset{\diagup\text{CF}_3}{\underset{\diagdown\text{CF}_2}{\big\|}}
\tag{35}
$$

$$
\textbf{(40b)} \qquad\qquad\qquad\qquad \textbf{(41)}
$$

Ketal formation is observed when a pentaoxophosphorane is reacted with HFA (*251*).

2. Reactions Involving P—N Bonds

The reaction of the aminophosphanes t-Bu_2PNH_2 and F_2PNH_2 results in the formation of hexafluoroacetoneimine and phosphoric acid derivatives with and without addition of HFA. With aminodifluorophosphane, the expected formation of a 1,3,2-dioxaphospholane (see Section III,A,4) has been observed (*263*).

Phosphoric acid ester amides suffer insertion into the N—H bond (*170*). Dehydration of **42** with trifluoroacetic acid anhydride and triethylamine yields the ketimide **43**. Decomposition of **42** at 170°C leads to recovery of the starting material.

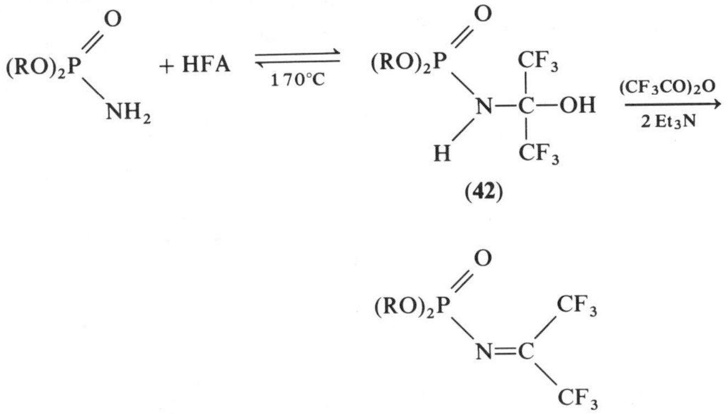

(36)

(R = Me, Et, Pr, i-Pr, Bu)   (**43**)

Silyl group migration has been observed in the reaction of silyl-substituted aminophosphanes to yield iminophosphoranes **44**. A mechanism has been discussed (*198, 201*).

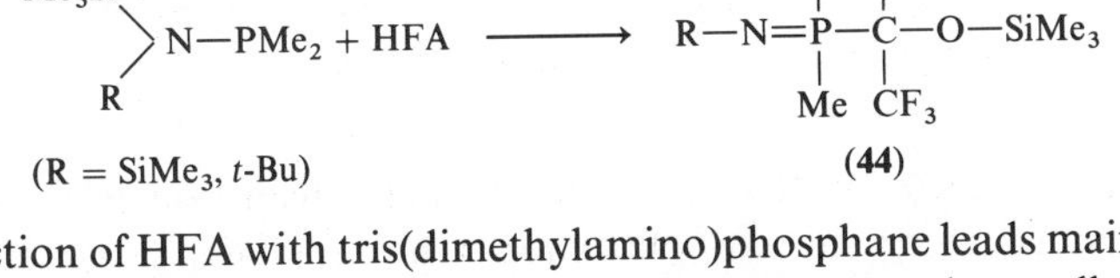

(37)

(R = $SiMe_3$, t-Bu)   (**44**)

The reaction of HFA with tris(dimethylamino)phosphane leads mainly to the formation of tris(dimethylamino)difluorophosphorane and a smaller amount of hexamethylphosphoric acid triamide, although indirect evidence for the formation of 1,3,2-dioxaphospholanes (Section III,A,4) has been found (*217*).

An oxaphosphirane **45** is accessible from HFA and a silylated amino-iminophosphane via [2 + 1] cycloaddition (*230*).

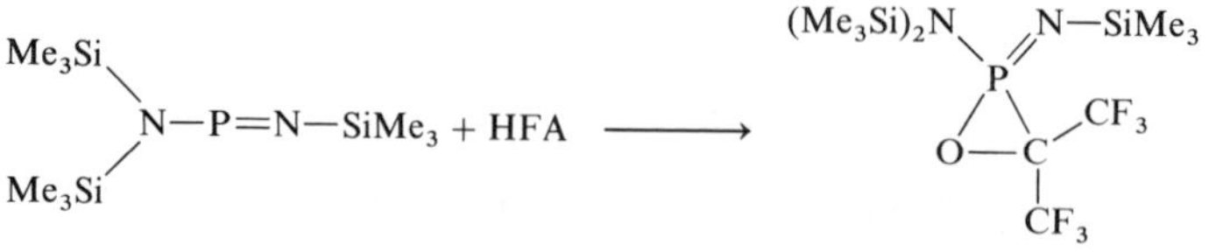

$$\begin{array}{c} Me_3Si \\ \diagdown \\ N-P=N-SiMe_3 + HFA \\ \diagup \\ Me_3Si \end{array} \longrightarrow \quad \begin{array}{c} (Me_3Si)_2N \quad N-SiMe_3 \\ \diagdown \diagup \\ P \\ \diagup \diagdown \\ O-C \diagdown CF_3 \\ | \\ CF_3 \end{array} \qquad (38)$$

(45)

A series of oxaazaphospetidines **46**, **49**, and **51** has been synthesized, starting from primary and secondary P(III) amines. The reaction proceeds by hydrogen migration with intermediate formation of iminophosphoranes **47** (*91, 263*), from activated iminophosphoranes (*246, 250*) and cyanates and thiocyanates (*91, 176, 239*), with preceding [3 + 2] cycloaddition involving formation of a phosphorus–nitrogen double bond. Although pseudohalides of phosphorus are treated in a subsequent section, we include these examples in this section due to their analogy to the present topic.

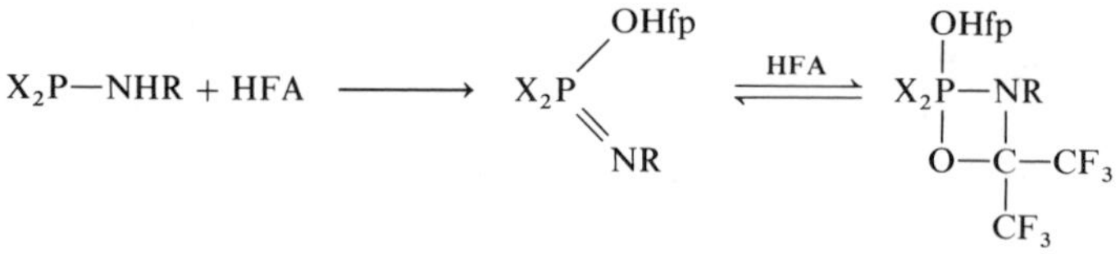

$$X_2P-NHR + HFA \longrightarrow X_2P\begin{array}{c} OHfp \\ \diagup \\ \diagdown \\ NR \end{array} \xrightleftharpoons{HFA} \begin{array}{c} OHfp \\ | \\ X_2P-NR \\ | \ | \\ O-C-CF_3 \\ | \\ CF_3 \end{array}$$

(47)   **(46)**

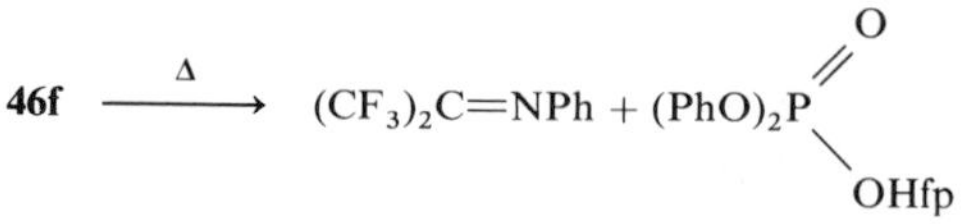

$$\textbf{46f} \xrightarrow{\Delta} (CF_3)_2C=NPh + (PhO)_2P\begin{array}{c} \diagup\!\!\!\!\diagup O \\ \diagdown \\ OHfp \end{array} \qquad (39)$$

(48)   **(33k)**

46, 47	a, b	c, d	e	f
R	H, Me	Me, *t*-Bu	H	Ph
X	OHfp	F	Pfp/$_2$	OPh
Reference	*106a*	*106a*	*263*	*91*

The X-ray structures of **46a** and **e** show slight distortion from trigonal-bipyramidal geometry at the phosphorus atom (*106a, 263*). The axial positions are occupied by oxygen atoms; the oxygen–phosphorus distances are relatively short due to the electron-withdrawing effect of the CF_3 groups (*263*).

Thermal decomposition of **46f** at 130°C yields **48**, **33k**, and **47f** in reversal of its formation (91).

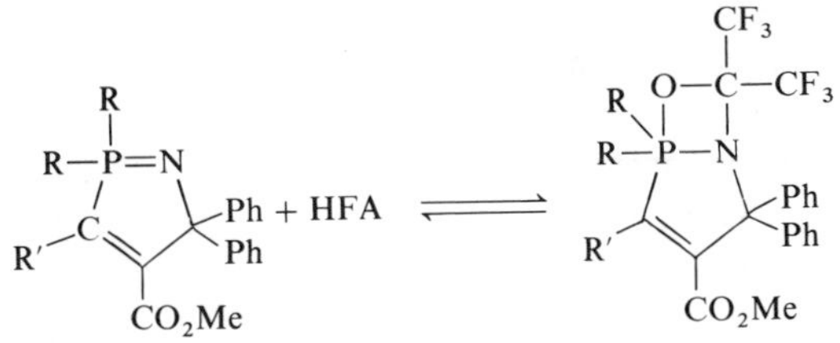

$$ \tag{40} $$

49	a	b	c
R	Ph	Me	Ph
R'	H	CO$_2$Me	CO$_2$Me

The adducts **49a–c** are crystalline solids, which in solution are in equilibrium with their precursors (246, 250). The X-ray structure of **49b** has been reported (250).

Similar bicyclic phosphoranes **51** are formed in the reactions of phosphorus isocyanates and isothiocyanates with HFA (91, 176, 239) via intermediate formation of oxazaphospholinones and -thiones **50**.

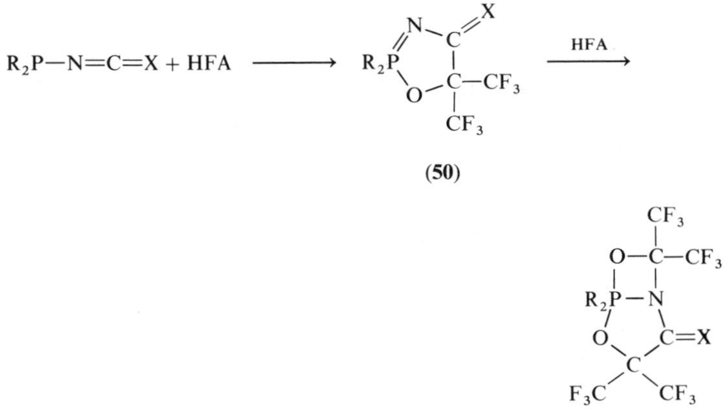

$$ R_2P-N=C=X + HFA \longrightarrow \quad (50) \quad \xrightarrow{\text{HFA}} $$

$$ \tag{41} $$

50, 51	a	b	c	d	e	f
R	OEt	OEt	OPh	NCO	NCS	OMe
X	O	S	O	O	S	O
Reference	91	91	91	239	239	166

Pudovik and co-workers (*166*) found the reaction [Eq. (41)] to stop at the stage of the five-membered ring **50f** when excess dimethoxycyanatophosphane reacts with HFA. The same bicyclic system (**51g**, R = Ph, X = S) has been found in the reaction of the $Hg(SCN)_2$—HFA adduct **192** (Section V, F) with diphenylchlorophosphane (*241, 242*).

The reaction of monoammonium perfluoropinacolate with 2-amino-1,3,2-dioxaphospholane (**52**) provides a spirocyclic phosphorane **53** with hydrogen attached to phosphorus, with evolution of ammonia occurring (*263*).

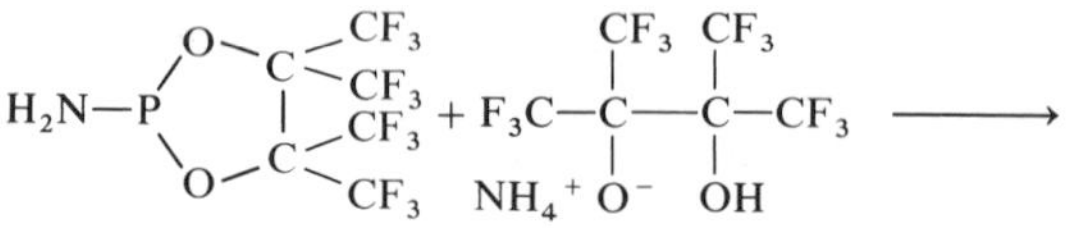

(**52**)

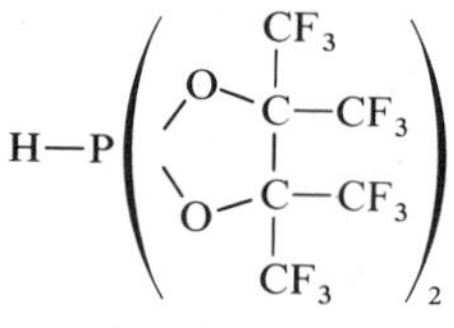

(42)

(**53**)

P-substituted derivatives **53a–g** have been obtained by metathesis of phosphorus(V) dihalides and dilithium perfluoropinacolate (see Section III,A,6). A similar compound, in which one ring bears methyl groups, has been prepared in an analogous fashion (*32*).

3. 1,3,4-Dioxaphospholanes

In contrast to trivalent phosphorus compounds with activated bonds, where insertion occurs, phosphanes bearing only alkyl, aryl, alkoxy, aryloxy, secondary amino, thio, and halogeno ligands are oxidized by HFA. Earlier work was reviewed by Ramirez (*216*) in 1970 and Hellwinkel (*139*) in 1972. The following scheme [Eq. (43)] shows all reactions occurring in the phosphane/HFA system.

The primary intermediates (dipolar 1:1 adducts **54**) can add another molecule of HFA to form five-membered rings. 1,3,4-λ^5-Dioxaphospholanes **56** are the kinetically favored products from reactions of trivalent phosphorus compounds with HFA, but normally rearrange below ambient temperatures to yield the thermodynamically stable 1,3,2-λ^5-dioxaphospholanes **57**.

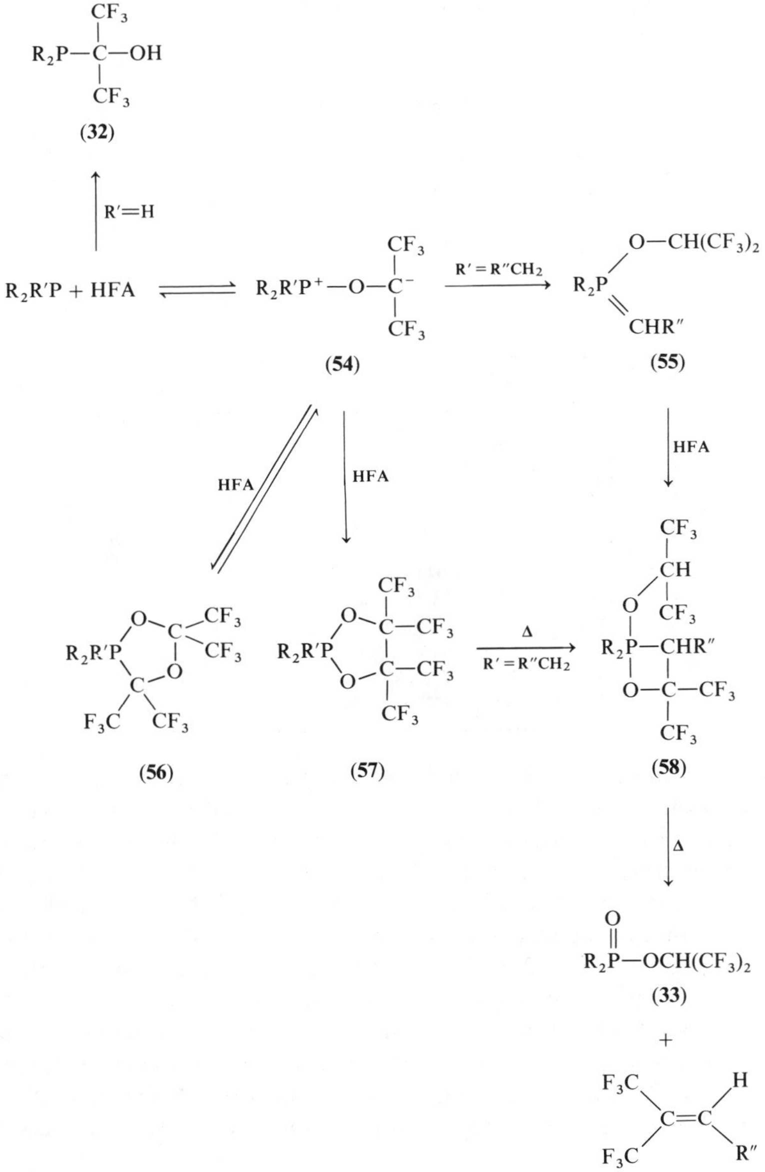

$$R_2P-\underset{\underset{CF_3}{|}}{\overset{\overset{CF_3}{|}}{C}}-OH$$

(32)

$\uparrow R'=H$

$$R_2R'P + HFA \rightleftharpoons R_2R'P^+-O-\underset{\underset{CF_3}{|}}{\overset{\overset{CF_3}{|}}{C}}^- \xrightarrow{R'=R''CH_2} R_2P\overset{O-CH(CF_3)_2}{\underset{CHR''}{\diagdown}}$$

(54)   (55)

HFA   HFA   HFA

(56)   (57) $\xrightarrow[R'=R''CH_2]{\Delta}$ (58)

$$R_2\overset{\overset{O}{\|}}{P}-OCH(CF_3)_2$$

(33)

+

$$\overset{F_3C}{\underset{F_3C}{\diagup}}C=C\overset{H}{\underset{R''}{\diagdown}}$$ (43)

A series of stable heterocycles **56** is listed in Table I.

TABLE I

1,3,4-λ^5-DIOXAPHOSPHOLANES STABLE AT ROOM
TEMPERATURE

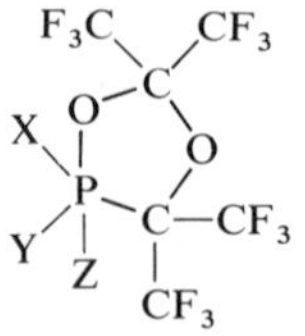

(56)

56	X	Y	Z	Reference
a	Me	Me	F	*117, 226*
b	Me	Me	Cl	*117*
c	Me	F	F	*226*
d	Et	Et	Cl	*278*
e	Pr	Pr	Cl	*9, 278*
f	Ph	CMe$_2$CMe$_2$CH$_2$		*89, 204*
g	Ph	SCH$_2$CH$_2$O		*90*
h	OPh	OCH$_2$CH$_2$O		*32*
i	Cl	CMe$_2$CHMeCMe$_2$		*32*
k	Me	NMeC(O)NMe		*282, 283*
l	NMe$_2$	o-C$_6$H$_4$CH$_2$O		*74*

The fluoro-substituted compounds **56a** and **c** are the only examples where all three isomers **56**, **57**, and **58** are formed together under identical conditions (*226*). Compounds **56a–e** are converted into the corresponding oxaphosphetanes **58** by gentle heating. The isomeric five-membered rings **57** are also formed in the case of **56a**, and **c**; **56f** and **h** behave similarly. The spirophosphoranes **56g** and **56k** show remarkable stability: Decomposition occurs only at 150°C (*90, 283*). The structure of **56g** with two oxygen atoms occupying the axial positions has been confirmed by X-ray diffraction (*89*). Compound **56i** slowly regenerates the starting materials at room temperature (*32*). Ligand exchange reactions with **56c** yield derivatives otherwise inaccessible (*9, 278*). Compound **56m** can be synthesized from diethylchlorophosphane and cesium perfluoropropylate at room temperature [Eq. (44)], as well

$$Et_2PCl + 2CsOCF(CF_3)_2 \longrightarrow \quad\quad + CsCl + CsF \quad (44)$$

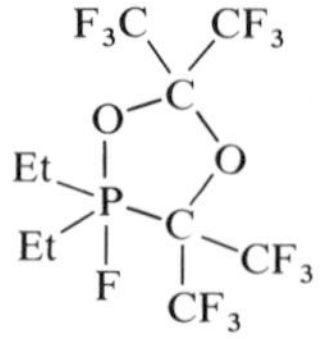

(56m)

as by halogen exchange in **56d** with CsF (*278*). The *n*-propyl derivative **56e** slowly rearranges at ambient temperature to form **59** (*278*).

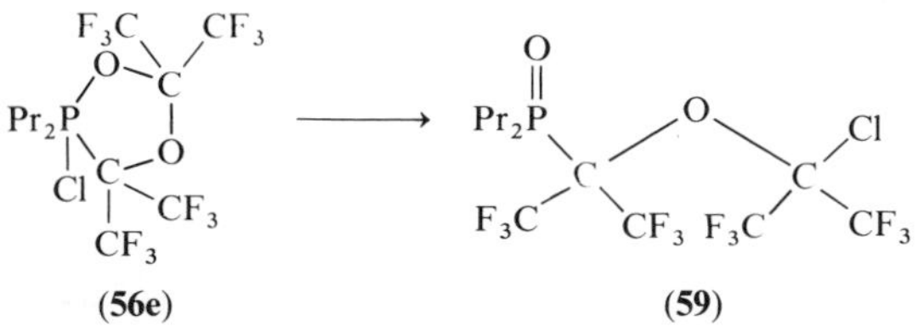

(45)

(56e)   (59)

4. 1,3,2-Dioxaphospholanes

The majority of the publications in the field of oxidative addition of HFA to tertiary phosphanes deal with the formation and properties of 1,3,2-λ^5-dioxaphospholanes **57** (*60, 61*). These compounds are generally stable when the α carbon atoms do not bear hydrogen atoms. Heterocycles with α hydrogen atoms are stable up to about 70°C; at this temperature the isomeric phosphetanes are formed. The syntheses of some sterically hindered phosphoranes require drastic conditions.

The compounds **57** (*60, 61*) formed according to Eq. (43) are listed in table II; the substituent Z normally occupies the second axial position, and the perfluoropinacolyl bridge has—with exceptions—axial-equatorial conformation. Spirocyclic phosphoranes **60** are shown in Table III.

The reaction of HFA with 2,2,3,4,4-pentamethyl-λ^3-phosphetane has been investigated (*204*). A crystalline 2:1 adduct has been obtained. NMR studies indicate two isomers of the dioxaphospholane **60e** relative to the position of the 3-methyl group. Compound **60e** slowly decomposes to yield diastereomeric phosphites **62** with the same isomeric composition. Further action of HFA affords **60o** (*204*).

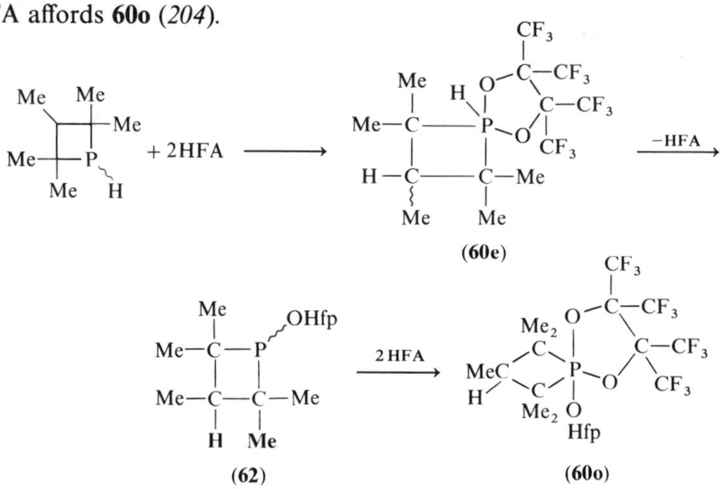

(60e)

(62)   (60o)

(46)

TABLE II

1,3,2-λ^5-DIOXAPHOSPHOLANES

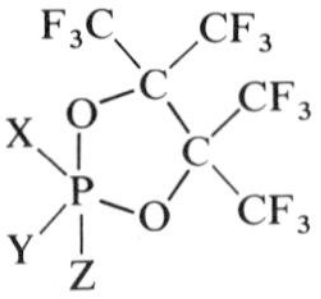

$$F_3C \quad CF_3$$

(57)

57	X^a		Y		Z	References
	i. X	**=**	**Y**	**=**	**Z**	
a–j	Me, Et, Bu, Ph, Tpo, OMe, OEt,					97, 104, 110–1
	OPh, OCH$_2$CF$_2$CF$_2$H					151, 162, 21
						221–223, 24
						259, 262
	ii. X	**=**	**Y**	**≠**	**Z**	
k–m, be, bf		Me			Ph, F, OSiMe$_3$	222, 226, 277
					Cl, N$_3$	
n, o		Et			Ph, OMe	97, 218, 222, 2
p, q		t-Bu			Cl, OHfp	75
r–u		Ph			OEt, OHfp,	97, 151, 221
					Hfp, Cl	
v, w		OMe, Oi-Pr			F	88, 119
x, y		OEt, NEt$_2$			Cl	278
	iii. X	**≠**	**Y**	**=**	**Z**	
z–ai	Me, Et, t-Bu,				F	75, 88, 92,
	Ph, OMe, NEt$_2$,					115, 116, 11
	NAll$_2$, NTmc$_2$,					121, 232
	N(t-Bu)SiMe$_3$,					
	N(SiMe$_3$)$_2$					
aj–am	Me				Ph, Tpo, OMe, OHfp	97, 218, 226, 2
an–ao	Et				Ph, OMe	97, 218
ap–as	Ph				OMe, OBuO,	97, 151, 212,
					OHfp, OPh	218, 222, 22
at–au	NMe$_2$, NEt$_2$				OHfp	107
av	C(CF$_3$)$_2$OP(O)(OEt)$_2$				OEt	106
	iv. X	**≠**	**Y**	**≠**	**Z**	
aw	Me		—(CH$_2$)$_3$—		OSiMe$_3$	166
ax, ay	Me		t-Bu		OHfp, Cl	68
az	Me		OEt		Cl	184
ba	t-Bu		OHfp		F	57
bb, bc	Ph, C$_6$H$_4$-p-Me		NEt$_2$		F	68
bd	OEt		NEt$_2$		Cl	184
bg–bi	t-Bu		NEt$_2$		F, Cl, OHfp	107a

^{a} Tpo, 4-oxy-3,3,5,5-tetramethylpiperidyl; All, allyl; Tmc, 2,2,6,6-tetramethylcyclohexyl.

TABLE III

SPIROCYCLIC PHOSPHORANES[a]

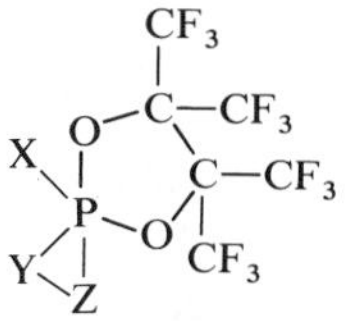

(60)

60	X	Y–Z	Reference
a	Ph	$CH_2CH_2CH_2$	204
b	Ph	$CH_2CMe_2CH_2$	204
c, d	Ph, C_6H_4-p-Br	$CMe_2CH_2CMe_2$	89, 144
e–t	H, Me, i-Pr, CH=CMe_2, t-Bu, 2-C_4H_4O, Ph, C_6H_4-p-Br, C_6H_4-p-OH, Cl, OHfp, OPh, NMe_2, N(CH_2)_4, N(i-Pr)_2, NMePh	$CMe_2CHMeCMe_2$	89, 204, 205, 266
u, v	OPh, SPh	OCH_2CH_2O	32
w–aa	OPh, OC_6H_4-p-Br, NMe_2, SPh, SePh	$OCMe_2CMe_2O$	32, 159
ab–ad	OPh, OC_6H_4-p-Br, NMe_2	o-OC_6H_4O	32
ae–ag	Ph, OPh, NMe_2	o-OC_6H_3ClO	32
ah	Ph	$OC(CF_3)_2C(CF_3)_2O$	231
ai	OPh	OCH_2CH_2NMe	33
aj–al	OPh, NMe_2, SPh	$NMeCH_2CH_2NMe$	32, 33, 217
am	OC_6H_4-p-Br	SCH_2CH_2O	33
an	OC_6H_4-p-Br	o-SC_6H_4O	33
ao	OPh	$O(CH_2)_3O$	33
ap	OC_6H_4-p-Br	$O(CH_2)_3NMe$	33

[a] Y–Z member of a ring, Z axial.

In addition to the compounds **57**, **60**, and **61** listed in Tables II–IV, further 1,3,2-λ^5-dioxaphospholanes are accessible via halogen exchange reactions (*106, 115, 229, 230, 232, 263, 263a, 277*) (*vide infra*), oxidative addition of α,β-diketones to λ^3-dioxaphospholanes (*32*), hydrolysis of silylamino compounds with HCl to yield unsubstituted amines (*121*), and thermal elimination of Me_3SiF from the silylaminofluoro derivatives **57ah** and **57ai** (*115, 118*).

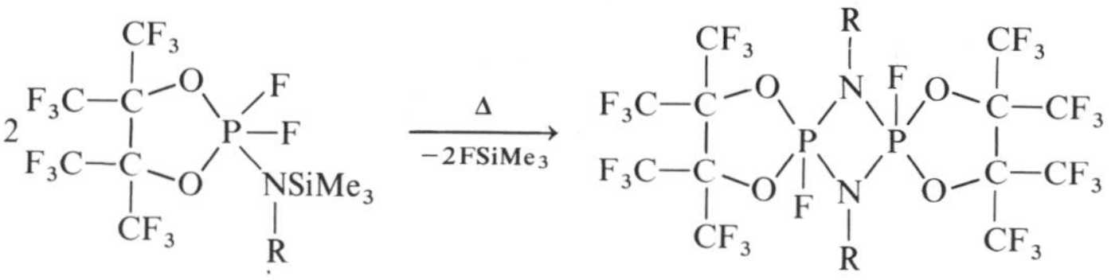

(57ah, 57ai)   (60aq, R = t-Bu; 60ar, R = SiMe_3)

$$\tag{47}$$

TABLE IV
PHOSPHORANE CAGE COMPOUNDS

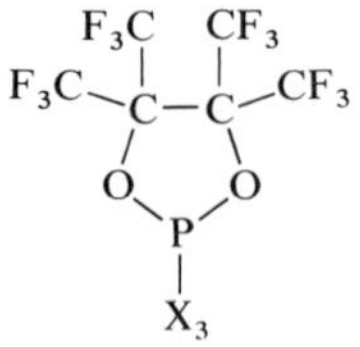

(61)

PX$_3$			Reference
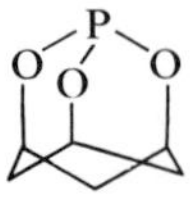 (61a) 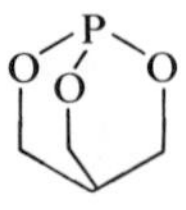	(61b) 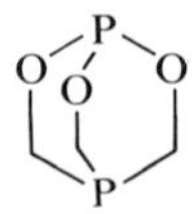	(61c)	*219, 223, 268, 269*

The X-ray structure of **60ar** has been reported and shows all rings having axial-equatorial conformation. This geometry is retained through pseudorotation processes, as shown from NMR studies (*118*). The chlorinated *t*-butyl-substituted derivative **60as** has been obtained similarly (*263a*) (*vide infra.*). Two isomers of the tricyclic system **60at**, in which phosphorus and nitrogen bear methyl groups, arise from photochemical nitrogen evolution via a tetracoordinated intermediate from azide **57bf** (*19a*).

The action of SOCl$_2$ on the bridged diphosphorane **57aw** yields a novel spirocyclic system (**60au**). The structure has been determined by X-ray analysis (*247*). All axial positions are shown to be occupied by oxygen atoms.

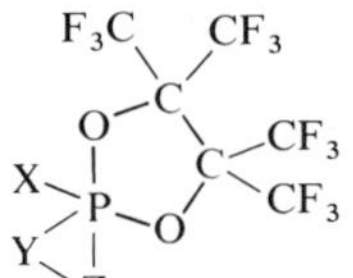

(60)

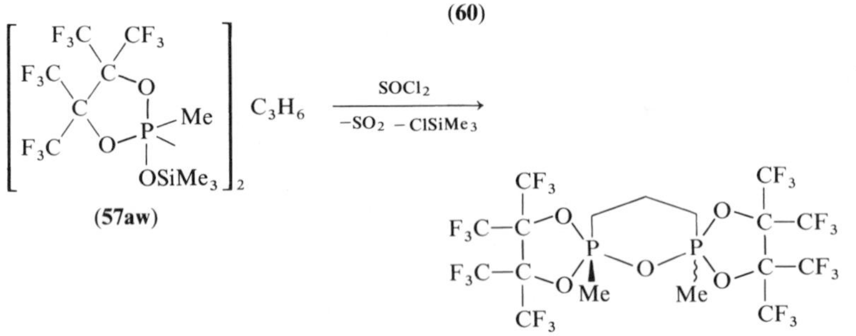

$$\left[\text{57aw structure} \right] C_3H_6 \xrightarrow[-SO_2 \ -ClSiMe_3]{SOCl_2} \text{(60au)} \tag{48}$$

(57aw)

(60au)

Four-coordinated phosphorus atoms are found in **63a** (*230*) [Eq. (49)] and **63b** (*110*) [Eq. (50)].

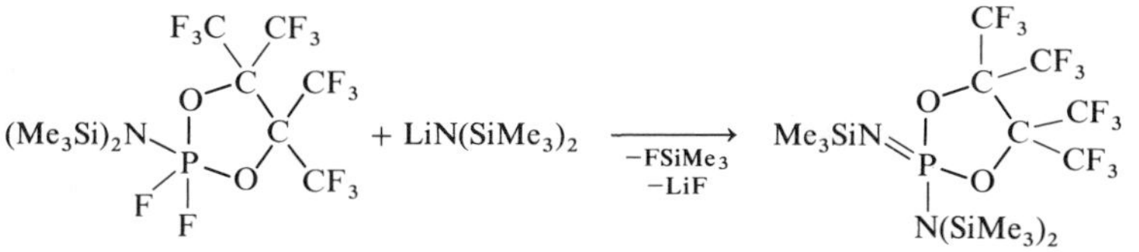

$$(Me_3Si)_2N-\underset{(57ah)}{\underset{F}{\overset{}{P}}}\cdots + LiN(SiMe_3)_2 \xrightarrow[-LiF]{-FSiMe_3} Me_3SiN\cdots\underset{(63a)}{\overset{}{P}}\cdots N(SiMe_3)_2 \qquad (49)$$

Hydrolysis of **57g** yields the phosphoric acid ester **63b**, which on exposure to air forms the orthophosphoric acid **57bj**. Prolonged heating of **57bj** affords perfluoropinacol (*110*).

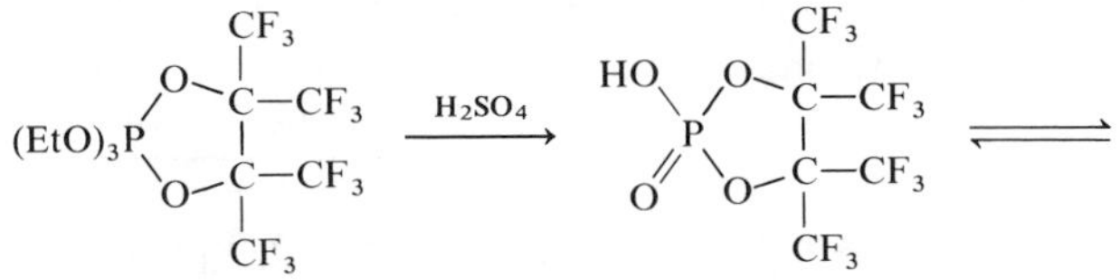

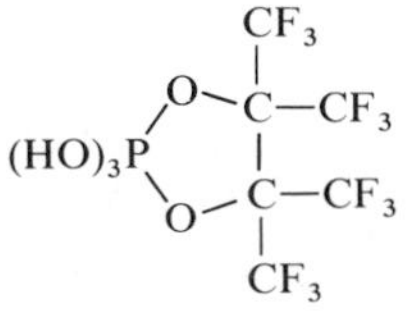

$$(EtO)_3P\underset{(57g)}{\cdots} \xrightarrow{H_2SO_4} \underset{(63b)}{\cdots} \rightleftharpoons$$

$$(HO)_3P\underset{(57bj)}{\cdots} \qquad (50)$$

The hydrolysis of various 1,3,2-dioxaphospholanes has been studied (*151, 212*) and is found to be a multistep process having a preliminary equilibrium with negative heat of reaction. Hydrolysis of **57m** proceeds via ring opening (*277*).

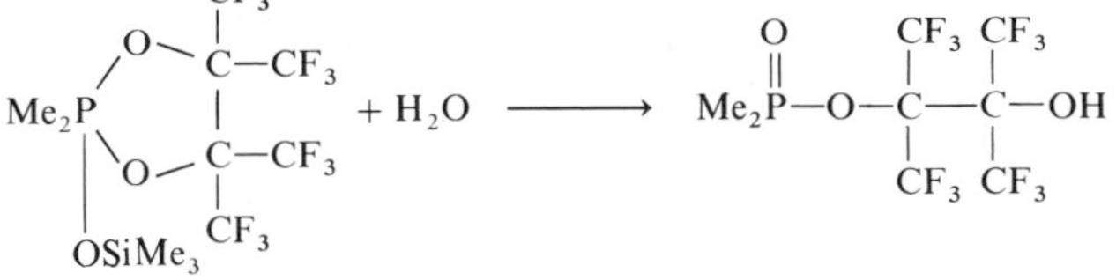

$$Me_2P\underset{\substack{| \\ OSiMe_3}}{\cdots} + H_2O \longrightarrow Me_2\overset{O}{\overset{||}{P}}-O-\underset{\underset{CF_3}{|}}{\overset{\overset{CF_3}{|}}{C}}-\underset{\underset{CF_3}{|}}{\overset{\overset{CF_3}{|}}{C}}-OH \qquad (51)$$

$$\underset{(57m)}{} \qquad \underset{(64)}{}$$

Different behavior has been observed on heating dixoaphospholanes. With a few exceptions compounds bearing hydrogen on a carbon atom adjacent to phosphorus rearrange to 1,2-oxaphosphetanes (see Section III,A,5). At 160°C, **57d** (X = Y = Z = Ph) dissociates into the starting materials (*220*). Compound **57g** (X = Y = Z = OEt) loses one molecule of HFA at 165°C and forms several decomposition products (*220*)

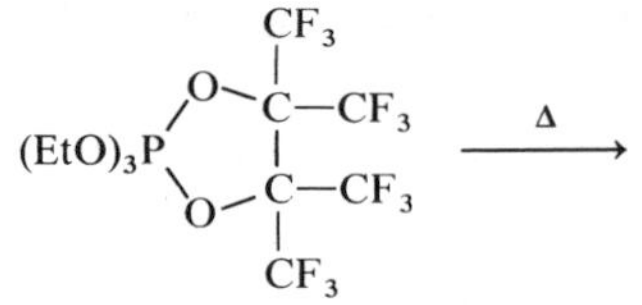

$$(EtO)_3P \underset{O}{\overset{O}{\diagdown}} \begin{array}{c} CF_3 \\ | \\ C-CF_3 \\ | \\ C-CF_3 \\ | \\ CF_3 \end{array} \xrightarrow{\Delta}$$

(57g)

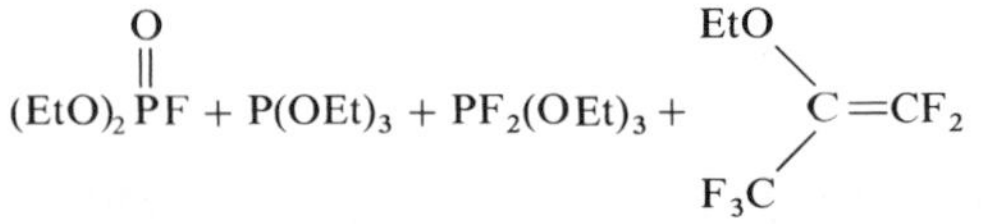

$$(EtO)_2 \overset{O}{\overset{||}{P}}F + P(OEt)_3 + PF_2(OEt)_3 + \underset{F_3C}{\overset{EtO}{\diagdown}}C=CF_2 \quad (52)$$

Pseudorotation processes have been investigated by means of ESR in the case of the paramagnetic Tpo derivatives **57e** and **57ak** (*262*) and dynamic ^{19}F NMR studies. Values of $\Delta G^{\ddagger}$ have been determined for a number of dioxaphospholanes and are strongly dependent on substitution. Small alkyl and amino substituents (also cyclic) provide low isomerization barriers ($\Delta G^{\ddagger} \approx 40$ kJ mol^{-1}); the CF$_3$ groups become equivalent at $<0°C$ (*159, 204, 220, 223, 266*). Steric hindrance and conformational rigidity make pseudoro­tation more difficult: in a series of spiro- and polycyclic derivatives (*60, 61*) $\Delta G^{\ddagger}$ values of >90 kJ mol^{-1} have been found, as manifested in the nonequivalence of the CF$_3$ groups at 180°C (*32, 89, 204, 223, 268, 269*). Exchange of O and S in bridging phenylene ligands in **60ac,an** has no effect on the $\Delta G^{\ddagger}$ value (*33*). ^{13}C NMR spectroscopy of difluorodioxaphospholanes **57z**, **57ab–ad**, **57ai**, and **74g** have been reported (*116*).

In contrast to the trigonal-bipyramidal geometry of most derivatives, the spirophosphoranes **60d** and **60k** have a square-planar structure, as seen from X-ray analysis (*144*).

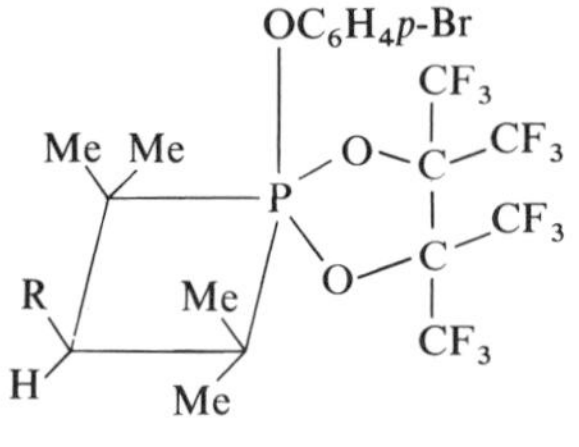

$$\begin{array}{c} OC_6H_4p\text{-Br} \\ | \\ CF_3 \\ | \\ \text{Me Me} \quad O\diagdown C - CF_3 \\ \diagup \ P \diagdown \quad | \\ R \quad Me \quad O\diagup C - CF_3 \\ \diagup \quad | \\ H \quad Me \quad CF_3 \end{array}$$

(R = H, Me)   (60d, k)

5. 1,2-Oxaphosphetanes

According to Eq. (43) the betaine **54** has a second possibility of stabilization. This is hydrogen migration from an α carbon to the carbonyl carbon bearing the negative charge with formation of a P=C double bond. The so-formed species (**55**) cannot be compared to stable Wittig reagents, which normally (cf. Section III,A,9) are not reactive toward HFA. No examples in which these intermediates (55) have been isolated are given in the literature. [2 + 2] Cycloaddition with HFA yields 1,2-λ^5-phosphetanes 58 with the oxygen ring atom always in an axial position.

A second access to these four-membered ring systems is the action of alkyl halides on the adduct of diphenylphosphane with HFA (32b) [Eq. (53)] (96).

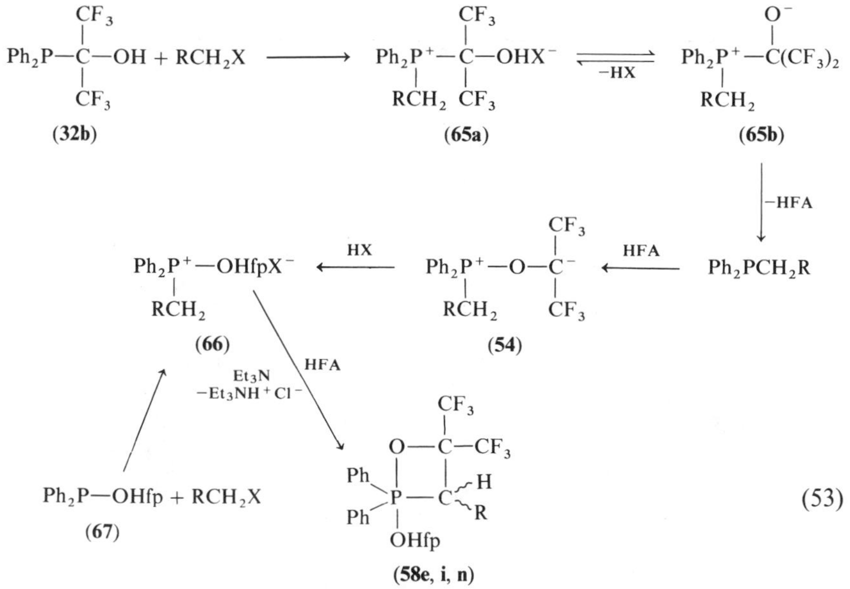

(53)

58	e	l	n
R =	H	Me	Ph
X =	I	I	Br

The extremely stable phosphonium salts 66 have been synthesized via the alternate route using diphenyl phosphite 67, which is accessible from diphenylchlorophosphane and hexafluoroisopropanol, thus proving the last step of the reaction (96).

Phosphetanes available either from phosphanes and HFA or by thermal ring contraction of the corresponding dioxaphospholanes 56, 57, and 60 are listed in Table V.

With X ≠ Y and R ≠ H, formation of isomers has been observed; equilibration occurs on heating. Pseudorotation involves a trigonal-bipyramidal intermediate with the four-membered ring in the diequatorial position (221).

TABLE V

1,2-λ^5 − Oxaphosphetanes

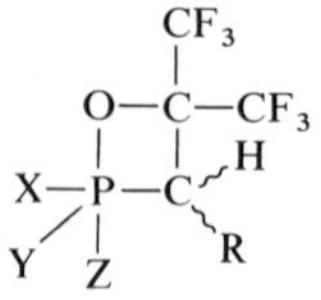

$$\begin{array}{c}
CF_3 \\
| \\
O-C-CF_3 \\
|\quad|\quad/H \\
X-P-C \\
/\ |\quad\backslash \\
Y\ \ Z\qquad R
\end{array}$$

(58)

58	X	Y	R	Reference
i.		**Z = OHfp**		
a	Me	Me	H	*69, 221*
b	Me	Ph	H	*69*
c	Me	OHfp	H	*226*
d	*t*-Bu	OHfp	H	*75, 92*
e	Ph	Ph	H	*69*
f	OMe	OMe	H	*69*
g	OHfp	OHfp	H	*226*
h	Et	Et	Me	*69, 218, 221*
i	Et	Ph	Me	*69, 218, 221*
j	Et	OMe	Me	*69, 218*
k	Et	OHfp	Me	*9*
l	Ph	Ph	Me	*69, 221*
m	*n*-Bu	*n*-Bu	*n*-Pr	*69*
n	OMe	OMe	Me	*69, 218*
	Y	Z	R	
ii.		**X = OHfp**		
o, p	Me	F, Cl	H	*117, 226*
q	*t*-Bu	Cl	H	*75, 92*
r, s	CMe$_2$CHMeCMe$_2$		H, Ph	*204*
t	F	F	H	*226*
u	Et	Cl	Me	*9*
v	OMe	OMe	Me	*69, 218*

Two fused four-membered rings are the main feature of the compounds **68a–c** (*89, 204*).

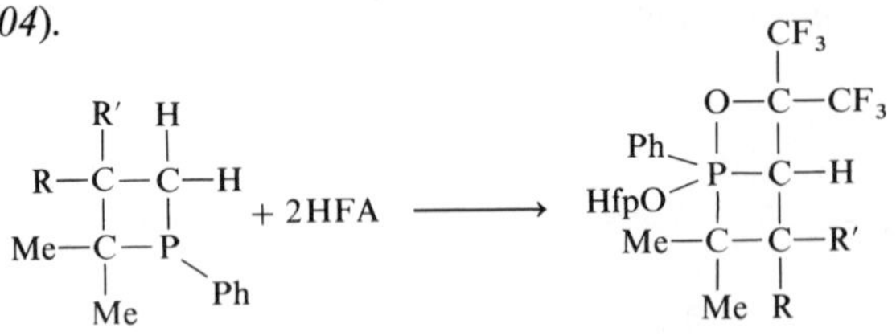

$$\begin{array}{c}
R'\ \ H \\
|\quad\ | \\
R-C-C-H \\
|\quad\ | \\
Me-C-P \\
|\quad\ \backslash \\
Me\quad Ph
\end{array}
\ + 2\,HFA \longrightarrow
\begin{array}{c}
CF_3 \\
| \\
O-C-CF_3 \\
Ph\diagdown\ |\quad| \\
\quad\ P-C-H \\
HfpO\diagup\ |\quad| \\
Me-C-C-R' \\
|\quad\ | \\
Me\ \ R
\end{array}
\qquad (54)$$

(R, R′ = H, H; H, Me; Me, Me)   (68)

Compound **68c** is available from thermal rearrangement of dioxaphospholane **60k** (*204*).

Decomposition of the insertion product **32b** of HFA with diphenylphosphane [Eq. (28)] yields, in addition to the oxidation products **33b**, **36**, and **37**, tetraphenyldiphosphane and two cyclic compounds **57r** and **69** (*97*).

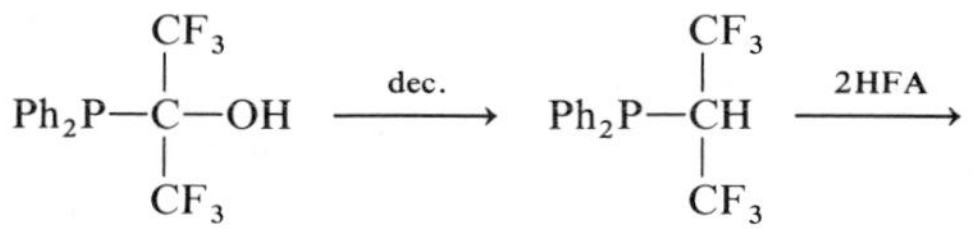

$$
\underset{\textbf{(32b)}}{\underset{\underset{CF_3}{|}}{\overset{\overset{CF_3}{|}}{Ph_2P\!-\!C\!-\!OH}}} \xrightarrow{\text{dec.}} \underset{\underset{CF_3}{|}}{\overset{\overset{CF_3}{|}}{Ph_2P\!-\!CH}} \xrightarrow{2\,HFA}
$$

$$\qquad(55)$$

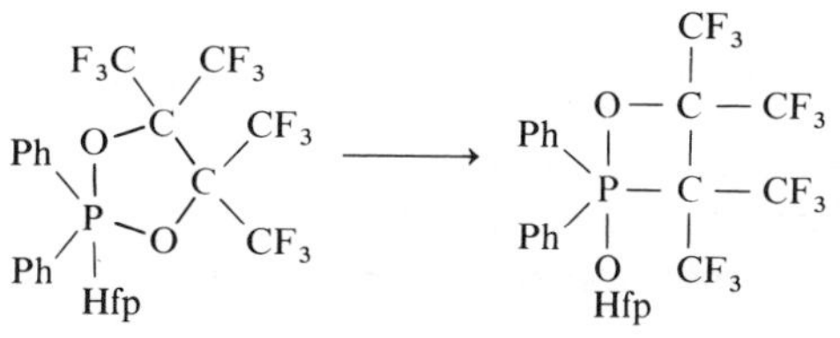

(57r) (69)

The 1,2-λ^5-oxaphosphetanes **58** and **68** can be considered stable intermediates of the Wittig reaction. Pyrolysis of the parent compounds yields the corresponding olefins and oxophosphoranes, respectively, which have been reported in several publications (*204, 220, 221, 226*).

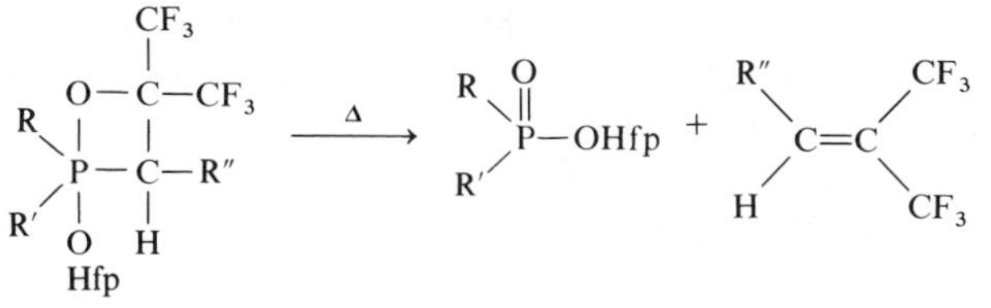

$$\qquad(56)$$

(58) (33)

Kinetics of formation and decomposition of **58** have been investigated (*221*). Pyrolysis of **68a** yields an ω-unsaturated phosphonous ester derivative **70** (*204*).

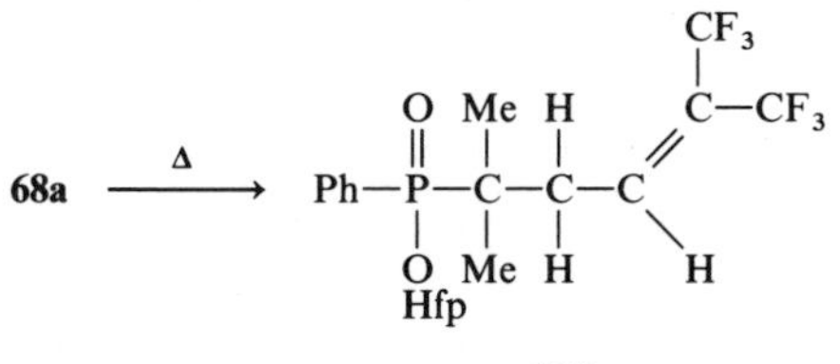

$$\qquad(57)$$

(70)

The tris-OHfp substituted oxaphosphetane **58g** is thermally stable up to 190°C (*226*).

Alcoholysis of **58g** has been studied (*218*). NMR studies show that the exchange of equatorial groups proceeds via a six-coordinated intermediate. Isomerization of equatorial substituents X and Y to R occurs at 120-140°C. Only one ^{19}F NMR signal for **58v** is observed at room temperature (*218*), which suggests that pseudorotation processes are rapid on the NMR time scale.

The X-ray structure of **58l** (X = Y = Ph, Z = OHfp, R = Me) has been reported. The two oxygen atoms are in axial positions (*183*).

6. Reactions Involving Phosphorus–Halogen Bonds

Only two examples are known in which HFA inserts into a phosphorus–halogen bond. While PF_2Cl, PF_3, and POF_2Br are inert even at elevated temperatures, insertion into the P—X bond is observed in PF_2Br and PF_2I (*177*).

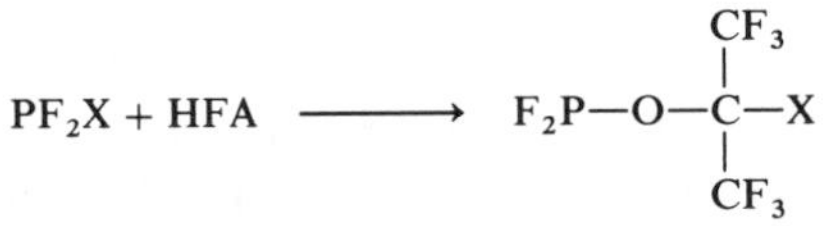

$$PF_2X + HFA \longrightarrow F_2P-O-\underset{\underset{CF_3}{|}}{\overset{\overset{CF_3}{|}}{C}}-X \qquad (58)$$

$$(X = Br, I) \qquad\qquad (71)$$

In contrast to reactions of other ketones with PCl_5, a temperature of 230°C is required to form 2,2-dichlorohexafluoropropane. The high activation energy might be due to the difficulty of coordinating the phosphorus atom by the oxygen atom of HFA (*100*).

Attempted oxidative addition of a urea-bridged diphosphane according to Eq. (43) with HFA leads to the spirophosphorane **56k** (*282, 283*).

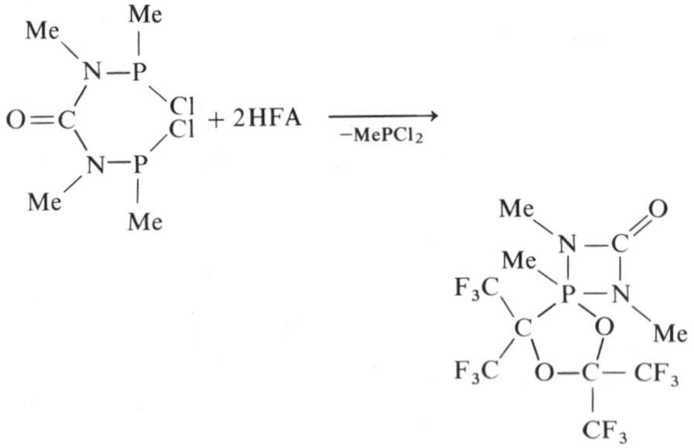

$$(59)$$

(56k)

Pyrolysis of the urea yields a 1,3,2-diazaphosphetanone, which is the actual precursor for the oxidation. The X-ray structure of **56k** shows nitrogen and oxygen to occupy the axial positions (*283*).

Most reactions of halophosphorus compounds do not involve direct attack of HFA. However, the reduced species dilithio perfluoropinacolate, accessible from HFA and lithium metal, is used as a precursor for λ^3- and λ^5-dioxaphospholanes [Eq. (60)] (Table VI).

λ^3-Dioxaphospholanes show no inversion at the phosphorus atom, which can be seen in the nonequivalence of the CF_3 groups (*273*). λ^3-Dioxaphospholanes **73** can be oxidized with halogens to form λ^5-dioxaphospholanes **74** (*228*). Exchange of the halogens in **73a–c** with suitable reaction partners (e.g., $LiNH_2$) yields substituted λ^3-phospholanes [e.g., **73g** (*263*)] (see also Section III,A,2).

Interestingly, POF_3 and $POBr_3$ yield not the tetracoordinated oxophosphoranes, but rather the pentacoordinated trihalogeno-λ^5-dioxaphospholanes, which is in sharp contrast to the formation of the chloro derivative **74i** (*228*).

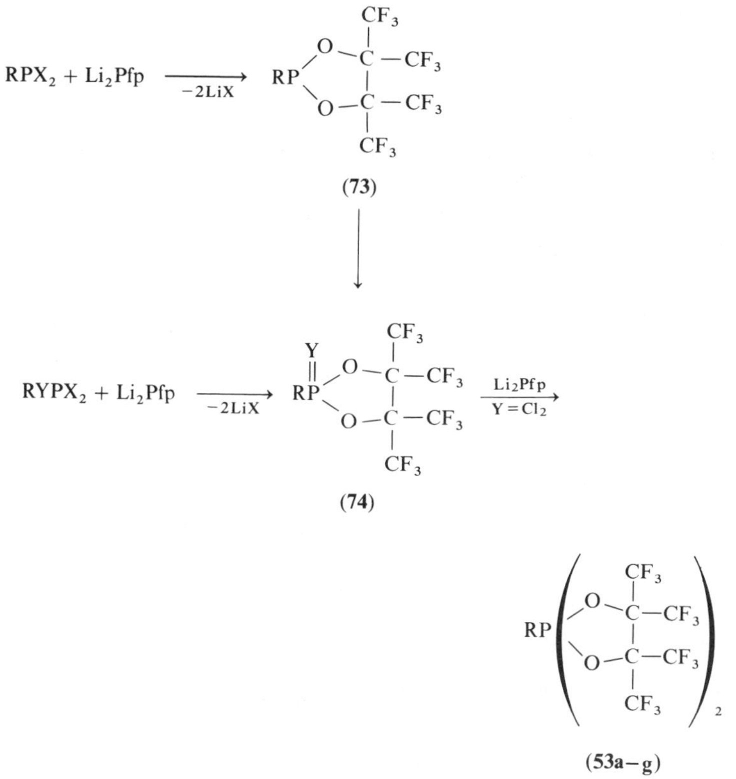

$$RPX_2 + Li_2Pfp \xrightarrow{-2LiX} \textbf{(73)}$$

$$RYPX_2 + Li_2Pfp \xrightarrow{-2LiX} \textbf{(74)} \xrightarrow[Y=Cl_2]{Li_2Pfp}$$

$$(60)$$

$$\textbf{(53a–g)}$$

TABLE VI

1,3,2-DIOXAPHOSPHOLANES BY METATHESIS WITH
PERFLUOROPINACOLATE

Compound	R		Reference
73a–c	F, Cl, Br		*228*
73d	OEt		*279*
73e	Ph		*64*
73f	CF$_3$		*273*
73g	NH$_2$		*263, 263a*
73h–l	NMe$_2$, NEt$_2$, NH-t-Bu, NHSiMe$_3$, N(SiMe$_3$)$_2$		*263a*
		Y	
53, 74	R		
a–f	NH$_2$, NMe$_2$, NEt$_2$, NH-t-Bu, NHSiMe$_3$, N(SiMe$_3$)$_2$	2Cl	*263a*
g	Ph	2F	*231*
h	OHfp	2Et	*9*
i	Cl	O	*228*
j–l (R = Y)	F, Cl, Br		*228*

Thermolysis of **74d** yields a tricyclic system **60as** (see Section III,A,4). The derivatives **74a, e**, and **f** with two leaving groups on the nitrogen atom form the phosphazene derivative **74m** on prolonged standing (*263a*) or heating in a sealed tube at 75°C (*263*).

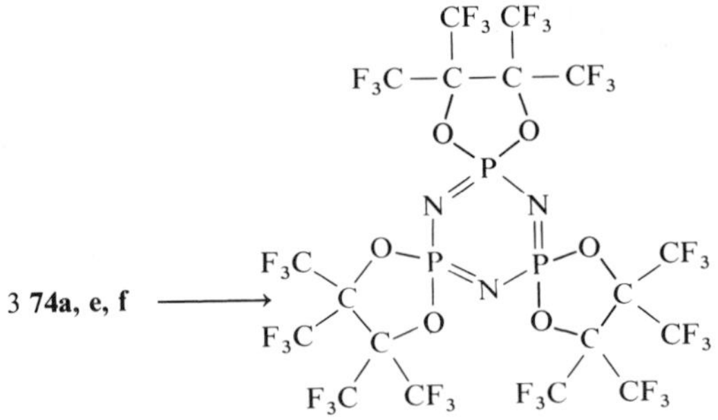

$$3\ \mathbf{74a, e, f} \longrightarrow \tag{61}$$

(**74m**)

The trifluoro-substituted derivative **74j** is accessible via fluorine exchange in **57ah** with PF$_5$ (*229*).

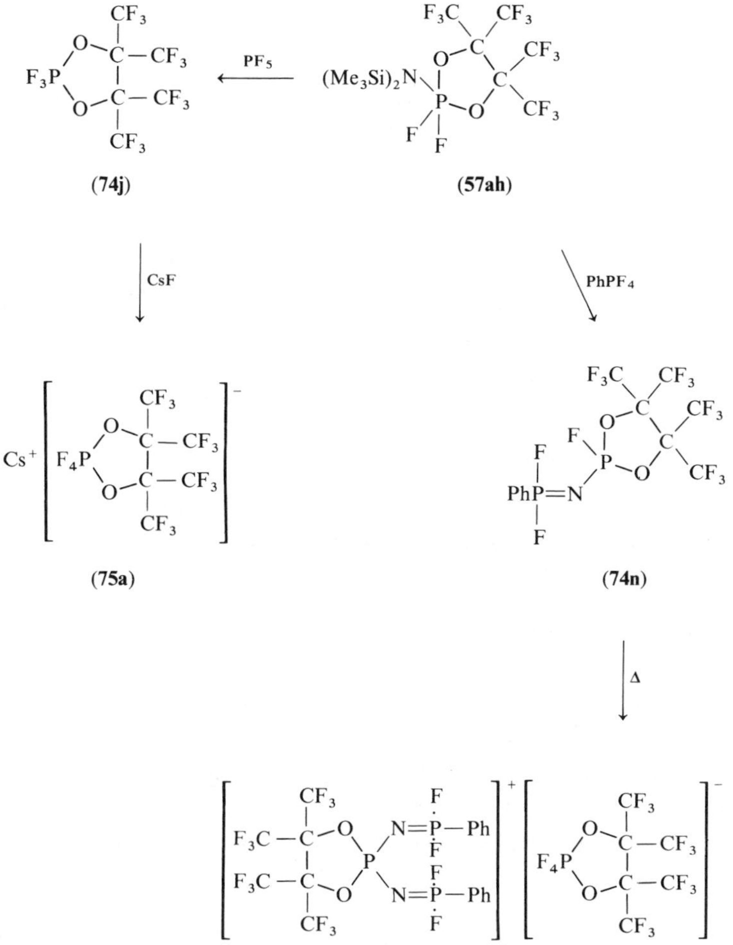

$(74j)$ $\xleftarrow{\;PF_5\;}$ $(Me_3Si)_2N$ ... $(57ah)$

$(74j)$ $\xrightarrow{\;CsF\;}$ $(75a)$ $\quad Cs^+$

$(57ah)$ $\xrightarrow{\;PhPF_4\;}$ $(74n)$

$(74n)$ $\xrightarrow{\;\Delta\;}$ $(75b)$ ... (62)

On treatment with CsF, **74j** yields the salt **75a** (*119*). The same anion together with a cation containing a dioxaphospholane ring **75b** has been synthesized from thermolysis of the aminoiminodiphosphorane **74n** (*229*). Compound **74j** forms a stable donor–acceptor complex with trimethylphosphane (*119*). Reduction of **74j** with trimethylsilyldiphenylphosphane affords the λ^3-dioxaphospholane **73a**, which serves as a phosphane ligand in the molybdenum complex **76** (*229*). The phosphane ligands are in cis-positions.

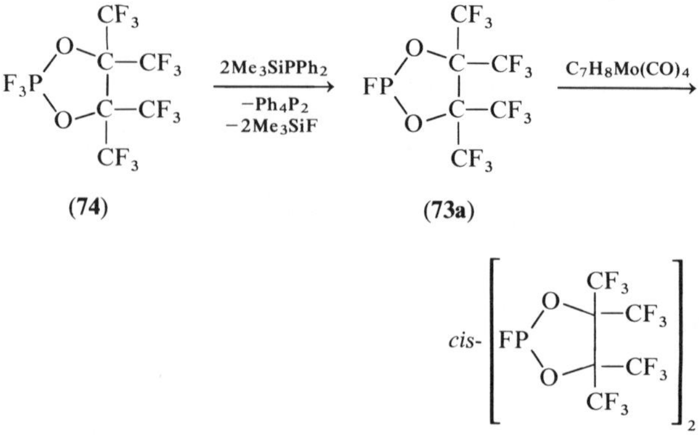

$$
\begin{array}{ccc}
(74) & (73a) &
\end{array}
$$

$$
cis\text{-}\left[\,FP \overset{O}{\underset{O}{\diagup}}\overset{CF_3}{\underset{CF_3}{\diagdown}}\,\right]_2 Mo(CO)_4 \qquad (63)
$$

$$(76)$$

7. Reactions with P—P Bonds

Only one report deals with insertion into a P—P bond. Tetrafluoro-diphosphane adds one molecule of HFA to form **77** (*30*).

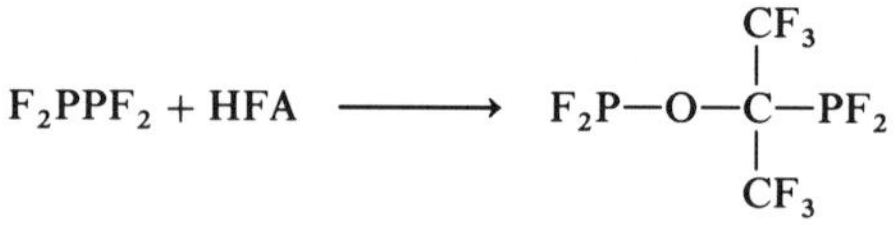

$$
F_2PPF_2 + HFA \longrightarrow F_2P-O-\underset{\underset{CF_3}{|}}{\overset{\overset{CF_3}{|}}{C}}-PF_2 \qquad (64)
$$

$$(77)$$

Röschenthaler and co-workers (*120, 122*) have investigated the action of excess HFA on tetramethyldiphosphane. The resulting 1,2-oxaphosphetanes **58c,o,w** and their relative conformations have been studied with NMR.

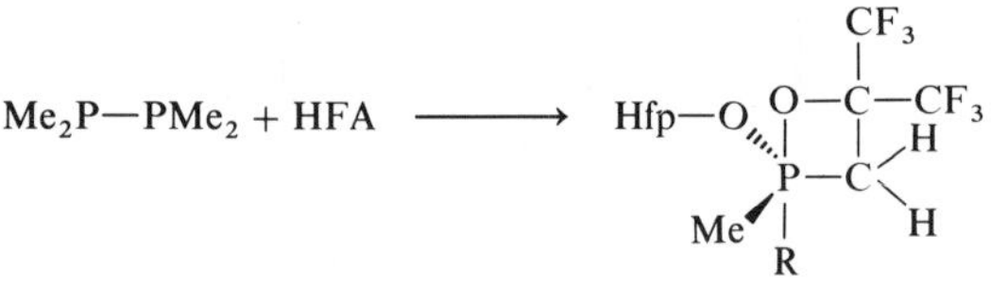

$$
Me_2P-PMe_2 + HFA \longrightarrow \text{Hfp}-O \qquad (65)
$$

$$[R = F, OHfp, OC(CF_3)_2PMe_2] \qquad (58c,o,w)$$

Attempted reaction of the cyclic diphosphane 1,3,4,5-tetramethyl-1,3,4-λ^3,5-λ^3-diazadiphospholidinone with HFA did not result in oxidative addition; the complex mixture has not been separated (*283*).

8. Insertion and Addition Reactions with Pseudohalides

Depending on the substituents, a variety of reaction types have been observed in the reaction of HFA and pseudohalides of phosphorus(III). Most reactions do not proceed in the absence of a basic catalyst, of which triethylamine has proved the best. This catalyst is probably responsible for a nitrile–isonitrile equilibrium in solution; the formation of five-membered rings has also been found with organic isonitriles (188).

Insertion into two P—C bonds and isonitrile cycloaddition occur with phosphorus tricyanide (240).

$$P(CN)_3 + 4HFA \longrightarrow \left(NC-\underset{CF_3}{\overset{CF_3}{C}}-O \right)_2 -P-N=C \qquad (66)$$

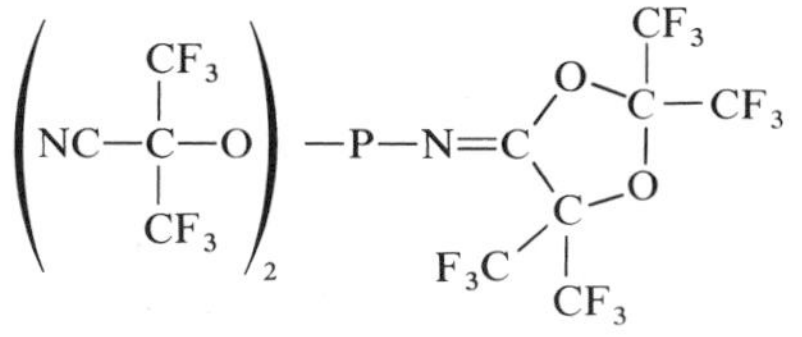

(78)

Substitution of one cyano group with a trifluoromethyl group and reaction with excess HFA lead to the formation of a chiral phosphane. The X-ray structure determination (79) shows only one enantiomer to be present in the solid state (240).

$$F_3C-P(CN)_2 + 5HFA \longrightarrow$$

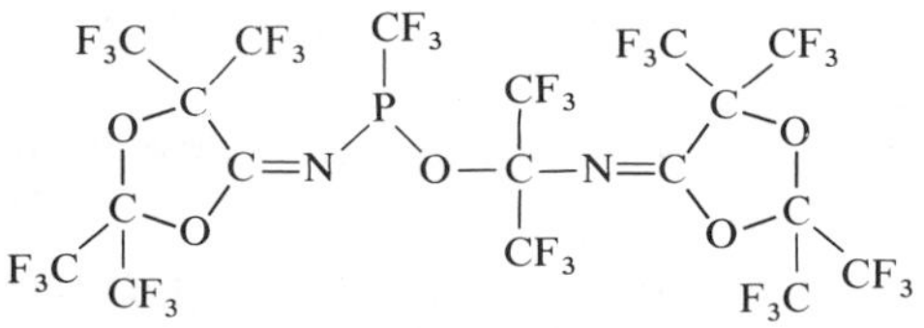

(67)

(79)

Again, an entirely different product geometry has been found in the reaction of phenyldicyanophosphane with HFA. X-Ray structure analysis of 80 shows the two cyclic oxygen atoms to occupy the axial positions (241).

$$PhP(CN)_2 + 6HFA \longrightarrow$$

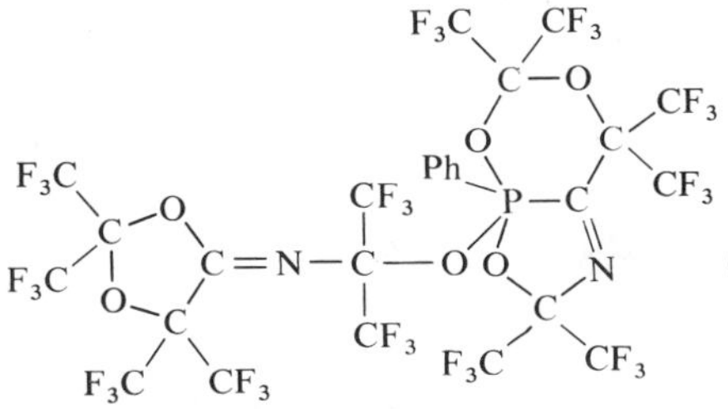

(68)

(80)

Triisothiocyanatophosphane reacts with HFA, with complete isomerization of all ligands and formation of three dioxazine rings, in the presence of triethylamine as a catalyst (240).

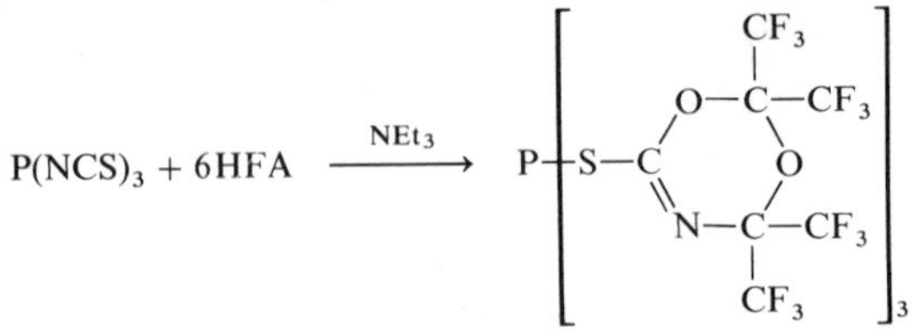

$$P(NCS)_3 + 6HFA \xrightarrow{NEt_3} \quad (69)$$

(81a)

Uncatalyzed reactions of isocyanato- and isothiocyanatophosphanes [Eq. (41)] have been mentioned in Section III,A,2. The structure assignment of **81a** has been made on the basis of analogy with the homologous arsenic compound (**81b**), the X-ray structure of which has been determined.

9. Reactions with P—C Multiple Bonds and C—C Multiple Bonds Attached to Phosphorus

Wittig reactions have been found in several cases when HFA reacts with substituted methylenephosphoranes (209, 213, 245, 249), by analogy with the

$$Ph_3P{=}CHR + HFA \longrightarrow Ph_3PO + (CF_3)_2C{=}CHR \quad (70)$$

(82)

[R = C(O)Me (209); CN, CO_2Me, COPh (245); CH_2SiMe_3 (249);
SR′ (R′ = Me, Ph, CH_2Ph) (213)]

thermal decomposition of some 1,2-oxaphosphetanes (**58**) [Eq. (70)] (see Section III,A,5). Similarly, a phosphonium salt as a precursor of a steroid Wittig reagent yields partially fluorinated desmosterol (141).

The intermediate of a Wittig reaction has been isolated (26) and structurally characterized (60) from the reaction of HFA with a carbodiphosphorane.

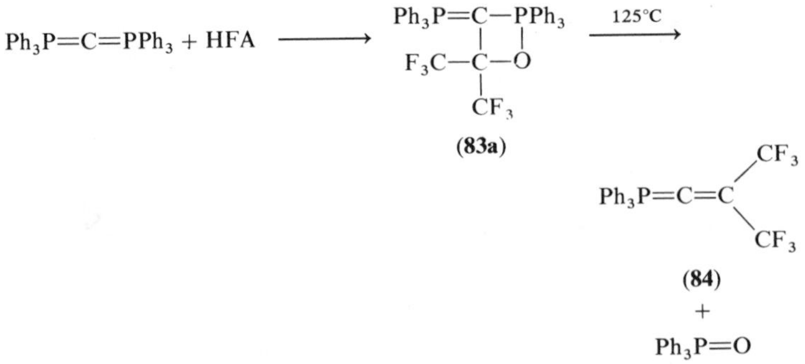

$$Ph_3P{=}C{=}PPh_3 + HFA \longrightarrow \qquad \xrightarrow{125°C}$$

(83a)

$$Ph_3P{=}C{=}C\begin{array}{c}CF_3\\ \\CF_3\end{array} \quad (71)$$

(84)

+

$$Ph_3P{=}O$$

The P—C bonds of **83** are essentially equivalent, suggesting contribution of polar forms. An extremely long P—O bond in contrast to **58l** indicates open chain mesomeric structures, stable due to electron delocalization (*60*). At 125°C a Wittig reaction occurs, and **84** is formed (*26*). The same type of [2 + 2] cycloaddition has been found with a series of *P*-(chloro)alkylidene-phosphoranes. The oxaphosphetanes **83b-d** thermally eliminate hydrogen chloride to yield vinyloxophosphoranes (*165a*).

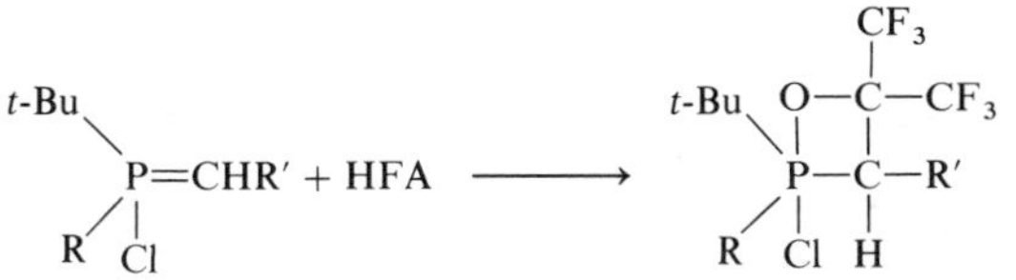

$$
\begin{array}{c}
t\text{-Bu} \\
\diagdown \\
\hspace{1em} P{=}CHR' + HFA \\
\diagup \; | \\
R \;\; Cl
\end{array}
\qquad\longrightarrow\qquad
\begin{array}{c}
\hspace{3em}CF_3 \\
\hspace{3em}| \\
t\text{-Bu}\;\; O{-}C{-}CF_3 \\
\diagdown \;| \;\;| \\
\hspace{1.5em}P{-}C{-}R' \\
\diagup \; | \;\;| \\
R \;\; Cl \;\; H \\
\text{\textbf{(83b-d)}}
\end{array}
\qquad (71a)
$$

	b	c	d
R	*t*-Bu	Et₂N	Et₂N
R'	H	H	Me

No reaction has been observed with compounds containing a carbon–phosphorus triple bond (*290*).

In analogy to the formation of **51** (Section III,A,2), diphenylvinyl-phosphane yields a bicyclic system **85** (*91*).

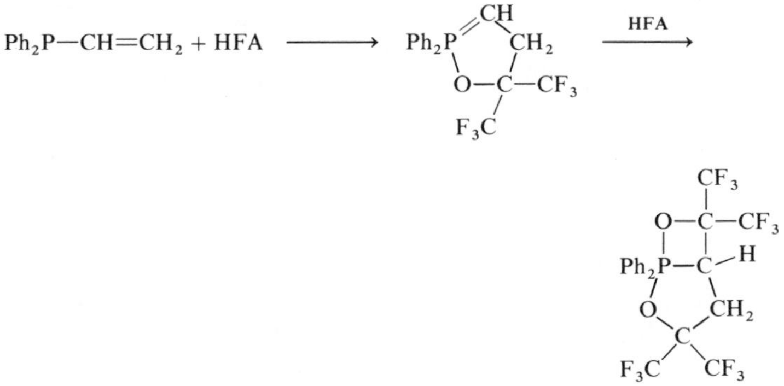

$$
Ph_2P{-}CH{=}CH_2 + HFA \longrightarrow
\begin{array}{c}
\hspace{1em}CH \\
Ph_2P \diagup \;\;\diagdown CH_2 \\
\diagdown \hspace{2.5em}| \\
\hspace{0.5em}O{-}C{-}CF_3 \\
\hspace{1.5em}| \\
\hspace{1em}F_3C
\end{array}
\xrightarrow{\;HFA\;}
$$

$$
\begin{array}{c}
\hspace{2em}CF_3 \\
\hspace{2em}| \\
\hspace{1em}O{-}C{-}CF_3 \\
\hspace{1em}| \;\;| \\
Ph_2P{-}C{\diagdown}^{H} \\
\hspace{1.5em}| \hspace{2em} \\
\hspace{0.8em}O \;\;\;\; CH_2 \\
\hspace{1.5em}\diagdown C \diagup \\
\hspace{1em}F_3C \;\;\; CF_3 \\
\text{\textbf{(85)}}
\end{array}
\qquad (72)
$$

A similar dehydrated ring system **86** is accessible from 1-phenylethinyl-2,2,3,4,4-pentamethylphosphetane (*10*). A mechanism [Eq. (73)] has been reported. The X-ray structure of **86** shows the two oxygen atoms to occupy the axial positions (*10*).

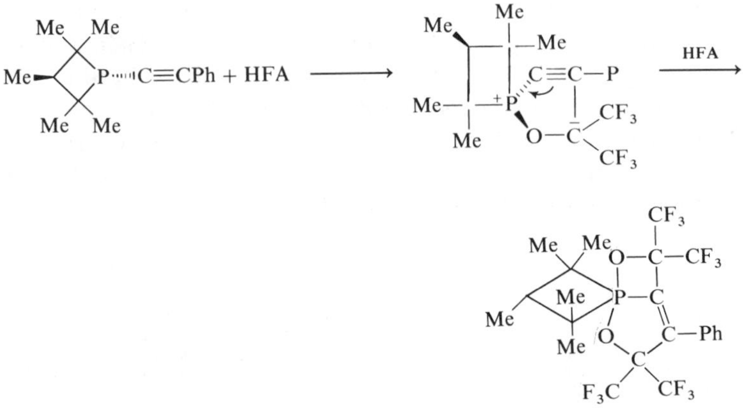

(73)

(86)

B. Arsenic and Antimony

1. Insertion and Addition Reactions

Only a few reports deal with reactions of arsenic and antimony compounds with HFA. Several reports describe insertion of HFA into As—H bonds (43, 72, 155). In contrast to the heavier group IV elements, insertion leads to the formation of 2-arsanoperfluoropropanols 87. This difference can be explained by assuming nucleophilic attack by the arsenic lone pair on the highly electrophilic carbonyl carbon.

$$R_{3-n}AsH_n + HFA \longrightarrow R_{3-n}AsH_{n-1}-\overset{\overset{\displaystyle CF_3}{|}}{\underset{\underset{\displaystyle CF_3}{|}}{C}}-OH \qquad (74)$$

(87)

[R = Me, n = 3 (43), 2, or 1 (72); R = Ph, n = 1 (155)]

Reaction of two molecules of HFA with methylarsane probably yields a 1,4,2,3-dioxadiarsenane (88) with release of hexafluoroisopropanol; similar compounds have been found with aldehydes (72).

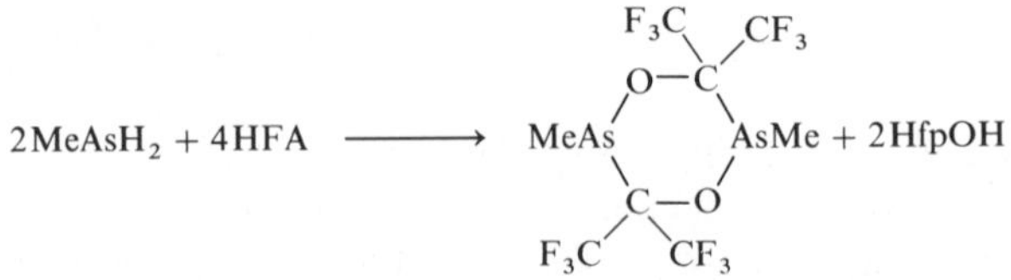

$$2MeAsH_2 + 4HFA \longrightarrow \text{MeAs} \overset{\displaystyle F_3C \diagdown \diagup CF_3}{\underset{\displaystyle F_3C \diagup \diagdown CF_3}{\overset{\displaystyle O-C}{\underset{\displaystyle C-O}{\big|}}}} \text{AsMe} + 2HfpOH \qquad (75)$$

(88)

No formation of λ^5-dioxaarsolanes has been observed, probably due to the relative instability of the oxidation state ($+V$) of arsenic compared to phosphorus (259).

Tetramethyldiarsane and trimethylarsane form unstable, probably dipolar, 1:1 complexes as suggested by ^{19}F NMR spectroscopy. In the former a four-membered ring is formed with the two arsenic atoms acting as donor and acceptor sites (72). Tris(dimethylamino)stibane reacts with HFA with insertion into all three Sb—N bonds (83).

2. Reactions with Pseudohalide Functions

Triethylamine has been employed as a catalyst in the following reactions [Eqs. (76)–(78)].

A 14-membered ring has been obtained in moderate yield from excess tricyanoarsane and HFA (233).

$$2As(CN)_3 + 12HFA \longrightarrow \quad (89) \quad (76)$$

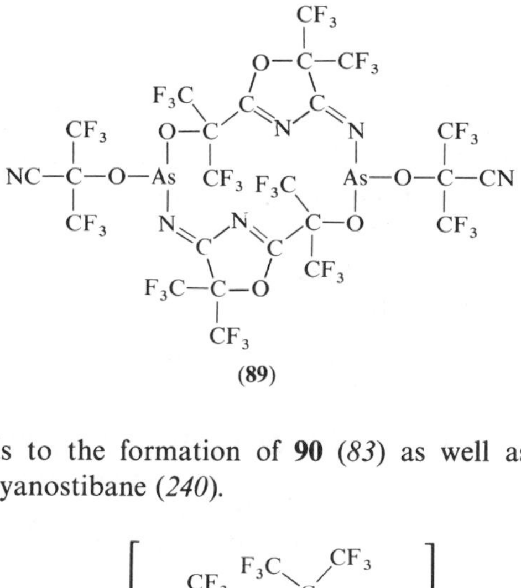

However, excess HFA leads to the formation of **90** (83) as well as the analogous reaction with tricyanostibane (240).

$$E(CN)_3 + 9HFA \longrightarrow \quad (90) \quad (77)$$

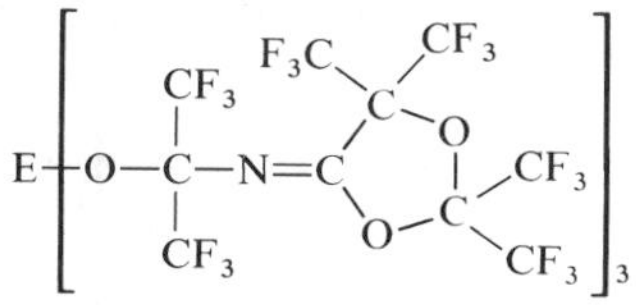

(E = As, Sb)

As in the case of cyanophosphanes (Section III,A,8), isomerization of the nitriles to the isonitriles is the initial step. While triisocyanatoarsane proved to be inert to HFA (176), the expected product **81b** with triisothiocyanato-arsane has been isolated (240). The structure of **81b** is reported.

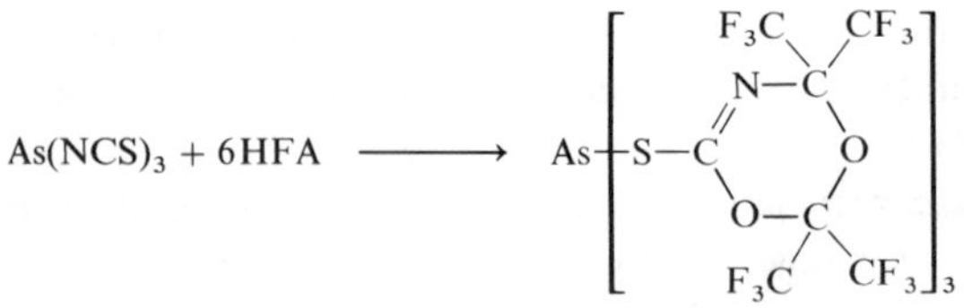

$$\text{As(NCS)}_3 + 6\,\text{HFA} \longrightarrow \text{As}\left[\text{S}-\text{C}\underset{\text{O}-\text{C}}{\overset{\text{N}-\text{C}}{<}}\text{O}\right]_3 \qquad (78)$$

(81b)

IV. Reactions of HFA with Compounds of Nitrogen and Group VI Elements

A. OXYGEN AND NITROGEN

Since the reactions of oxygen- and nitrogen-containing molecules with HFA are very similar, and several molecules contain both atoms which could serve as active centers, they are treated under one heading. Early work has been reviewed by Krespan and Middleton (*168*) and Gambaryan *et al.* (*111*).

1. Insertion into E—H Bonds

The exothermic reaction of HFA with a stoichiometric amount of water leads to the formation of a stable crystalline hydrate **91**; excess water affords a liquid sesquihydrate **91a** of unknown structure (*191*).

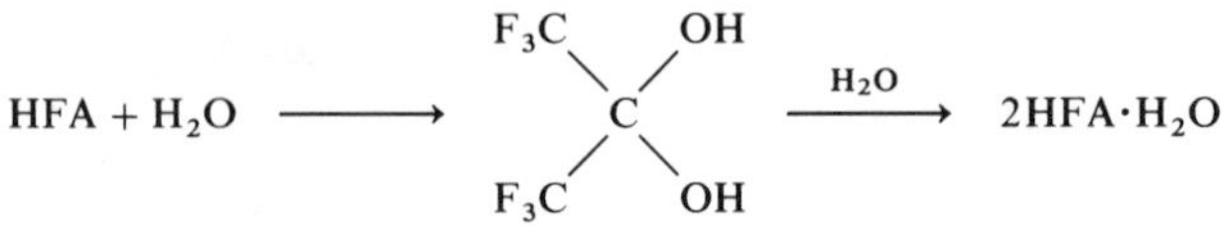

$$\text{HFA} + \text{H}_2\text{O} \longrightarrow \underset{\text{F}_3\text{C}}{\overset{\text{F}_3\text{C}}{>}}\text{C}\underset{\text{OH}}{\overset{\text{OH}}{<}} \xrightarrow{\text{H}_2\text{O}} 2\,\text{HFA}\cdot\text{H}_2\text{O} \qquad (79)$$

(91)   **(91a)**

Despite its rather high acidity ($pK_a = 6.58$), **91a** is an excellent solvent for polymers like polyamides, -esters, -acetals, and -ols (*185*). Hydrolysis of the anionic species **113** (*vide infra*) in the presence of tetraphenylphosphonium chloride yields the tetrameric dianion **92**.

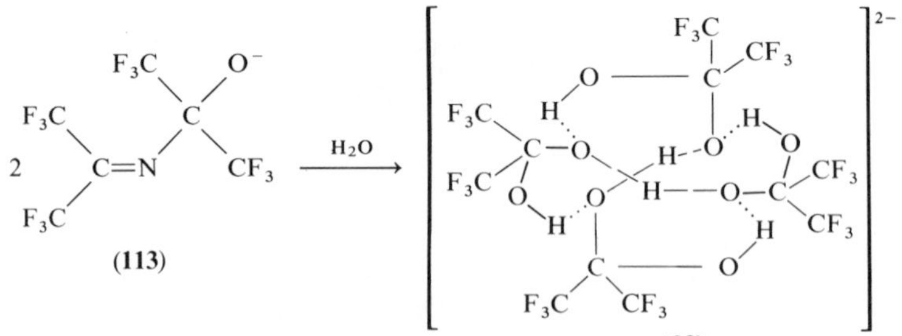

$$2 \; \underset{\text{F}_3\text{C}}{\overset{\text{F}_3\text{C}}{>}}\text{C}=\text{N}\underset{\text{CF}_3}{\overset{\text{C}\;\overset{\text{F}_3\text{C}\quad\text{O}^-}{<}}{}} \xrightarrow{\text{H}_2\text{O}} \left[\;\right]^{2-} \qquad (80)$$

(113)   **(92)**

The X-ray structure of **92** (*237*) shows fourfold symmetry with four asymmetrically bridging hydrogen atoms and two symmetrical bridges in the core of the molecule.

In analogy with Eq. (79), HFA forms unstable hemiketals **93a** which are converted into stable ketals **93b** with diazomethane (*163*) or dialkyl sulfates (*185*). The ketals **93b** are almost stable in 2 M HCl, only slight decomposition occurring.

$$
\text{HFA + ROH} \longrightarrow
\underset{(93a)}{
\begin{array}{c}
\text{F}_3\text{C} \quad \text{OH} \\
\diagdown \diagup \\
\text{C} \\
\diagup \diagdown \\
\text{F}_3\text{C} \quad \text{OR}
\end{array}}
\xrightarrow[\text{(R'O)}_2\text{SO}_2]{\text{CH}_2\text{N}_2}
\underset{(93b)}{
\begin{array}{c}
\text{F}_3\text{C} \quad \text{OR'} \\
\diagdown \diagup \\
\text{C} \\
\diagup \diagdown \\
\text{F}_3\text{C} \quad \text{OR}
\end{array}}
\tag{81}
$$

^{19}F NMR spectra of HFA adducts with compounds containing active hydrogen atoms like alcohols **93a**, amines **95**, and thiols **126** have been reported (*173*). HFA also forms unstable ketal esters with a series of carboxylic acids. The equilibrium has been investigated by means of ^{19}F NMR spectroscopy (*206*).

In a similar fashion HFA inserts into the O—H and O—M (M = Li, Na) bond of peroxides (*11, 12, 57*). The hydroperoxides **94a** can be metallated with MH (*11*).

$$
\text{ROOH + HFA} \longrightarrow
\underset{(94a)}{
\begin{array}{c}
\text{CF}_3 \\
| \\
\text{ROO}-\text{C}-\text{OH} \\
| \\
\text{CF}_3
\end{array}}
\xrightarrow{\text{MH}}
\underset{(94b)}{
\begin{array}{c}
\text{CF}_3 \\
| \\
\text{ROO}-\text{C}-\text{OM} \\
| \\
\text{CF}_3
\end{array}}
\tag{82}
$$

$$
[\text{R = H, HOC(CF}_3)_2, \text{ }t\text{-Bu, CF}_3\text{C(O)}] \qquad\qquad (\text{M = Li, Na})
$$

These compounds are strong oxidizers, but less flammable than nonfluorinated peroxides.

Hemiaminals are available from HFA and amines. A series of similar reactions has been carried out with ammonia (*187, 189*), amines (*189*), aliphatic, aromatic (*258*), and fluoroaliphatic acid amides (*169*). The hemiaminals **95** can be dehydrated with phosphorus oxychloride in pyridine to form 2-hexafluoropropaneimines **96** (*187*).

$$
\text{HFA + NH}_2\text{R} \longrightarrow
\underset{(95)}{
\begin{array}{c}
\text{F}_3\text{C} \quad \text{NHR} \\
\diagdown \diagup \\
\text{C} \\
\diagup \diagdown \\
\text{F}_3\text{C} \quad \text{OH}
\end{array}}
\xrightarrow{\text{POCl}_3}
\underset{(96)}{
\begin{array}{c}
\text{F}_3\text{C} \\
\diagdown \\
\text{C}=\text{NR} \\
\diagup \\
\text{F}_3\text{C}
\end{array}}
\tag{83}
$$

Hexafluoroacetoneazine (47, 276a) and hexafluoroacetonebis(trifluoromethyl)hydrazone (55) have been synthesized similarly. Dehydration has been achieved with phosphorus oxychloride (47) or oleum (276a) for the ketazine and oleum for the hydrazone. S-Arylsulfinamides also react only via cleavage of an N—H bond (46). Derivatives of **96a** are available from (phenylimino)triphenylphosphorane via aminals **97** (293, 295).

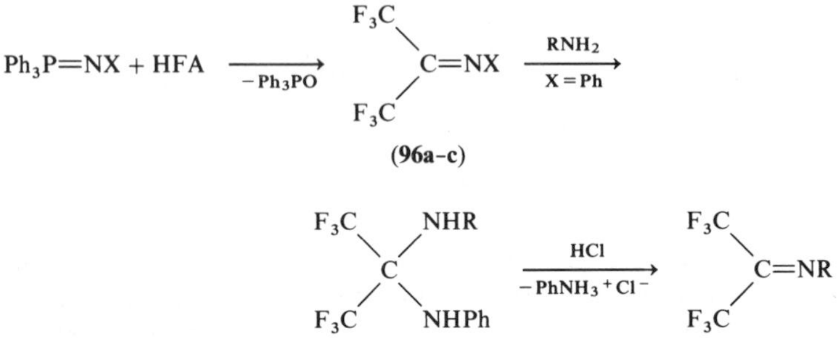

$$Ph_3P=NX + HFA \xrightarrow[-Ph_3PO]{} \underset{F_3C}{\overset{F_3C}{>}}C=NX \xrightarrow[X=Ph]{RNH_2}$$

(96a–c)

$$\underset{F_3C}{\overset{F_3C}{>}}C\underset{NHPh}{\overset{NHR}{<}} \xrightarrow[-PhNH_3{}^+Cl^-]{HCl} \underset{F_3C}{\overset{F_3C}{>}}C=NR \quad (84)$$

[X = H, Ph, C(O)Ph; R = H, i-Bu, CH$_2$PH] **(96)**

Compound **96a** (R = H) is also obtained from thermal decomposition of **97a** at 180°C or from the unsubstituted phosphoraneimine (294). Corresponding imides are found to react analogously (294). The same type of reaction has also been found with N-arylimino sulfoxides. Arylhexafluoropropaneimines are formed with HFA under the catalytic influence of CsF with evolution of SO$_2$ (296).

HFA reacts with trifluoroacetyl nitrite in the presence of potassium fluoride to yield potassium trifluoroacetate and heptafluoroisopropyl nitrite (171).

N,N,N',N'-Tetramethyl-p-phenylenediamine reacts with HFA to form a blue charge-transfer complex. Though no ESR signal has been observed, one-electron transfer seems to be likely (84).

Pyridines and pyridine N-oxides are attacked by HFA in ortho positions in the presence of lithium-2,2,5,5-tetramethylpiperidide. The intermediate formation of a zwitterionic species in the case of the pyridines has been postulated (265).

Aldol condensation has been observed with acetone and acetophenone (255).

Photochemical reactions of HFA with perfluorinated carbon–oxygen compounds have been reported (271, 272). HFA serves as a mild source of CO in the reaction with bis(trifluoromethyl) peroxide (271) to yield bis(trifluoromethyl) carbonate; with perfluoromethyl oxalate, CF$_3$ radicals are the reactive species to yield perfluoromethyl acetate (272).

2. Reactions with Pseudohalides

Like ordinary ketones, HFA forms a cyanohydrin **98** (*165*) with (*193*) and without (*210*) catalysis by bases like piperidine.

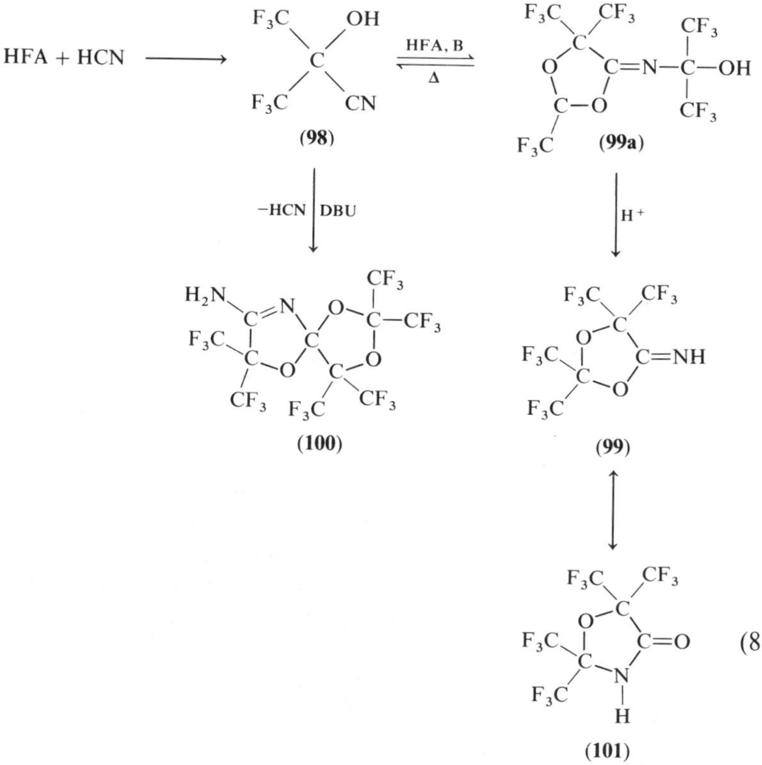

With excess HFA a dioxolaneimine **99a** is formed from a typical isonitrile reaction mentioned in previous sections. Acidification yields the unsubstituted five-membered ring **99**, which is the parent compound of alkylated dioxolanes **99b–e** available from isonitriles (R = Me, Et, *t*-Bu, *c*-C$_6$H$_{11}$) (*111, 188*). Whether the imine structure **99** or the ketone structure **101** has to be assigned to the hydrolysis product is not known. The Chapman rearrangement has been proved in the following system [Eq. (86)] (*190*). Treatment of **98** with strong bases like 1,8-diazabicyclo[5.4.0]undec-7-ene (DBU) yields a spirocyclic compound **100** with elimination of HCN; the structure of **100** has been derived by spectroscopic methods (*193*). The reaction of **98** with HFA × H$_2$O (**91**) in sulfuric acid produces a 1,3-dioxolan-(4)one (*112*).

With sodium cyanide the anion of hexafluorocyanohydrin **98a** is formed. The structure of **98a** has been confirmed by etherification with dimethyl sulfate (*190*). Further action of HFA or hexafluoroacetoneimine on **98a** leads to the formation of ionic 1,3-oxazolidines **101a** and **101b** (*190, 192*).

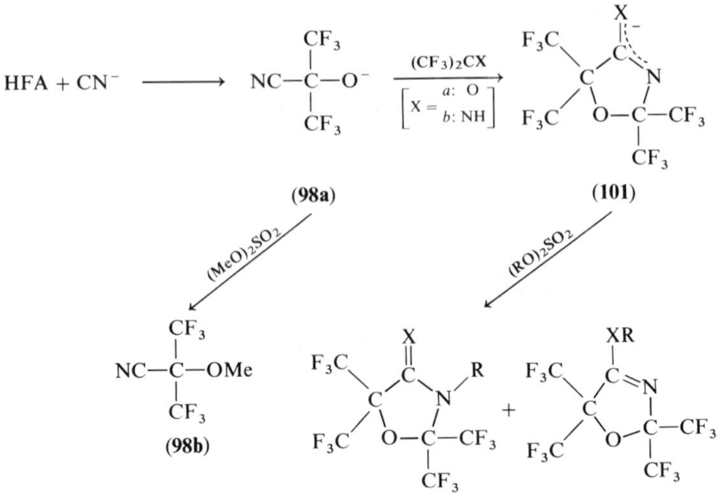

$$R \stackrel{\text{Me}}{=} X = O^{101}\underset{d}{\overset{c}{\text{X}}} = NH^{e}_{f}$$

Compound **101a** is generated from an unstable intermediate by a Chapman rearrangement. Its structure has been confirmed unambiguously by reaction with dimethyl sulfate (*190*), whereby both isomers **101c** and **101e** are formed. The X-ray structure of **101a** is in agreement with these observations (*237*). The neutral compounds **101d** and **101f** are strong acids.

HFA has been found to insert into one of the acidic C—H bonds of malodinitrile (*186*).

HFA forms an adduct **102** with hydrogen cyanate that decomposes above 0°C. Storage over a long period produces oxadiazinedione **103** and loss of one molecule of HFA (*143*).

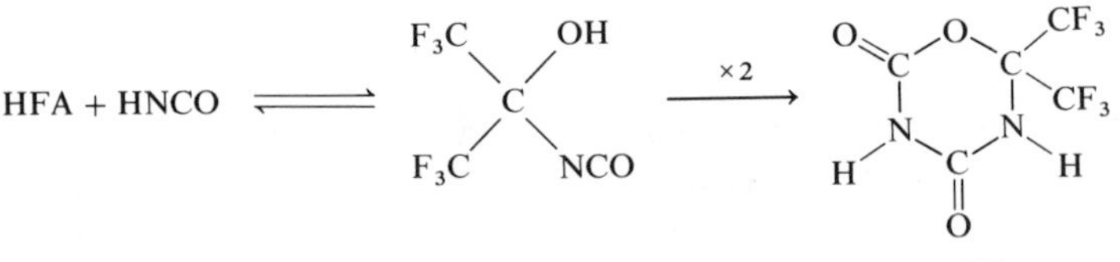

Mercaptodicyanamides react with two molecules of HFA to form six-membered heterocycles **104b–d**. Derivative **d**, a recrystallizable solid, has been isolated. Pyrolysis yields triazinones **105b–d** via Chapman rearrangement with evolution of HFA. The unsubstituted six-membered ring **105a** is also available, but no indication of intermediate **104a** has been found (*147, 291*).

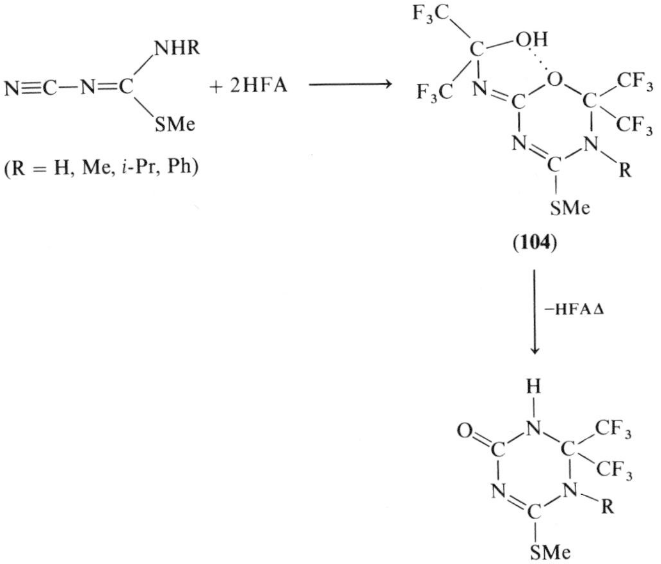

(104)

$$-HFA \triangle$$

(105)

$$(88)$$

A substituted diazinedione **106** has been synthesized from HFA and cyanoacetamide (*146, 147*) in the presence of pyridine.

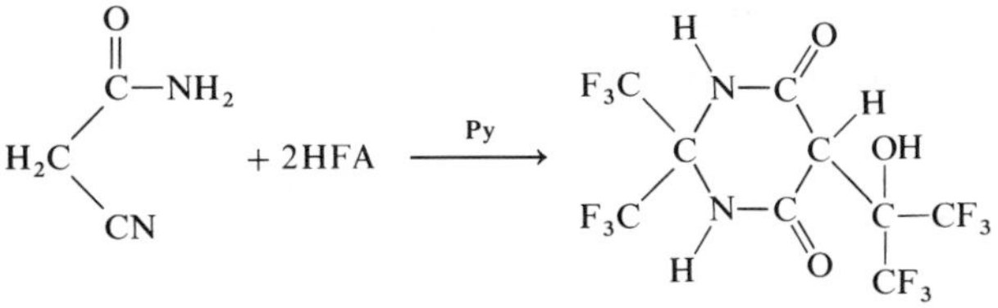

(106)

$$(89)$$

The pyridine-catalyzed reaction of cyanogen chloride with HFA yields the perhalogenated dioxazine **107** (*270*).

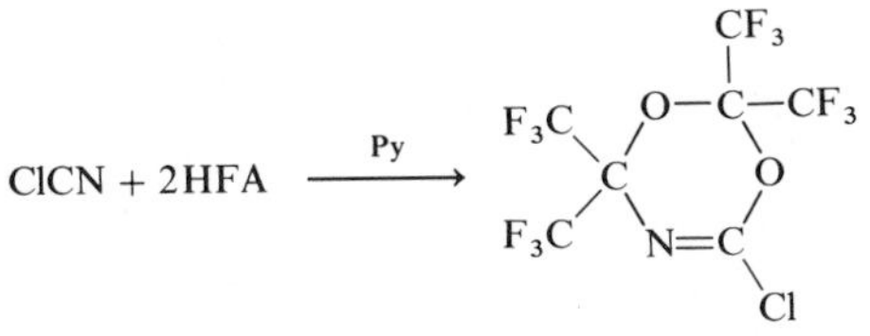

$$ClCN + 2HFA \xrightarrow{Py}$$

(107)

$$(90)$$

The acid-catalyzed reaction of benzonitrile proceeds with formation of 1,3,5-oxadiazine **108a** (*256*).

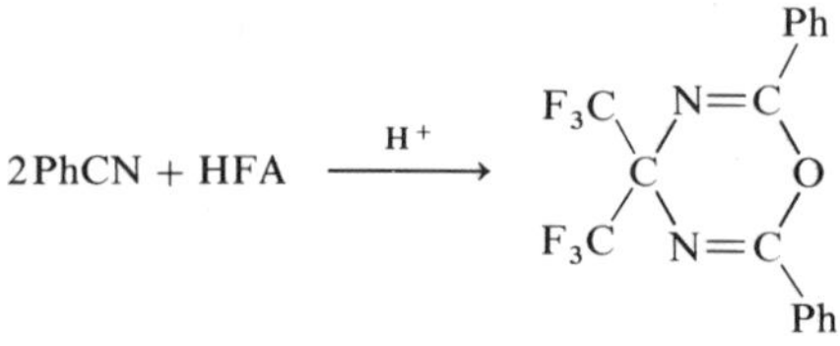

$$2 PhCN + HFA \xrightarrow{H^+}$$

(91)

(**108a**)

Amino-substituted derivatives **108b–d** are formed via 1,3-oxazetines **109b–d** from substituted cyanoamides and HFA (*53, 53a, 140*).

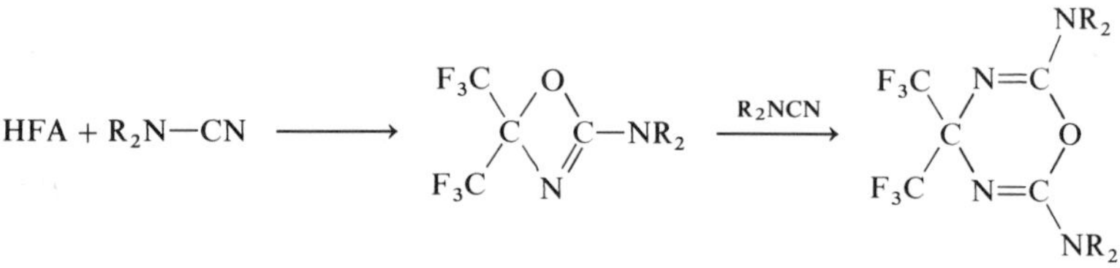

$$HFA + R_2N-CN \longrightarrow \qquad \xrightarrow{R_2NCN}$$

(92)

[R = Me, Et, (CH$_2$)$_{5/2}$]   (**109**)   (**108b–d**)

Amidines (*51, 52*), guanidine (*79*), and biguanidine (*80*) react with HFA with elimination of water and formation of oxadiazines **110**.

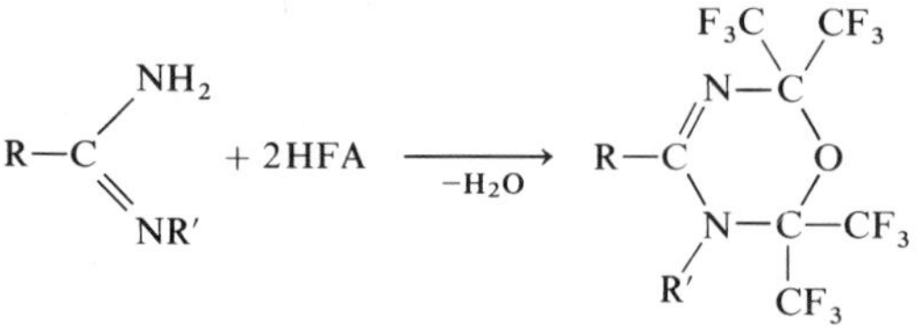

(93)

(**110**)

	a	b	c	d	e
R =	NH$_2$	(NH)$_{1/2}$	Ar	Al	Ar
R' =	H	H	Ar, Al, H	Ar	Ar

The dehydrating agent in the case of the guanidines is HFA itself; phosphorus oxychloride has to be used to effect ring closure in the case of the amidines (*52*). An intermediate 1:1 adduct **111** can be isolated, which, in addition to the formation of **110d** and **110e**, on dehydration reacts with the aromatic system to yield 3,4-dihydrochinazolines **112** (*52*).

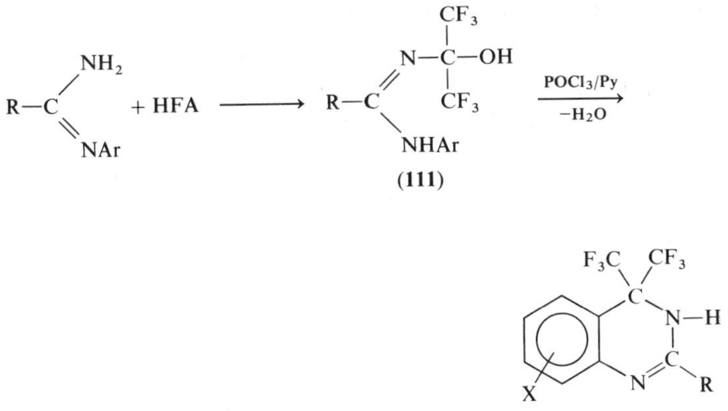

$$(94)$$

(111)

(112)

Thermal elimination of CO_2 and formation of a perfluorinated ketimine have been found when pentafluorophenyl isocyanate was treated with HFA at 150°C in dimethylformamide (82), in analogy to the reaction with imino sulfoxides (296) mentioned earlier.

With evolution of CO_2 or COS, respectively, cyanate and thiocyanate anions react with HFA to form an ionic hexafluoropropaneimine 113, whose crystal structure shows the anion to be dimeric in the solid state (237).

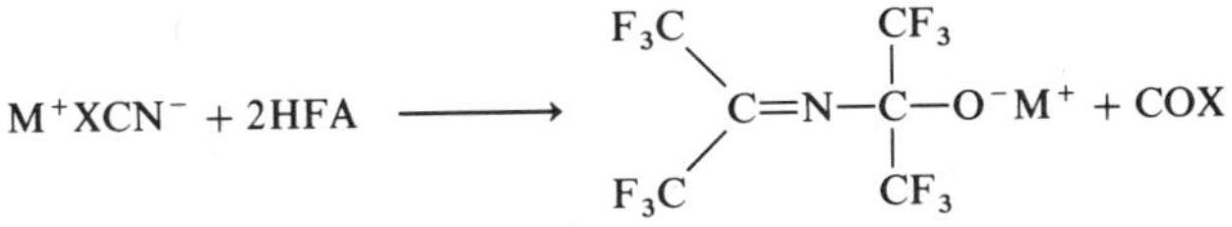

$$M^+XCN^- + 2HFA \longrightarrow \quad \underset{F_3C}{\overset{F_3C}{>}}C=N-\underset{CF_3}{\overset{CF_3}{\underset{|}{C}}}-O^-M^+ + COX \qquad (95)$$

(X = O, S; M = K, Na)   (113)

A by-product (114, 7 % yield) has been found in the HFA/NaSCN reaction. The X-ray structure shows the elimination of all oxygen atoms and cleavage of some C—C bonds (238).

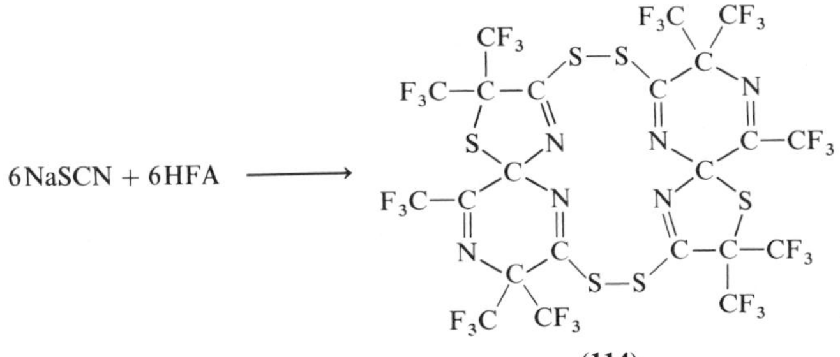

$$6 NaSCN + 6 HFA \longrightarrow \qquad (96)$$

(114)

3. Cyclizations Not Involving Pseudohalides

The reaction of perfluorinated nitrosoalkanes with substituted diazomethanes has been found to produce oxaziridines (81, 243). Insertion of HFA into the O—N bond yields 1,3,4-dioxazolidines **115a** and **115b** (81, 273, 274).

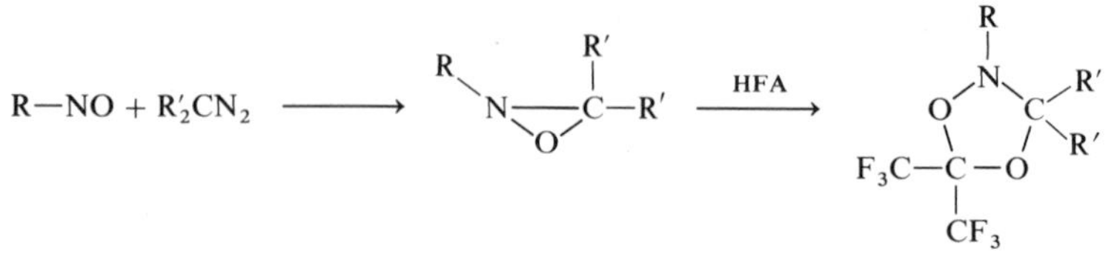

$$R-NO + R_2'CN_2 \longrightarrow \underset{O}{\overset{R}{\underset{}{N}}}{-}\overset{R'}{\underset{}{C}}{-}R' \xrightarrow{HFA} \quad (97)$$

[**a**: R = t-C$_4$F$_9$, R' = Ph (81); **b**: R = R' = CF$_3$ (273, 274)] (115)

Stable Δ^3-1,3,4-oxadiazolines **116** have been synthesized from HFA and disubstituted diazomethanes (252). Thermal elimination of N$_2$ yields oxiranes **117b** if both substituents are aryl, and vinyl ethers **117a** in the case of methyl-substituted diazomethanes (252).

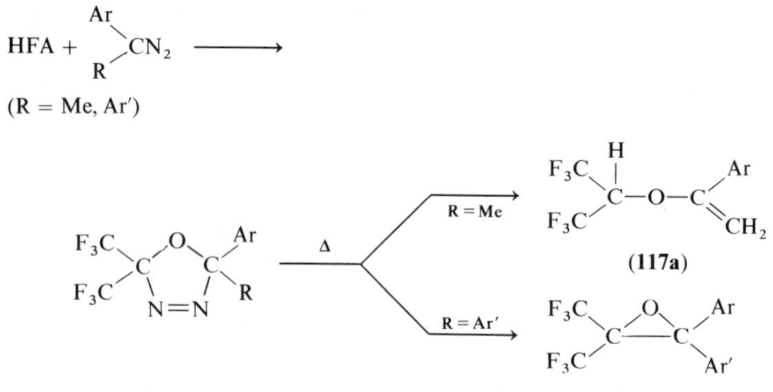

$$HFA + \overset{Ar}{\underset{R}{\diagup}}CN_2 \longrightarrow$$

(R = Me, Ar')

(116) (117a) (117b) (98)

The condensation products of benzoylamides with HFA have been reductively cyclized with anhydrous tin dichloride to yield oxazoles **118** (50a).

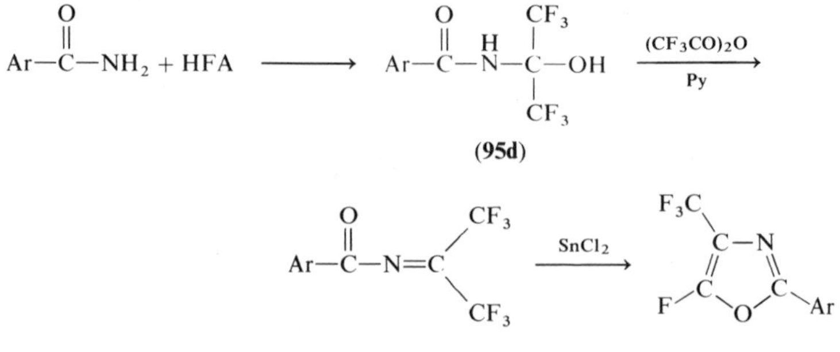

$$Ar-\overset{O}{\overset{\|}{C}}-NH_2 + HFA \longrightarrow Ar-\overset{O}{\overset{\|}{C}}-\overset{H}{\underset{CF_3}{\overset{|}{N}}}-\overset{CF_3}{\underset{|}{C}}-OH \xrightarrow[Py]{(CF_3CO)_2O}$$

(95d)

$$Ar-\overset{O}{\overset{\|}{C}}-N{=}C\overset{CF_3}{\underset{CF_3}{\diagup}} \xrightarrow{SnCl_2} \qquad (98a)$$

(96d) (118)

1,3,4-Dioxazolines **119** are available from fluorinated oximes via insertion of HFA into the O—H bond and subsequent thermal elimination of hydrogen fluoride with triethylamine or potassium fluoride (*275, 276*)

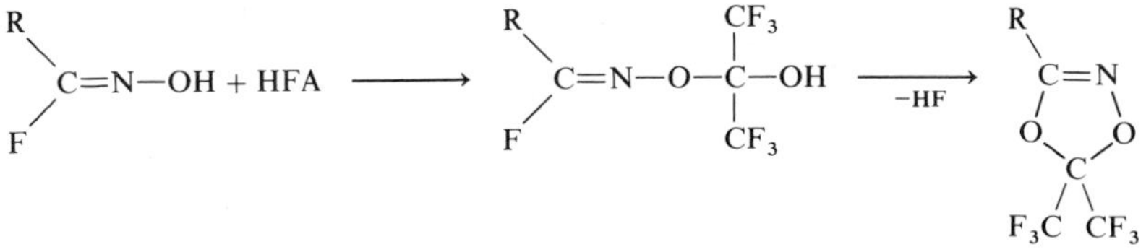

$$
\begin{array}{c}
R \\
\diagdown \\
\diagup \quad C{=}N{-}OH + HFA \\
F
\end{array}
\longrightarrow
\begin{array}{c}
R \quad\quad CF_3 \\
\diagdown \quad\quad | \\
\quad C{=}N{-}O{-}C{-}OH \\
\diagup \quad\quad | \\
F \quad\quad CF_3
\end{array}
\xrightarrow[-HF]{}
\begin{array}{c}
R \\
\diagdown \\
C{=}N \\
O \diagdown \quad \diagup O \\
C \\
\diagup \diagdown \\
F_3C \quad CF_3
\end{array}
\quad (99)
$$

(R = CH₂CF₃, CH(CF₃)₂, CHFCF₃)   (**119**)

Dioxolanones are accessible from the reactions of oxalyl fluoride (*71*) and α-hydroxy acids (*285*) with HFA.

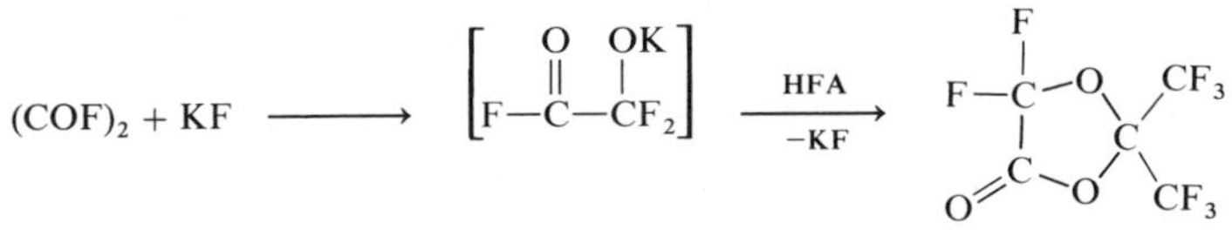

$$
(COF)_2 + KF \longrightarrow
\left[
\begin{array}{cc}
O & OK \\
\| & | \\
F{-}C{-}CF_2
\end{array}
\right]
\xrightarrow[-KF]{HFA}
\begin{array}{c}
F \\
| \\
F{-}C{-}O \quad CF_3 \\
\quad\quad \diagdown \diagup \\
\quad\quad C \\
O{=}C{-}O \quad CF_3
\end{array}
\quad (100)
$$

(**120a**)

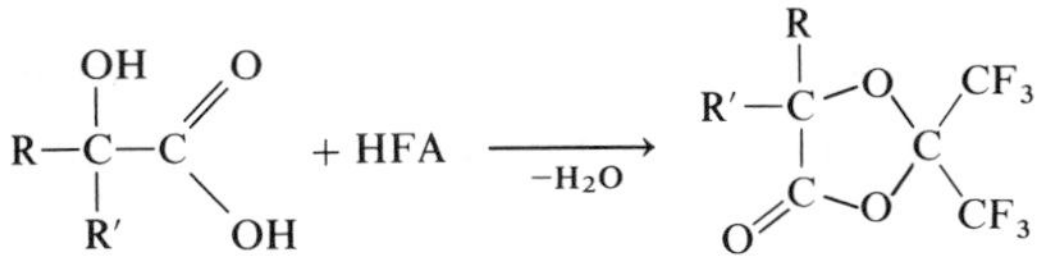

$$
\begin{array}{c}
OH \quad O \\
| \quad\quad \diagup\diagdown \\
R{-}C{-}C \\
| \quad\quad \diagdown \\
R' \quad\quad OH
\end{array}
+ HFA
\xrightarrow[-H_2O]{}
\begin{array}{c}
R \\
| \\
R'{-}C{-}O \quad CF_3 \\
\quad\quad \diagdown\diagup \\
\quad\quad C \\
O{=}C{-}O \quad CF_3
\end{array}
\quad (101)
$$

(**120b**)

Oxazolidinones containing the hexafluoroisopropylidene group have been synthesized from α-amino acids and HFA (*253, 286*). Elimination of hydrogen fluoride from difluoromethyldimethylamine with HFA, and addition of HFA, yield dioxolane **121a**, which can be converted with sulfuric acid into the fully fluorinated dioxolanone **120c**. The reaction has been described as proceeding via an aminofluoronitrene (*164*).

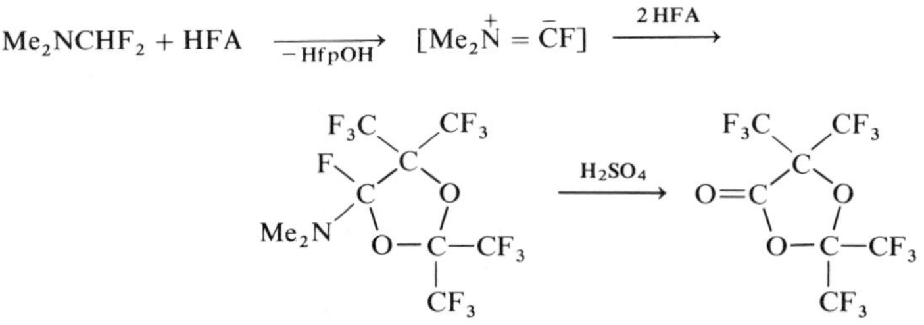

$$
Me_2NCHF_2 + HFA \xrightarrow[-HfpOH]{} [Me_2\overset{+}{N}{=}\overset{-}{C}F] \xrightarrow{2\,HFA}
$$

$$
\begin{array}{c}
F_3C \diagdown \quad \diagup CF_3 \\
F \diagdown \quad C \\
\quad C \diagdown O \\
Me_2N \diagup \diagdown \\
O{-}C{-}CF_3 \\
| \\
CF_3
\end{array}
\xrightarrow{H_2SO_4}
\begin{array}{c}
F_3C \diagdown \quad \diagup CF_3 \\
C \\
O{=}C \diagdown O \\
\quad \diagdown \\
O{-}C{-}CF_3 \\
| \\
CF_3
\end{array}
\quad (102)
$$

(**121a**)   (**120c**)

Alkoxy-substituted dioxolanes are available from orthoformic esters and HFA (36).

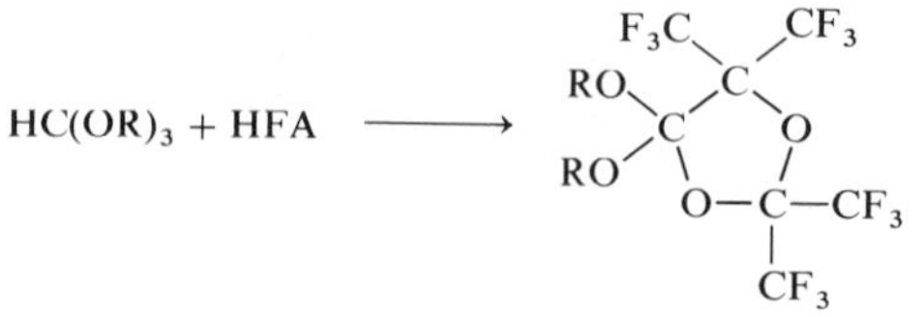

$$HC(OR)_3 + HFA \longrightarrow$$

(121b)

(103)

A Δ^3-Oxazolines (122) has been obtained from the acid-catalyzed reaction of diiminosuccinonitrile with HFA (20).

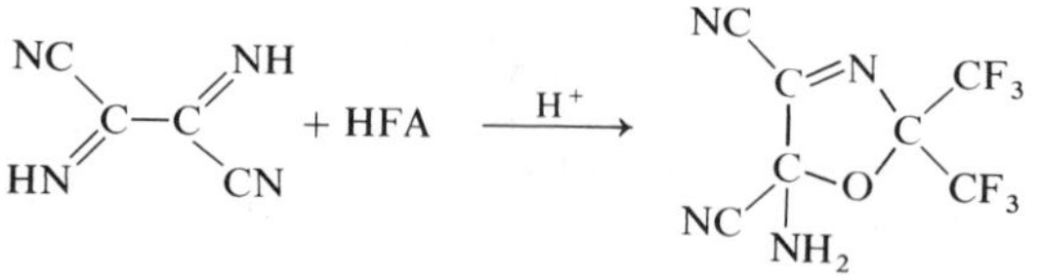

(122)

(104)

Cyclic carbodiimides react with HFA to form six- and four-membered rings 123 and 124, which have been investigated spectroscopically. Isolation was successful only in the case of the 10-membered ring 123b; the other two compounds are in equilibrium with their precursors (225).

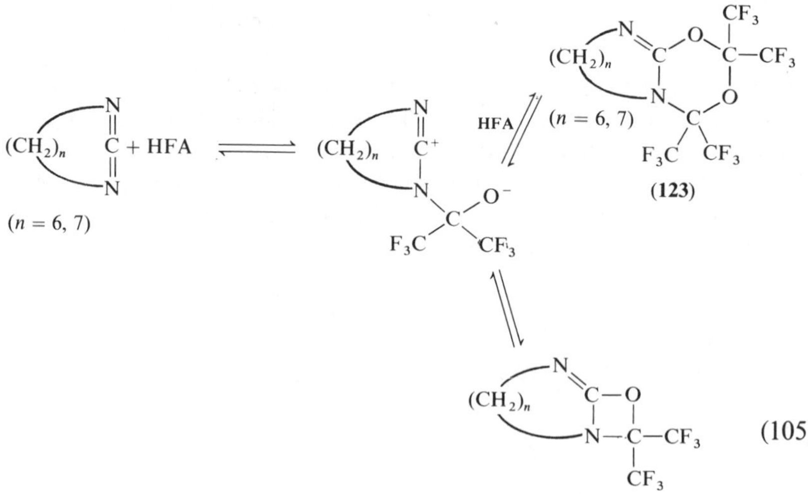

(123)

(105)

$(n = 6)$ (124)

Interestingly only the C–C double bond of ketenes (*111*) and ketimines (*284*) is attacked by HFA.

Carboxamides form rather unstable 1:1 adducts with HFA insertion into one N—H bond occurring as with urea (*202*). Addition of another molecule of HFA can only be effected by salt formation with pyridine. The addition is reversible. A stable product **125a** is obtained from urea in the presence of acetic acid anhydride (*257*). Thiourea forms the corresponding oxadiazin-thione **125b** with HFA (*257*).

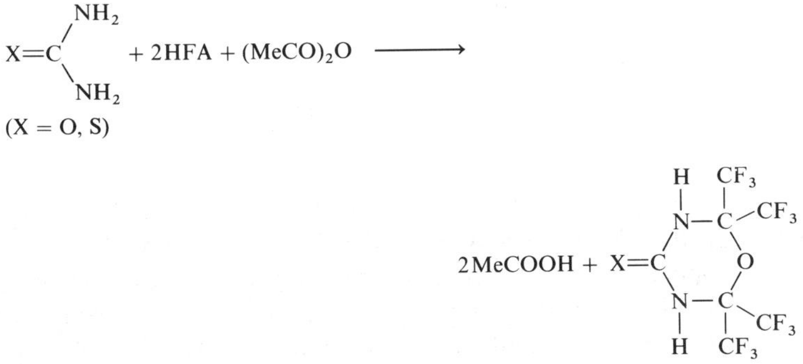

$$X{=}C\diagup^{NH_2}_{\diagdown NH_2} + 2HFA + (MeCO)_2O \longrightarrow$$

(X = O, S)

$$2MeCOOH + X{=}C \qquad (106)$$

(125)

B. SULFUR

1. Insertion Reactions

The reaction of HFA with hydrogen sulfide (*136*) and mercaptans (*103, 150*) yields hemimercaptals **126** at moderate temperatures in analogy to Eqs. (79), (81), and (84)–(87). Monothioacetic (*206*), trifluoroacetic (*218*), and benzoic (*206*) acids react similarly.

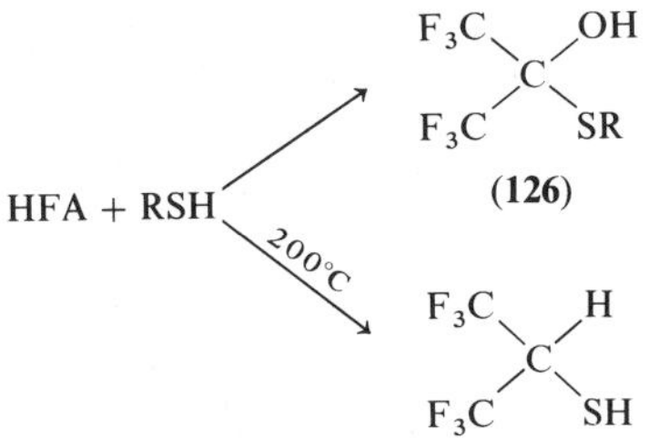

$$HFA + RSH \qquad (107)$$

$$F_3C\diagup^{OH}_{\diagdown SR} \quad C$$

(126)

200°C

$$F_3C\diagup^{H}_{\diagdown SH} \quad C$$

(127)

(R = H, Me, *n*-Pr, *i*-Pr, *t*-Bu, Ph, C$_6$F$_5$, MeCO, CF$_3$CO, PhCO)

Pyrolytic conditions favor the formation of the thiol **127** (*168*). Attempted dehydration of **126a** (R = H) with diethylamine and subsequent acidification yield hexafluoropropanol (*150*).

Carbon–sulfur bonds show a remarkable inertness toward attack of HFA. For example, thiirane, bis(trifluoromethyl) disulfide, and tetrafluoro-1,3-dithietane are not affected. However, thietane forms 1,2-oxathiane **128** (*195*).

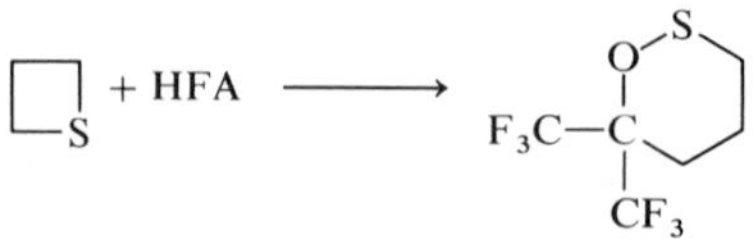

$$\square_S + HFA \longrightarrow F_3C-C(CF_3)... \tag{108}$$

(128)

Sulfides with α hydrogen atoms like dimethyl sulfide, dimethyl disulfide, and tetrahydrothiophene form diols under UV irradiation (*195*). In the tetrahydrothiophene-2HFA$_2$ adduct the hydroxyhexafluoropropyl groups are in trans positions and the trifluoromethyl groups show nonequivalence (*195*). Aromatic thio compounds like thiophenols (*17*), diaryl sulfides (*179*, *180*, *207*), and thiophene (*95*) add HFA in an ortho position, as does furan (*95*).

Reduction of the adduct **129** of HFA and a para-substituted diarylsulfane with potassium hydride yields a spirocyclic sulfur(IV) compound **130** (*180*).

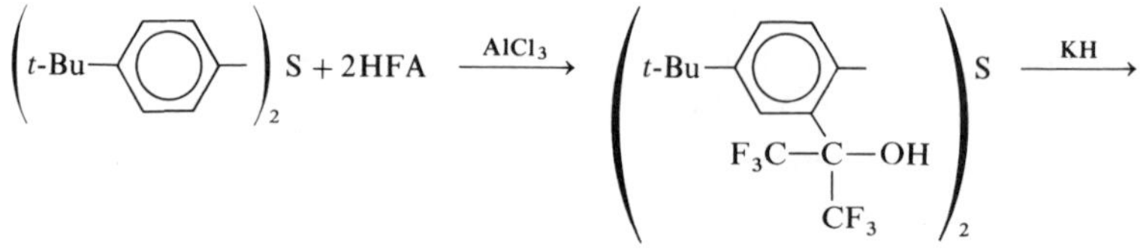

$$\left(t\text{-Bu}-\bigcirc-\right)_2 S + 2HFA \xrightarrow{AlCl_3} \left(t\text{-Bu}-\bigcirc- \right)_2 S \xrightarrow{KH}$$

with F$_3$C—C—OH and CF$_3$ groups

(129)

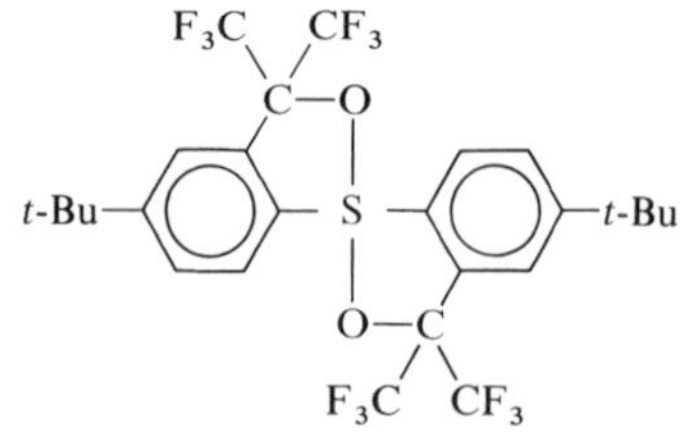

$$t\text{-Bu}-\bigcirc- S -\bigcirc- t\text{-Bu} \tag{109}$$

with C—O, F$_3$C CF$_3$ and O—C, F$_3$C CF$_3$

(130)

Multistep synthesis according to Eq. (110) affords sulfur anions **132a** and **132b** with unusual coordination via alcohol **131** (*207*). The spirocyclic anion **132b** has been estimated to be ~13 kJ/mol more stable than its open-chain isomer **132c**.

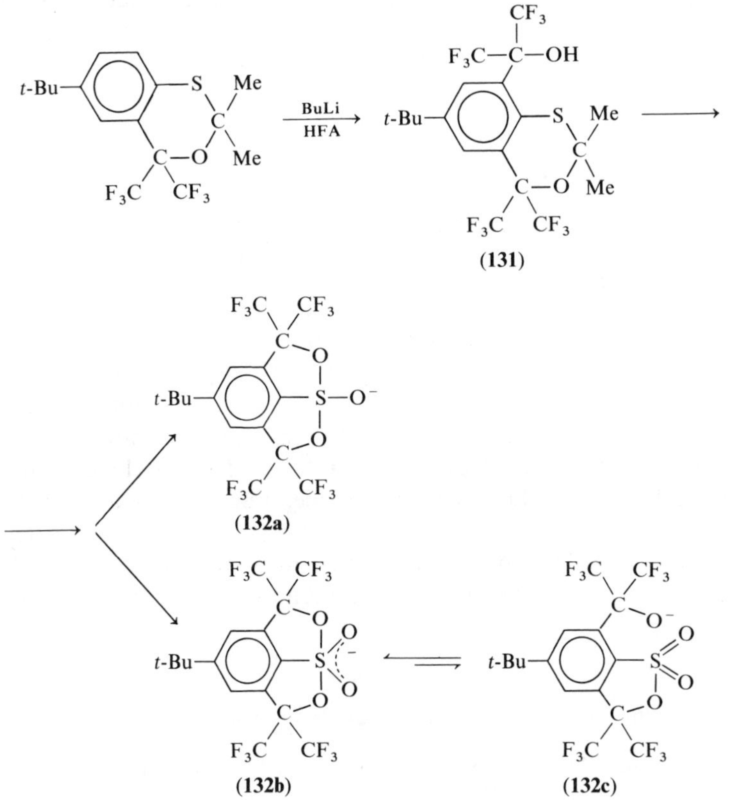

(110)

2. Dioxathiolanes

A λ^4-1,3,2-dioxathiolane **133a** is formed in the reaction of HFA with trichlorothiophosphorane (*195*).

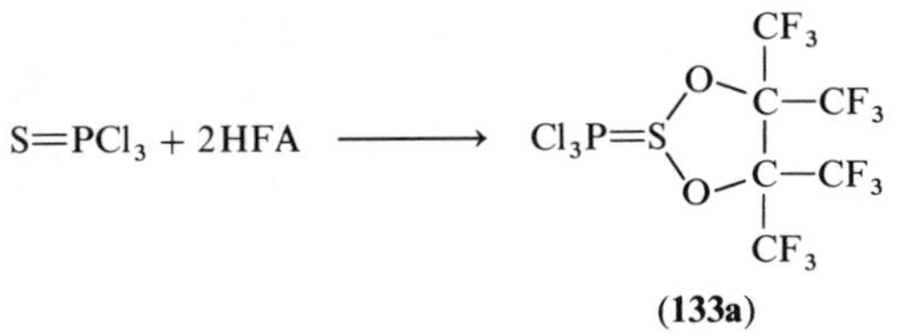

(111)

The reaction of sulfur chlorides (7, 64) and iminosulfur difluorides (64) with dialkali salts of perfluoropinacol is another synthetic route to 1,3,2-dioxathiolanes **133b–e**.

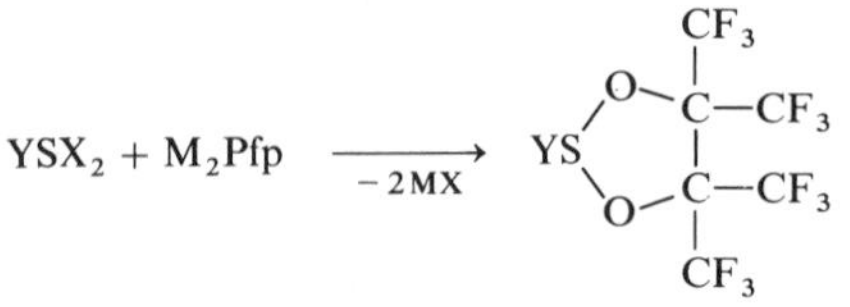

$$YSX_2 + M_2Pfp \xrightarrow[-2MX]{} \text{(133)} \tag{112}$$

133	b	c	d	e
Y	O	O$_2$	CF$_3$N	C$_2$F$_5$N
X	Cl	Cl	F	F
M	Li, Na	Li, Na	Li	Li
Ref.	7, 64	7, 64	64	64

In analogy with the disproportionation in the reaction of dichlorosulfane and sodium fluoride to yield sulfur tetrafluoride, sulfur dichloride reacts with two molecules of perfluoro pinacolate to yield a spirocyclic λ^4-bisthiadioxolane **133f** with elimination of elemental sulfur (64). Compound **133f** has also been obtained from the reaction of perfluoropinacol with dichlorosulfane in the presence of pyridine (18).

3. Reactions with X=S Double Bonds and Pseudohalides

Carbon disulfide has been found to be inert toward attack of HFA even at elevated temperatures (195). Thiocarboxamides react with two molecules of HFA. The intermediates with the likely structure **134** can be dehydrated with phosphorus oxychloride, pyridine (44), or trifluoroacetic acid anhydride (49) to yield Δ^4-1,3,5-oxathiazines **135**. A retro Diels–Alder reaction takes place with evolution of HFA when **135** is heated to 140°C (44, 45). The heterobutadienes **136** are in equilibrium with the thiazetes **137**, which are more stable at ambient temperature (45).

The stability of the strained four-membered ring is due to the trifluoromethyl groups. Compounds like P$_4$S$_{10}$ (49), P$_4$Se$_{10}$, Sb$_4$Te$_6$, In$_2$Te$_3$, and elemental Te (50) cause ring expansion. Six-membered heterocycles have also been reported (49).

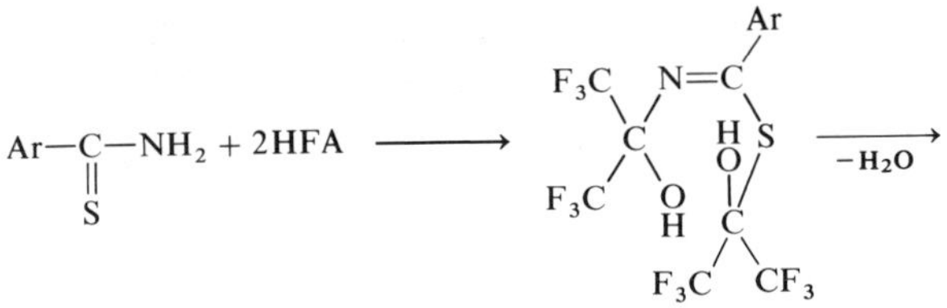

$$Ar-\underset{\underset{S}{\parallel}}{C}-NH_2 + 2HFA \longrightarrow (134) \xrightarrow{-H_2O}$$

(134)

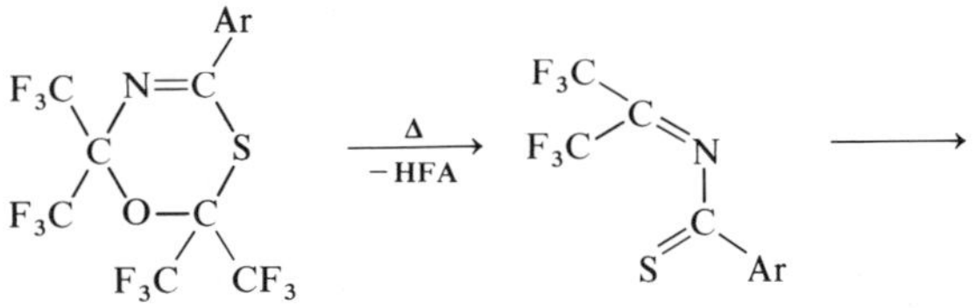

(135) $\xrightarrow[-HFA]{\Delta}$ (136) $\longrightarrow$

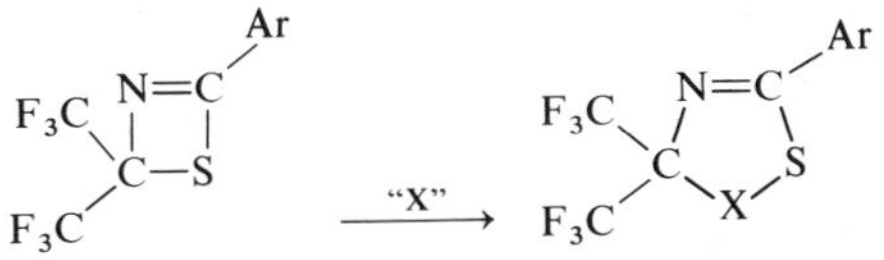

(135)   (136)

(X = S, Se, Te)   (137) $\xrightarrow{\text{"X"}}$ (138)   (113)

HFA inserts into an N—H bond of dithiooxamide with elimination of sulfur to yield monothiooxamide 139 (234). The X-ray structure of 139 shows the six atoms of the thiooxamide skeleton to be almost planar.

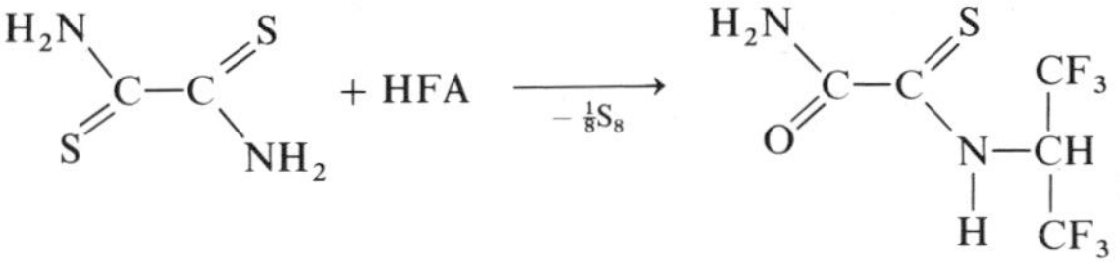

(139) + HFA $\xrightarrow{-\frac{1}{8}S_8}$ (139)   (114)

(139)

HFA has been reacted with a series of dicyanosulfanes NCS_nCN ($n = 1-4$) under the catalytic action of triethylamine. The monosulfane produces a red oil which shows extensive decomposition during attempts of purification (176). Spectral data suggest formation of a Δ^4-1,3,5-dioxazine ring system as has been found in the reactions of the homologous dicyanoselenane [Eq. (116)] and thiocyanogen with HFA, which results in compound 140. The X-ray structure has been reported. A remarkable feature is the stability of the sulfur–sulfur bond, which is not affected by HFA similarly to the reaction of dialkyl disulfides with HFA (*195*). Cleavage has been achieved with elemental chlorine (*235*).

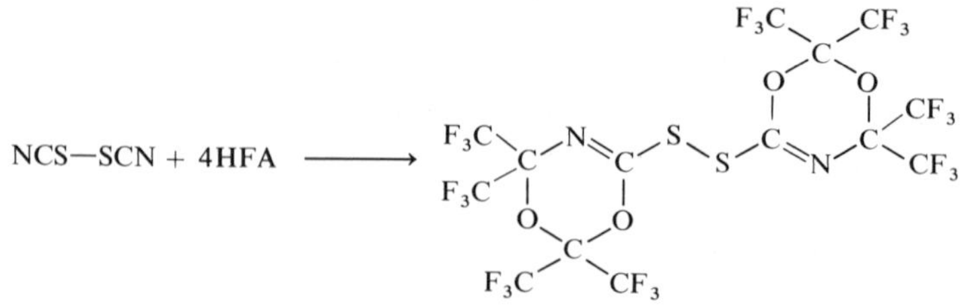

$$\text{NCS—SCN} + 4\text{HFA} \longrightarrow \quad (115)$$

(**140**)

The same product **140** is formed when the $Hg(SCN)_2$-HFA adduct **192** (Section V,F) is treated with elemental bromine (*241*).

The higher sulfanes ($n = 3, 4$) react in a similar manner; the products obtained aᵣe rather unstable and form **140** with elimination of sulfur (*142*).

C. SELENIUM AND TELLURIUM

Involvement of the catalyst has been observed in the reaction of excess HFA with dicyanoselenane. A bicyclic compound is formed with one dioxazine ring attached to selenium, the other ring being generated by complete dehydrogenation of an ethyl group (*240*). The X-ray structure of **141** has been reported.

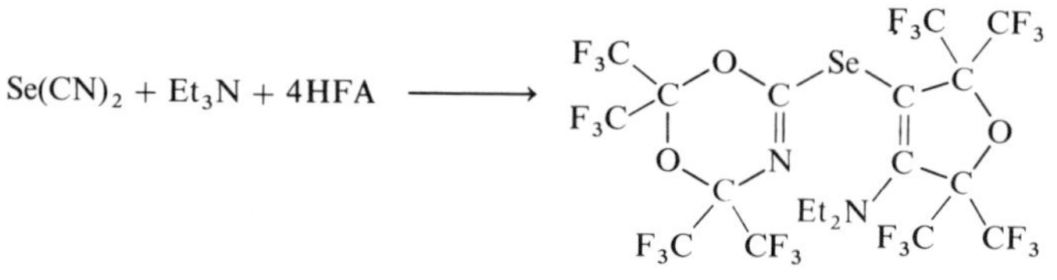

$$\text{Se(CN)}_2 + \text{Et}_3\text{N} + 4\text{HFA} \longrightarrow \quad (116)$$

(**141**)

A striking difference to the reactions with cyanate and thiocyanate [Eq. (95)] has been found with potassium selenocyanate. Elemental selenium is precipitated and cyanide ion is the reactive species in accordance with Eq. (86) (*237*).

Diselenazolines **142** are formed from selenourea or selenocarboxamides and HFA (*48*).

Compound **142** is a useful synthon for the synthesis of other selenium-containing heterocycles (*48, 49*).

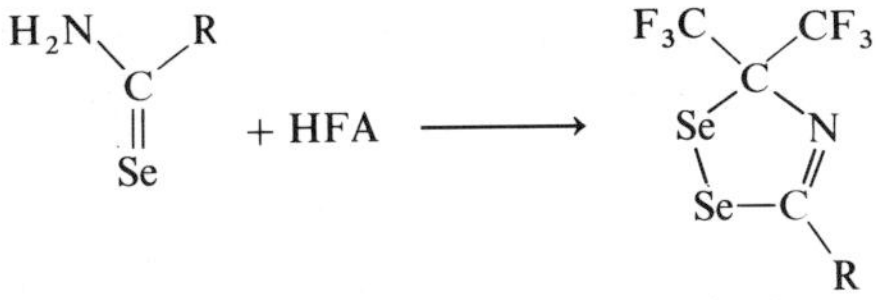

$$\underset{\substack{\|\\ \text{Se}}}{\overset{H_2N}{\underset{}{\underset{C}{\overset{R}{\diagdown}}}}} + \text{HFA} \longrightarrow \underset{\substack{\\ \text{Se}-C\\ \diagdown\\ R}}{\overset{F_3C \diagdown \diagup CF_3}{\underset{}{\underset{C}{\overset{}{}}}} \quad (117)$$

(R = NMe$_2$, Ar) (**142**)

A mechanism [Eq. (118)] has been proposed for the generation of diselenetane **143** from HFA and triphenylselenophosphorane at elevated temperatures (*214*).

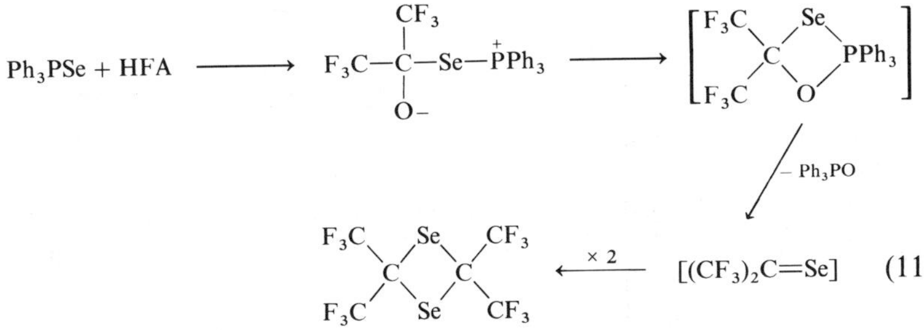

$$\text{Ph}_3\text{PSe} + \text{HFA} \longrightarrow \underset{\substack{|\\ O-}}{F_3C-\overset{\overset{CF_3}{|}}{C}-Se-\overset{+}{P}Ph_3} \longrightarrow \left[\underset{\substack{F_3C\\ }}{\overset{F_3C}{\diagdown}}\underset{O}{\overset{Se}{\underset{}{C}}}\overset{}{\diagdown}PPh_3 \right]$$

$$\Bigg\downarrow {\scriptstyle - \text{Ph}_3\text{PO}}$$

$$\underset{\substack{F_3C\\ \diagup}}{\overset{F_3C}{\diagdown}}\underset{Se}{\overset{Se}{\underset{}{C}}}\underset{\substack{\diagup\\ CF_3}}{\overset{CF_3}{\diagdown}} \quad \xleftarrow{\times 2} \quad [(CF_3)_2C{=}Se] \quad (118)$$

(**143**)

Tellurium tetrachloride reacts with disodium perfluoropinacolate to yield the spirocycle **144**, a homologue of the sulfur spirocycle **133f** (*7*).

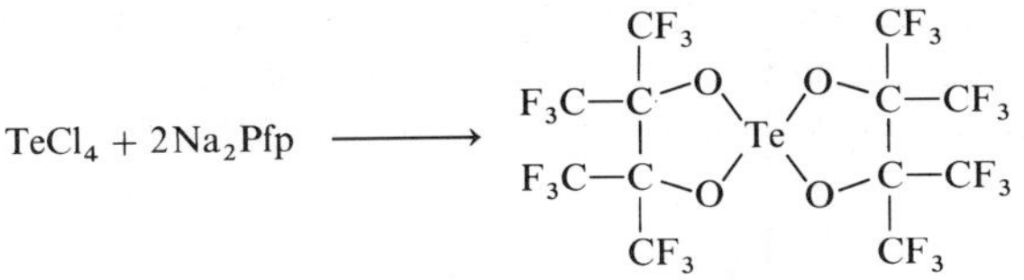

$$\text{TeCl}_4 + 2\,\text{Na}_2\text{Pfp} \longrightarrow \underset{\substack{CF_3\\ }}{\overset{CF_3}{\underset{}{}}} \quad (119)$$

(**144**)

V. Metal Complexes of HFA

Green, Stone, and co-workers have done considerable work on HFA metal complexes. Most of their publications, of which the earlier ones were reviewed by Stone in 1972 (*261*), deal with low-valent group VIIIB elements. Two kinds of reactions are generally observed.

1. Insertion into C—H bonds of the ligands as well as formation of metallacycles (i.e., metallaoxiranes and 1,2,4-metalladioxolanes) according to Eqs. (2)–(4), depending on the coordination sphere and reaction conditions. All metallaoxiranes and 1,2,4-metalladioxolanes are listed in Tables VII and VIII. Three-membered rings have been synthesized by reaction of metal

TABLE VII

3-Bis(trifluoromethyl)metallaoxiranes

Compound	M	L^a	n	L'^a	m	Method	Reference
145	Fe	C_4Me_4 CO	1	CO	1	a	31
158a	Ni	t-BuNC	2	L	2	a	124, 135
b		PhNC	2	L	2	a	135
c		COD	1	L	1	a	38, 40
d		PEt_3	2	L	2	a	39
e		$PMePh_2$	2	L	2	a	39
f		PPh_3	2	C_2H_4	1	a	15
a, d–l		L''	2	COD	1	b	38, 39, 41, 124, 135
163a	Ru	$PMePh_2$, CO	2 2	CO	1	a	54
163b		$P(OCH_2)_3CEt$ CO	2 2	CO	1	a	65
164	Os	PMe_2Ph, CO	2 2	CO	1	a	125
167a	Rh	$PMePh_2$, acac	2 1	CO		a	200
b		PPh_3	2				
		$NCH(CF_3)_2$	1			a	184
		Cl	1				

						PPh₃	2	a	63

Compound	M	Ligand	n	L′		n	a/b	Ref.
171a		PPh$_3$	1	PPh$_3$	2		a	63
b		NO	1					
		PMePh$_2$	2				a	61
c		CO	1					
		Cl	1					
		PPh$_3$	2	PPh$_3$			a	61
		CO	1					
		Cl	1					
173a	Pd	PEt$_3$	2	L	2		a	199
b		(PPh$_2$CH$_2$)$_2$	1	P(OPh)$_3$	2		b	94
c		PMePh$_2$	2	L	2		a	94, 199
d		PPh$_3$	2	(CF$_3$)$_2$CNH	1		b	94
e		P(OPh)$_3$	2	L	2		a	94
179a	Pt	COD	1	L	1		a	128, 129
		COD	1	C$_2$H$_4$	3		b	129
b		PEt$_3$	2	L	1		a	138
c		P(i-Pr)$_3$	2	L	1		a	138
d		PMePh$_2$	2		2		a	61
			2	PPh$_3$	2		b	40
e		(PPh$_2$CH$_2$)$_2$	1	PPh$_3$	2		b	40
f		PPh$_3$	2	L	2		a	61, 134
			2	COD	1		b	128, 129
			2	DBA	1		a	58, 59
			2	CF$_3$CN	1		b	29
g		P(OPh)$_3$	2	L	2		a	40

^{a} L″, t-BuNC, PEt$_3$, PMePh$_2$, PPh$_3$, P(OPh)$_3$, P(OCH$_2$)$_3$CEt, $\frac{1}{2}$(PPhCH$_2$)$_2$, $\frac{1}{2}$(C$_5$H$_4$N)$_2$, $\frac{1}{2}$(o-AsMe$_2$)$_2$C$_6$H$_4$; acac, acetylacetonate; DBA, dibenzalacetone.

TABLE VIII

3,3,5,5-Tetrakis(trifluoromethyl)-1,2,4-metalladioxolanes

Compound	M	L	n	L'	m	Method	Reference
160a	Ni	t-BuNC	2	L	2	a, b	*124, 135*
b		o-$(AsMe_2)_2C_6H_4$	1			b	*41*
166	Rh	PPh_3	1	PPh_3	1	a	*200*
		acac					
174a	Pd	t-BuNC[a]	2			a	*77, 93*
b		c-$C_6H_{11}NC$[a]	2			a	*77*
c		PEt_3	2	L	2	a	*199*
d		$(PPh_2CH_2)_2$	1			b	*94*
e		$PMePh_2$	2	L	2	a, b	*94, 199*
f		$P(OMe)_2Ph$	2	L	2	a	*94*
g		$P(OMe)_3$	2	L	2	a	*94*
h		$AsMe_2(CH_2Ph)$	2	L	2	a	*94*
180a	Pt	t-BuNC[a]	2			a	*105, 131*
b		COD	1	L	1	a	*128, 129*
			1	i-Pr	2	a	*40*
c		$PMePh_2$	2			b	*40*
d		$(PPh_2CH_2)_2$	1			b	*40*
e		$P(OMe)_3$	2	L	2	a	*40*

[a] Trimer $M_3(RNC)_6$.

complexes containing labile ligands with HFA (method a) and by ligand displacement, with the heterocycle remaining intact (method b) according to Eq. (120).

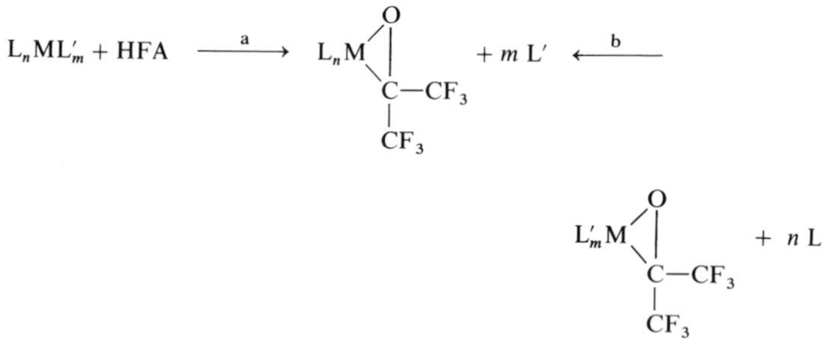

$$L_nML'_m + HFA \xrightarrow{\ a\ } L_nM \overset{O}{\underset{C-CF_3}{\big|}} \overset{}{\underset{CF_3}{\big|}} + m\,L' \xleftarrow{\ b\ }$$

$$L'_mM \overset{O}{\underset{C-CF_3}{\big|}} \overset{}{\underset{CF_3}{\big|}} + n\,L \qquad (120)$$

2. Similarly, five-membered rings can be synthesized by either addition of excess HFA to the complexes with exchange of ligands (method a) or by ring expansion of the corresponding metallaoxiranes (method b) [Eq. (121)].

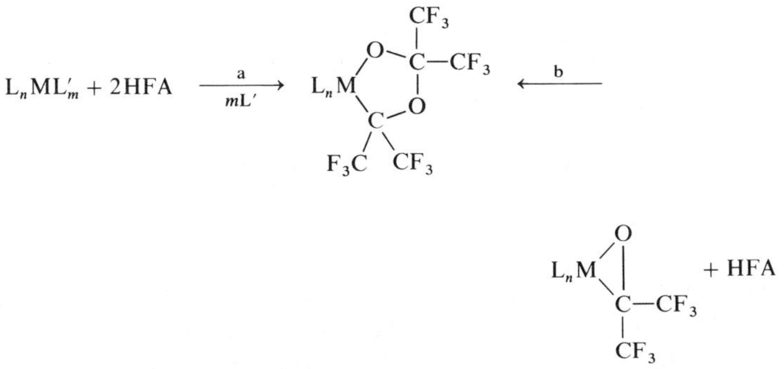

$$L_nML'_m + 2HFA \xrightarrow[mL']{a} L_nM \underset{F_3C \quad CF_3}{\overset{\underset{|}{CF_3}}{\overset{O-C-CF_3}{\underset{C-O}{|}}}} \xleftarrow{b}$$

$$L_nM \overset{O}{\underset{\underset{CF_3}{|}}{\overset{||}{C-CF_3}}} + HFA \qquad (121)$$

A. IRON, COBALT, AND NICKEL

The photochemical reaction of tetramethylcyclobutadieneiron tricarbonyl with HFA leads to the formation of three isomers of the complex **145**. The stereochemistry of these isomers has been investigated by NMR spectroscopy (*31*).

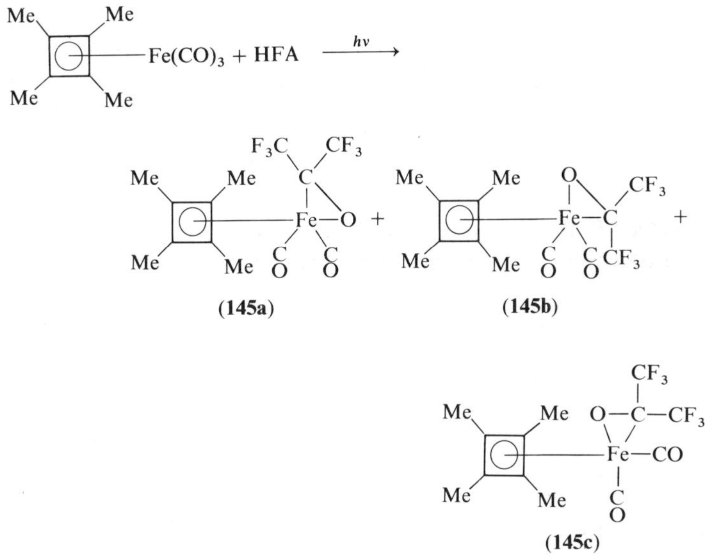

$$(145a) \qquad\qquad (145b) \qquad\qquad (145c) \qquad (122)$$

Ferrocenes react directly (180°C/15 days) or in the presence of $AlCl_3$ as catalyst (20°C/24 hours) to yield fluorinated 2-ferrocenylpropanols **146a–c** (*42*).

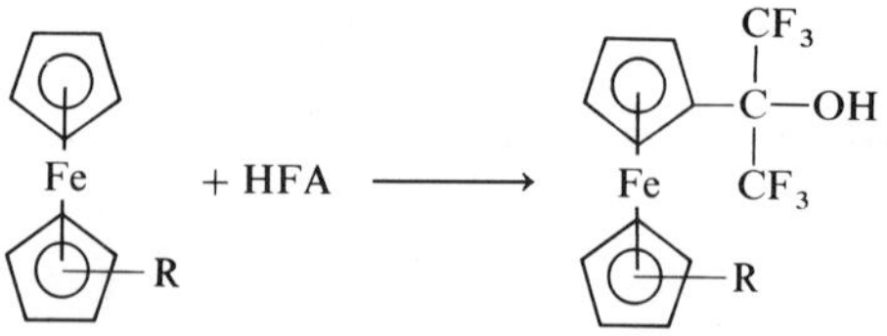

$$(123)$$

(R = H, COMe, COPh)   (146a–c)

In η^5,η^1-bis(cyclopentadienyl)iron dicarbonyl the σ-bonded ligand is easily attacked by HFA with conservation of the σ bond. Photochemical treatment of **147** affords **146d**, in which both cyclopentadienyl rings become π bonded (*76*).

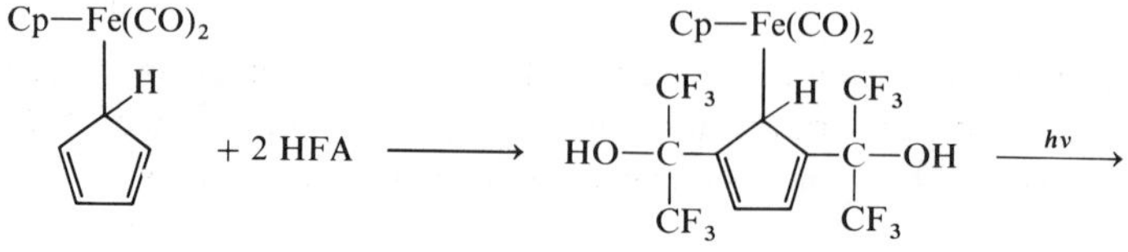

(147)

$$(124)$$

(146d)

High stereospecifity has been found in the reaction of HFA with tricarbonyl (η^4-cycloheptadiene)iron complexes (*126, 127*).

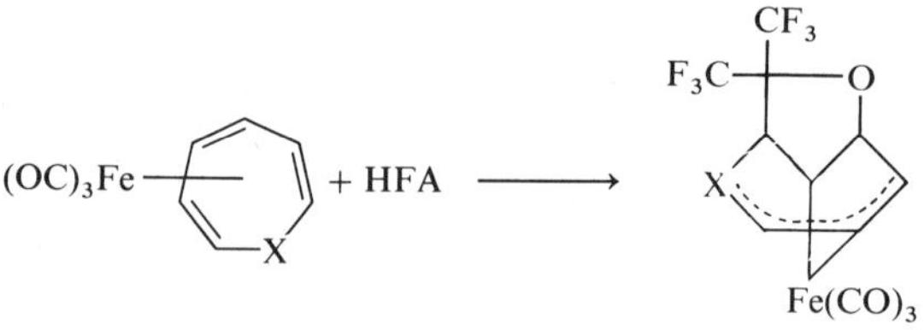

$$(125)$$

(X = CH$_2$, CO, NCO$_2$Me)   (148a–c)

Mainly oxo addition products **148a–c** are formed.

Iron complexes with hydrated furane ligands are available from appropriate alkynyl- and alkenyl-substituted iron compounds as precursors (*174*).

Cp(CO)$_2$Fe—CH$_2$—C≡C—R + HFA ⟶
(R = Me, Ph)

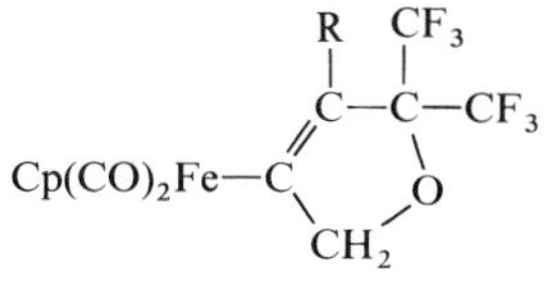

$$\text{Cp(CO)}_2\text{Fe—C} \underset{\text{CH}_2}{\overset{\displaystyle \overset{\text{R} \quad \text{CF}_3}{\overset{|\quad\; |}{\text{C—C—CF}_3}}}{\diagup\!\!\!\diagdown_{\!\!\text{O}}}} \qquad (126)$$

(149)

Cp(CO)$_2$Fe—CH$_2$—CH=CRR' + HFA ⟶
(R = Me, H; R' = Me, H, Me, Ph, Cl)

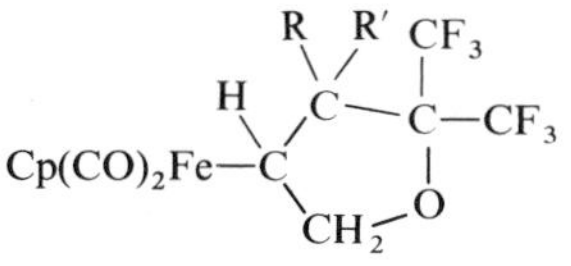

$$\text{Cp(CO)}_2\text{Fe—C} \qquad (127)$$

(150)

Photochemical decarbonylation of **149b** in the presence of PPh$_3$ affords a chiral metal center as reflected in the nonequivalence of the two CF$_3$ groups (*174*).

A σ-phenylethinyliron complex reacts with either one or two molecules of HFA to yield four- and six-membered rings **151** and **152**, respectively (*78*).

Cp(CO)$_2$Fe—C≡CPh + HFA ⟶

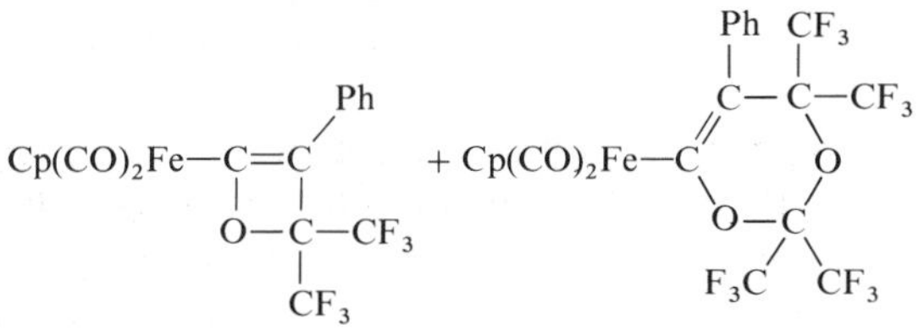

$$\text{Cp(CO)}_2\text{Fe—C=C} \quad + \quad \text{Cp(CO)}_2\text{Fe—C} \qquad (128)$$

(151)   **(152)**

Exchange of one CO group in the starting material by PPh$_3$ only leads to the formation of the oxetane (*78*). Once formed, the four-membered rings **151** do not undergo ring expansion.

Addition of HFA to substituted butadieneiron complexes includes the metal atom (*132, 133*).

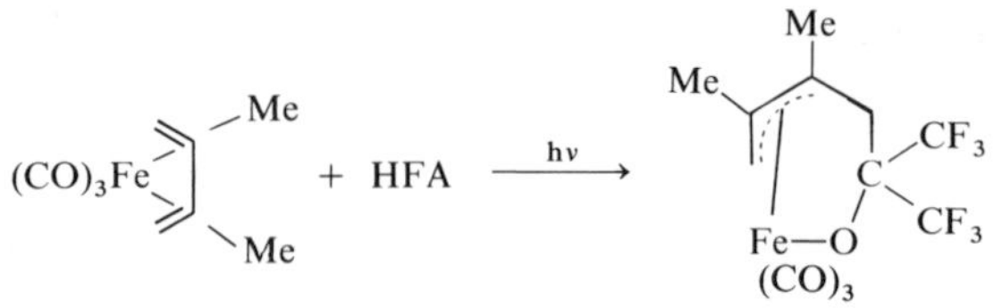

$$(CO)_3Fe \overset{Me}{\underset{Me}{\bigtriangleup}} + HFA \xrightarrow{h\nu} (153) \qquad (129)$$

While the unsubstituted butadiene complex does not react, the isoprene complex yields a 2:1 addition product **154** when trace amounts of Fe(CO)$_5$ are present (*132, 133*).

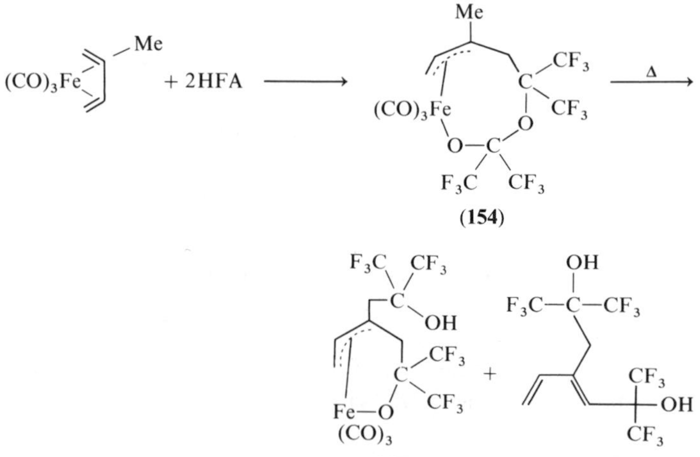

$$(CO)_3Fe \overset{Me}{\underset{}{\bigtriangleup}} + 2HFA \longrightarrow (154) \xrightarrow{\Delta}$$

$$(155) \quad + \quad (156) \qquad (130)$$

Heating **154** in hexane in a sealed tube affords the isomer **155** by insertion of one molecule of HFA into a C—H bond of the methyl group, and the diene **156** by cleavage of the metal bonds and migration of a hydrogen atom (*132, 133*).

Only two reports deal with the reactions of cobalt complexes with HFA. Insertion into the cobalt–hydrogen bond of a hydride complex affords a cobalt hexafluoroisopropylate (*136a*). An oxolene(2) is formed from an alkylcobalt compound and HFA (*66*). For mechanistic reasons the authors favor the depicted structure **157** over the isomeric oxolene (3) ring reported for the analogous iron complex **149** (*174*).

$$PyL_2Co-CH{=}C{=}CH_2 + HFA \longrightarrow (157) \qquad (131)$$

Potassium tetrafluorocobaltate(III), a mild fluorinating agent, leads to the formation of molecules like CF_4, COF_2, and CF_3COF (21).

Starting from $Ni(COD)_2$, a series of three-membered rings **158a, d–l** has been synthesized (see Table VIII) (38, 39, 41, 124, 135). Single-crystal X-ray diffraction studies of **158a,f** reveal that the complex is nearly planar, the nickel atom being equidistant from carbon and oxygen (68, 69).

While ring expansion of bis(t-butylisocyanide)hexafluoroisopropylidene-iminenickel with HFA affords two isomers **159a,b** in a 1:4 ratio, the inverse reaction of the HFA complex **158a** with the imine yields only **159b** (135), the structure of which has been determined (69).

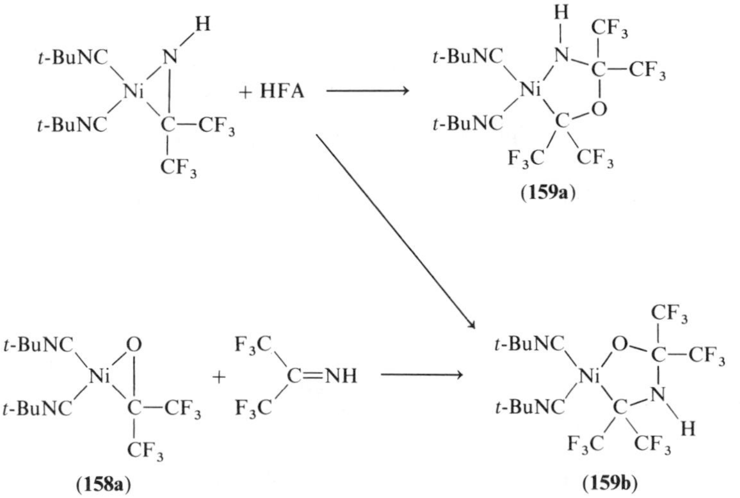

$$(132)$$

Only little evidence has been found for ring expansion of the analogous phenylisocyanide complex **158b** with HFA. The observation is in contrast to the reaction with hexafluoroisopropylideneimine (135). Compound **158a** reacts with tetrafluoroethane eliminating HFA and forming a perfluorinated niccolacyclopentane (135).

The five-membered nickel ring **160a** can be obtained in both ways [Eq. (121)] (124, 135); the diarsano-o-phenylene complex **160b** has been synthesized via ring expansion (41) (see Table VII).

At $-50°C$ peroxobis(t-butylisonitrile)nickel forms an explosive 1:1 adduct with HFA **161**, which has been assigned a five-membered ring structure by analogy with the complexes of peroxoplatinum compounds with CO_2 and CS_2. Excess HFA yields a labile 2:1 complex with no fluorine coupling in the ^{19}F NMR spectrum. Thus, structure **161a** has been proposed for this complex. Peroxide **161** easily transfers one oxygen atom to diethyl ether with ring contraction to form a niccoladioxetane **162** (124, 135).

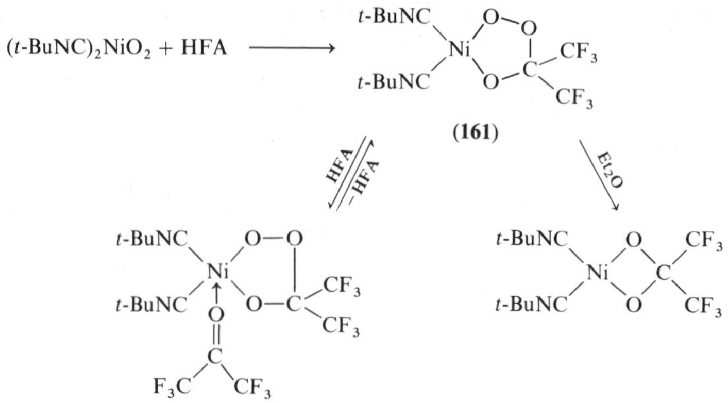

(161) (161a) (162) (133)

Similarly, dioxobis(triphenylphosphane)platinum reacts with HFA. Ring contraction has been achieved with triphenylphosphane. An unstable bis adduct which is believed to be a seven-membered ring has also been reported (*137*).

B. Ruthenium and Osmium

Only a few examples are known in which ruthenium and osmium compounds undergo addition reactions with HFA. While *trans*-$(PPh_3)_2$-$Ru(CO)_3$ does not yield stable products (*65*), *trans*-$(PMePh_2)_2Ru(CO)_3$ forms a complex in which the phosphane ligands are still in trans positions (*54*). The analogous osmium complex **164** (*125*) and the phosphite ruthenium complex **163b** show isomerization; the CO ligands are in trans positions. In the case of **163b**, the rearrangement has been proved by ^1H NMR spectroscopy (*65*).

(**163a**, L = $PMePh_2$) (134)

[L = $P(OCH_2)_3CEt$, M = Ru; **163b**,]
[L = PMe_2Ph, M = Os; **164**]

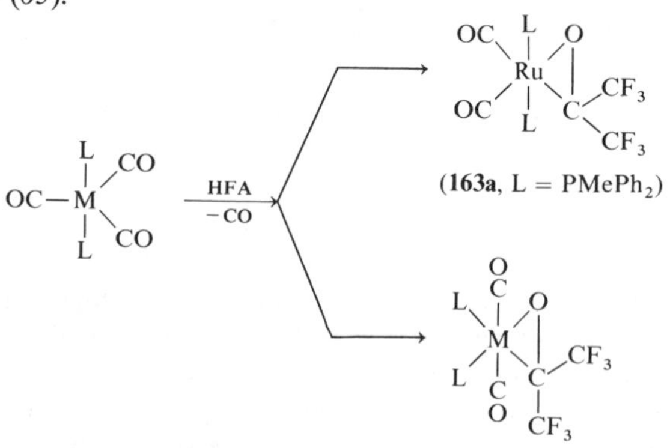

In analogy to ferrocenes [Eq. (123)], the cyclopentadienyl ring is attacked by HFA according to Eq. (135) (27).

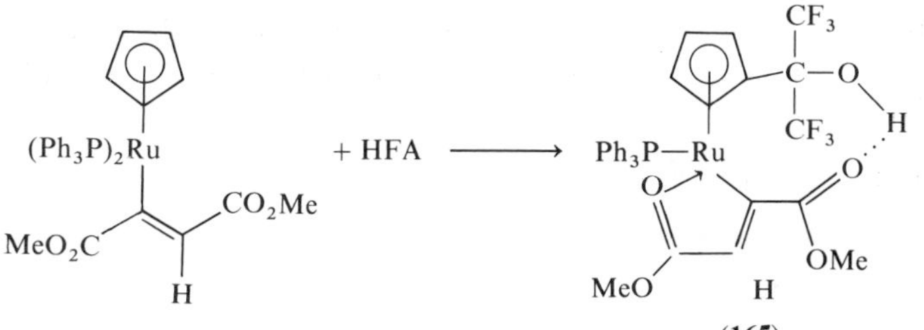

(135)

(165)

The loss of one molecule of PPh_3 in **165** is compensated by coordination to one carbonyl oxygen atom. The other one forms a hydrogen bond, as seen in the X-ray crystal structure of **165** (28, 215).

In analogy to the homologous iron complex **148c** [Eq. (125)], (η^4-N-methoxycarbonyl-^{1}H-azepine)ruthenium tricarbonyl reacts with HFA (127).

C. RHODIUM AND IRIDIUM

Probably due to steric effects, an interesting contrast in reactivity depending on the ligands has been found in the reaction of excess HFA with bisphosphanorhodium(+I) acetylacetonates. The bis(triphenylphosphane)-substituted complex yields a metalladioxolane **166** by losing one ligand. With methyldiphenylphosphane a metallaoxirane **167a** is formed (200). The ^{19}F NMR chemical shift of **167a** with trans configuration of the phosphane ligands shows a signal shifted 10 ppm to higher field in comparison to the related group VIIIA complexes **158**, **163**, **173**, and **179** (200).

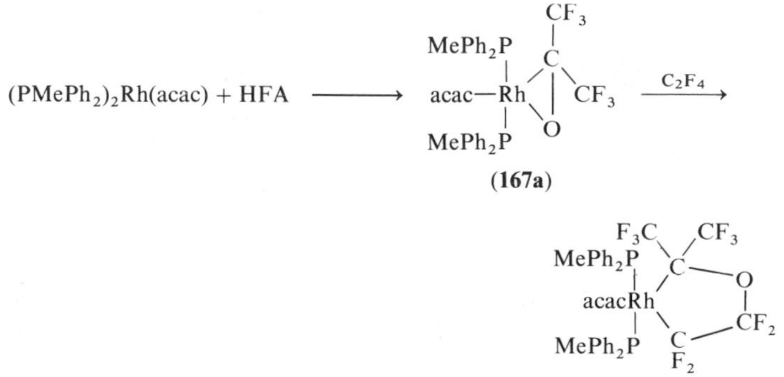

(167a)

(136)

(168)

Addition of tetrafluoroethylene to **167a** affords the 1,3-metallaoxolane **168** (*200*) in contrast to the corresponding reaction of the nickel complex **158a**, in which HFA is exchanged (*135*).

A similar extension of the coordination sphere at the rhodium atom has been found in the reaction of the nitrene complex **169** with HFA (*184*).

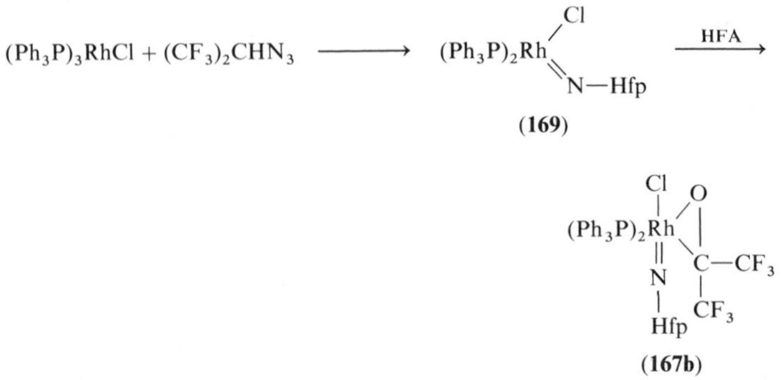

$$(Ph_3P)_3RhCl + (CF_3)_2CHN_3 \longrightarrow (Ph_3P)_2Rh \overset{\displaystyle Cl}{\underset{\displaystyle N-Hfp}{}} \xrightarrow{\text{HFA}}$$

(169)

$$(Ph_3P)_2Rh \overset{\displaystyle Cl \quad O}{\underset{\displaystyle \underset{\displaystyle Hfp}{N}}{\overset{\displaystyle \|}{}}} \overset{\displaystyle}{\underset{\displaystyle CF_3}{C-CF_3}} \qquad (137)$$

(167b)

Cyclopentadienylrhodium(+I) complexes with substituted butadiene ligands react thermally with HFA with oxidation of the metal atom and addition to the diene in analogy to Eq. (129). Two isomers are found with isoprene (*132, 133*). 1,3-Pentadienecyclopentadienylrhodium forms only one complex with HFA (*133*) and the dimethylbutadiene complex, in addition to oxidation, undergoes insertion of a HFA molecule into a C—H bond of the cyclopentadienyl ring (*132, 133*). Whereas higher temperatures are required to effect addition of HFA to the cyclopentadienyl complexes, the corresponding indenylbutadienerhodium complexes undergo insertion at room temperature (*56*). The 2,4-hexadiene complex undergoes addition at the indenyl system.

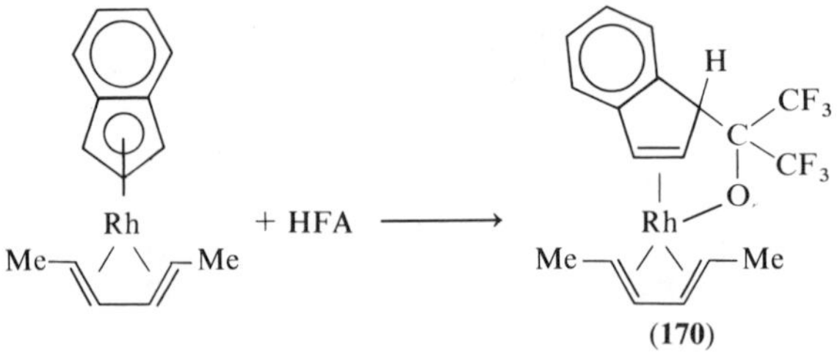

$$(138)$$

(170)

The X-ray structure of **170** has been reported (*56*). Generally only unsubstituted sp^2-hybridized carbon atoms of butadiene ligands are attacked by HFA (*133*). Different behavior has been found in the reactions of iridium complexes with HFA. With tris(triphenylphosphane)nitrosyliridium the geometry is retained [Eq. (139)] (*63*). However, the Vaska complexes lead to octahedral environments of the metal atom [Eq. (140)] (*61*). IR and NMR spectra suggest that the phosphane ligands are in trans positions and the chlorine atom is trans to oxygen (*61*).

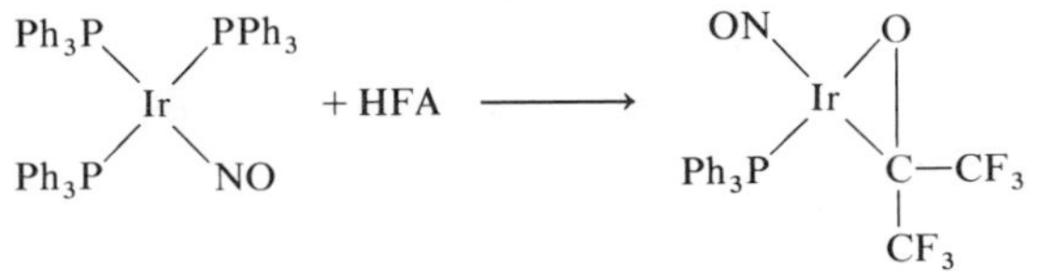

$$
\underset{\text{(171a)}}{}
$$

Ph$_3$P, PPh$_3$ ... Ir ... Ph$_3$P, NO   + HFA $\longrightarrow$ ON, O ... Ir ... Ph$_3$P, C—CF$_3$, CF$_3$ (139)

(171a)

Ph$_2$XP, CO ... Ir ... Ph$_2$XP, Cl   + HFA $\longrightarrow$ OC, PPh$_2$X, O ... Ir ... Cl, PPh$_2$X, C—CF$_3$, CF$_3$ (140)

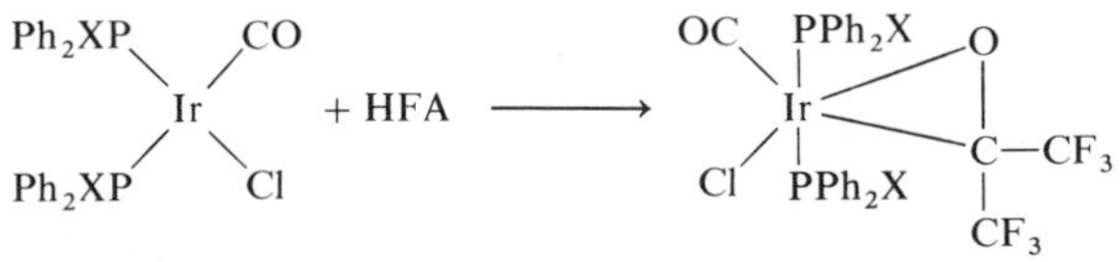

(X = Me, Ph)   **(171b, c)**

Mechanistic and kinetic studies have been carried out on some oxygenated Vaska-type complexes. Using ^{18}O-labeled complexes shows that insertion occurs into the Ir—O rather than into the O—O bond (*23*).

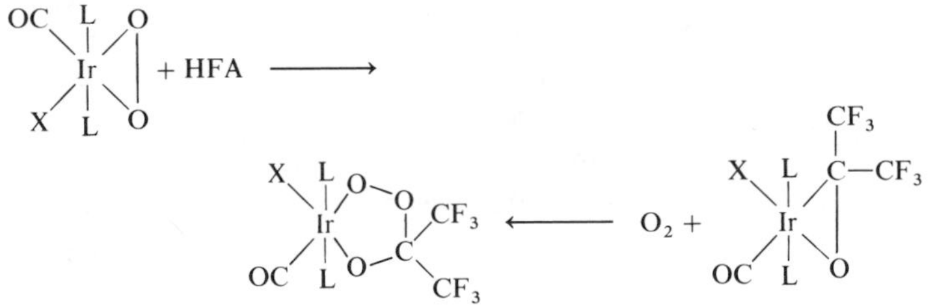

OC, L, O ... Ir ... X, L, O   + HFA $\longrightarrow$

X, L, O—O, CF$_3$ ... Ir ... OC, L, O—C, CF$_3$ $\longleftarrow$ O$_2$ +   X, L, C—CF$_3$ (CF$_3$) ... Ir ... OC, L, O   (141)

(172)   **(171)**

[L = PPh$_3$, X = Cl, Br, I; L = PMePh$_2$, P(C$_6$H$_4$-p-Me)$_3$, P(C$_6$H$_4$-p-OMe)$_3$, AsPh$_3$, X = Cl]

The reaction of the oxygen complexes with HFA proceeds at a rate two orders of magnitude faster than the addition of O$_2$ to the metallaoxirane (*23*).

D. PALLADIUM AND PLATINUM

While three-membered ring formation prevails in the iron and cobalt triad and with nickel complexes, more metalladioxolanes than metallaoxiranes are known with palladium and platinum (see Tables VII and VIII). Formation of the three- or five-membered rings is strongly dependent on stoichiometry, reaction conditions, and steric requirements of the ligands. While the bis(diphenylphosphano)ethane-substituted metallaoxirane **173b** undergoes ring expansion with excess HFA to yield **174d** and the corresponding imine, the bis(triphenylphosphane)-substituted palladaaziridine reacts with HFA with exchange and retention of the three-membered ring (**173d**). Interestingly, attempted exchange of the phosphite ligand in **173e** with methyldiphenylphosphane results in the loss of HFA (*94*).

An interesting addition reaction has been found with the isocyanide complexes of the nickel triad and dialkylamines [Eq. (142)]. The metal atom is oxidized, with formation of diaminocarbenes (*77*). The ligand cis to the metal-bonded oxygen is attacked. This is shown in the X-ray structure of **177b** (*197*).

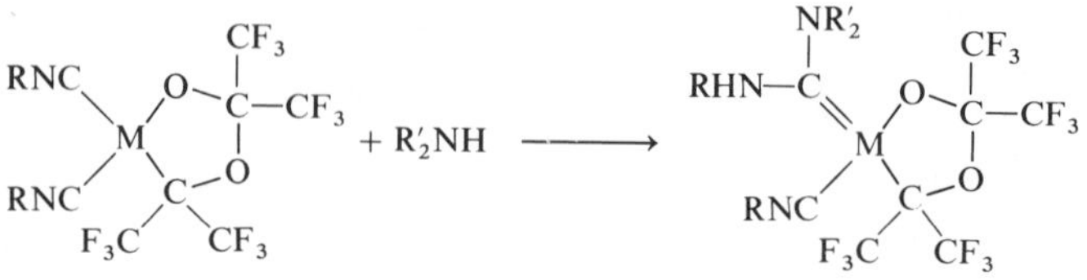

$$(142)$$

M	R	Compound
Ni	*t*-Bu, *i*-Pr	**160a,c**
Pd	*t*-Bu, *c*-C$_6$H$_{11}$	**174a,b**
Pt	*t*-Bu	**180**

M	R	R'	Compound
Ni	*t*-Bu, *i*-Pr	Et	**176a,b**
Pd	*t*-Bu, *c*-C$_6$H$_{11}$	Me, Et	**177a**
Pt	*t*-Bu	Me, Et	**178a,b**

Whereas bis(triphenylphosphane)platinaaziridine **181** reacts readily with HFA, the corresponding oxirane is inert toward ring expansion. This may be explained in terms of stronger π acceptor and weaker σ donor capacity of

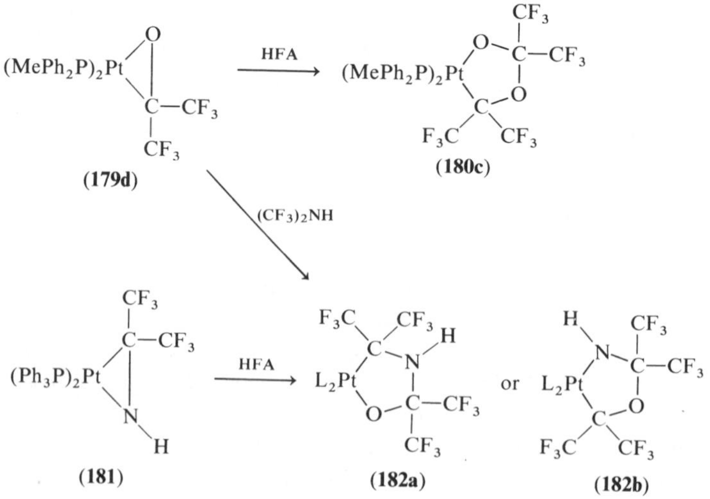

$$(143)$$

HFA than of the imine. Attack on the metal atom of one molecule of HFA and imine is a prerequisite for the formation of five-membered rings. The lack of electron density in the HFA adduct inhibits ring expansion. According to these observations the displacement of triphenylphosphane by the more nucleophilic methyldiphenylphosphane promotes dioxolane formation (40). Platinaoxazolidines are available in both ways. No structural assignment has been given for the mixed compounds **182a, b** (16).

Together with the formation of platinaoxirane **179b**, dioxaphospholane **57b** (R = Et) has been isolated (138). In contrast to the reaction of bis(triphenylphosphane)dibenzalacetoneplatinum in which platinaoxirane **179f** is formed, the analogous reaction of the bis(triethylphosphane)dibenzalacetone complex with HFA yields a product, which from spectroscopic data is assumed to arise from attack of HFA at the dienone ligand (59). The ^{195}Pt NMR spectrum of **179b** has been reported (123).

Like HFA, indanetrione reacts with tetrakis(triphenylphosphane)platinum to form a platinaoxirane, which readily adds one molecule of HFA to yield **183**. Further action of HFA affords a rather unstable seven-membered ring compound (**184**), which thermally loses either indanetrione or HFA to reform a platinadioxolane. Whether isomer **a** or **b** is formed cannot be deduced from ^{19}F NMR spectroscopy (145).

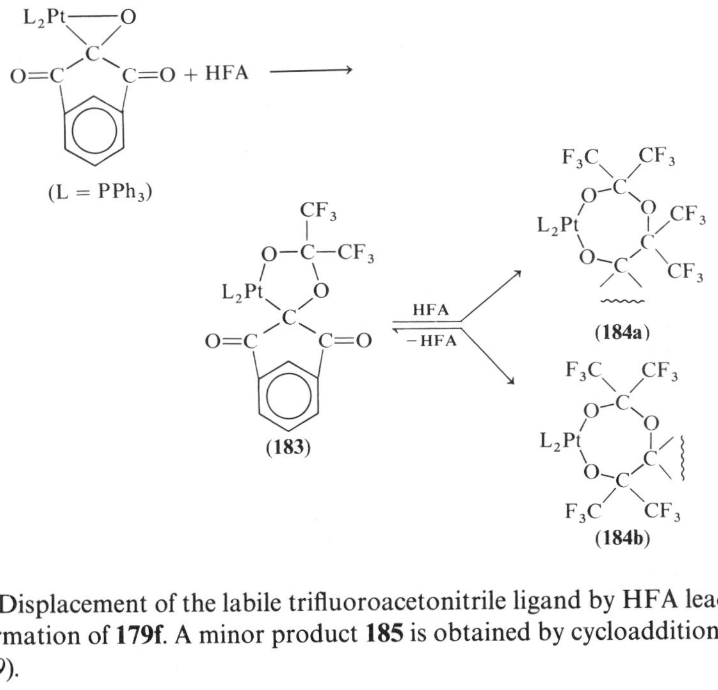

(144)

Displacement of the labile trifluoroacetonitrile ligand by HFA leads to the formation of **179f**. A minor product **185** is obtained by cycloaddition of HFA (29).

$$\text{(Ph}_3\text{P)}_2\text{Pt} \overset{\overset{\overset{\text{CF}_3}{|}}{\text{C}}}{\underset{\text{N}}{||}} + \text{HFA} \longrightarrow \text{(Ph}_3\text{P)}_2\text{Pt} \overset{\overset{\text{F}_3\text{C}\diagdown\diagup\text{CF}_3}{\text{C}}}{\underset{\text{O}}{|}} + \text{(Ph}_3\text{P)}_2\text{Pt} \overset{\overset{\overset{\text{CF}_3}{|}}{\text{C}=\text{N}}}{\underset{\text{O}-\underset{\underset{\text{CF}_3}{|}}{\overset{|}{\text{C}}}-\text{CF}_3}{}} \qquad (145)$$

$$\qquad\qquad\qquad\qquad\qquad (179f) \qquad\qquad\qquad (185)$$

Some novel structures have been found in the reaction between bis-(cyclooctadiene)platinum(0) and HFA, depending on the reaction conditions (128, 129).

$$\text{COD}_2\text{Pt} + \text{HFA}$$

$$\xrightarrow[\text{Et}_2\text{O}]{1:1} \quad \begin{array}{c} \overset{\text{CF}_3}{|} \\ \text{O}-\text{O}-\text{CF}_3 \\ |\quad\ | \\ \text{Pt}-\text{Pt} \\ \text{COD}\qquad\text{COD} \\ \textbf{(186)} \end{array}$$

$$\xrightarrow[\text{Et}_2\text{O}]{1:20} \quad \text{CODPt} \overset{\overset{\overset{\text{CF}_3}{|}}{\text{O}-\text{C}-\text{CF}_3}}{\underset{\underset{\text{F}_3\text{C}\diagup\diagdown\text{CF}_3}{\text{C}}}{|}}_{\text{O}} + \text{CODPt} \overset{\overset{\text{O}\diagdown\quad\diagup\text{CF}_3}{\text{C}-\text{CF}_3}}{\underset{\text{H}\blacktriangleright\quad\blacktriangleleft\text{H}}{}} \qquad (146)$$

$$\qquad\qquad\qquad\qquad\quad \textbf{(180b)} \qquad\qquad\qquad\qquad \textbf{(187)}$$

$$\xrightarrow[\text{C}_6\text{H}_6]{1:1} \quad \text{CODPt} \overset{\overset{\text{O}}{\diagdown}}{\underset{\underset{\text{CF}_3}{|}}{\overset{}{\text{C}-\text{CF}_3}}} \quad + \textbf{187}$$

$$\qquad\qquad\qquad\qquad \textbf{(179a)}$$

Whereas ligand exchange in **187** with triphenylphosphane and *o*-bis(dimethylarsano)phenylene proceeds with retention of conformation, ring contraction occurs with triphenylphosphane and **163** to form **179a**, which is also formed in the reaction between HFA and tris(ethylene)platinum in the presence of COD (129). The X-ray structures of **186** (128, 129) and **180b** have been reported (129).

$$\text{CODPt}\left(\overset{\overset{\text{H}\diagdown\quad}{\text{C}}}{\underset{\underset{\text{CH}_2}{||}}{|}}\overset{\overset{\text{O}}{||}}{\underset{\text{C}}{\diagup}}\text{Me}\right)_2 + \text{HFA} \longrightarrow \text{CODPt}\overset{\overset{\text{H}\diagdown\quad\diagup\text{H}\quad\overset{\text{O}}{||}}{\text{C}-\text{C}-\overset{|}{\text{C}}\diagdown\text{Me}}}{\underset{\text{O}-\underset{\underset{\text{CF}_3}{|}}{\overset{|}{\text{C}}}-\text{CF}_3}{}} \qquad (147)$$

$$\qquad\qquad\qquad\qquad\qquad\qquad\qquad\qquad \textbf{(188)}$$

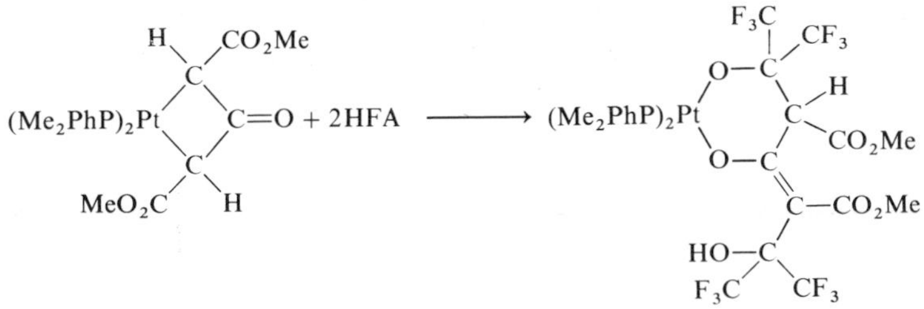

$$
\begin{array}{c}
\text{H} \diagdown \quad \diagup \text{CO}_2\text{Me} \\
\text{C} \\
(\text{Me}_2\text{PhP})_2\text{Pt} \diagup \quad \diagdown \text{C=O} + 2\,\text{HFA} \\
\diagdown \quad \diagup \text{C} \\
\text{MeO}_2\text{C} \diagup \quad \diagdown \text{H}
\end{array}
\quad\longrightarrow\quad
\begin{array}{c}
\text{F}_3\text{C} \quad \diagup \text{CF}_3 \\
\big| \\
\text{O—C} \\
\diagup \quad \diagdown \text{H} \\
(\text{Me}_2\text{PhP})_2\text{Pt} \quad \text{C} \diagdown \\
\diagdown \quad \diagup \text{CO}_2\text{Me} \\
\text{O—C} \\
\diagdown\diagdown \\
\text{C—CO}_2\text{Me} \\
\big| \\
\text{HO—C} \\
\text{F}_3\text{C} \quad \diagdown \text{CF}_3 \\
\textbf{(189)}
\end{array}
\qquad (148)
$$

Insertion into Pt—C bonds occurs with a π-bis(1-buten-3-one)platinum(0) complex with oxidation [Eq. (147)] (*130*) and a σ-bonded platina(II)cyclobutanone with total change of the coordination sphere at the metal atom [Eq. (148)] (*62*). The structure of **189** has been characterized by X-ray analysis (*62*).

E. GOLD

Only one reaction is known in which a gold compound reacts with HFA. (Triphenylphosphane)methylgold(I) forms a four-membered ring **190**, which is assigned a structure related to the diplatinum compound **186** (*196*).

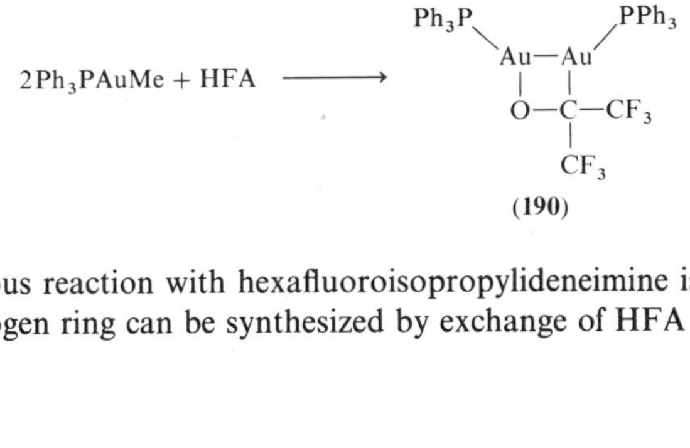

$$
2\,\text{Ph}_3\text{PAuMe} + \text{HFA} \quad\longrightarrow\quad
\begin{array}{c}
\text{Ph}_3\text{P} \diagdown \qquad \diagup \text{PPh}_3 \\
\text{Au—Au} \\
\big| \quad \big| \\
\text{O—C—CF}_3 \\
\big| \\
\text{CF}_3 \\
\textbf{(190)}
\end{array}
\qquad (149)
$$

The analogous reaction with hexafluoroisopropylideneimine is not successful; the nitrogen ring can be synthesized by exchange of HFA in **190** (*196*).

F. MERCURY

HFA reacts with trihalogenomethylphenylmercury compounds to form oxiranes **191**, probably via dihalocarbene intermediates (*248*).

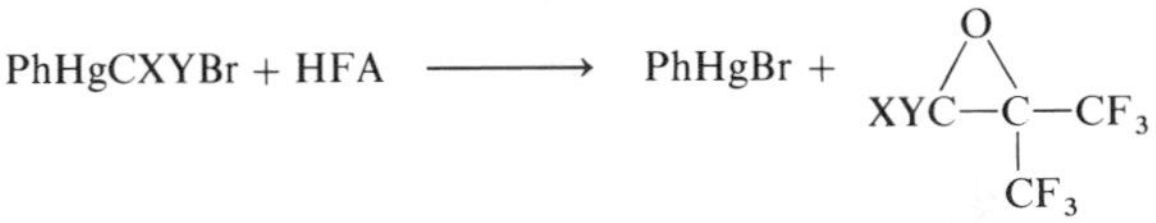

$$
\text{PhHgCXYBr} + \text{HFA} \quad\longrightarrow\quad \text{PhHgBr} +
\begin{array}{c}
\text{O} \\
\diagup \diagdown \\
\text{XYC—C—CF}_3 \\
\big| \\
\text{CF}_3 \\
\textbf{(191)}
\end{array}
\qquad (150)
$$

(X, Y = Cl, Br)

Mercury dithiocyanate reacts with HFA to form six-membered rings, in analogy to phosphorus, arsenic, and sulfur compounds (*vide supra*) (*241*).

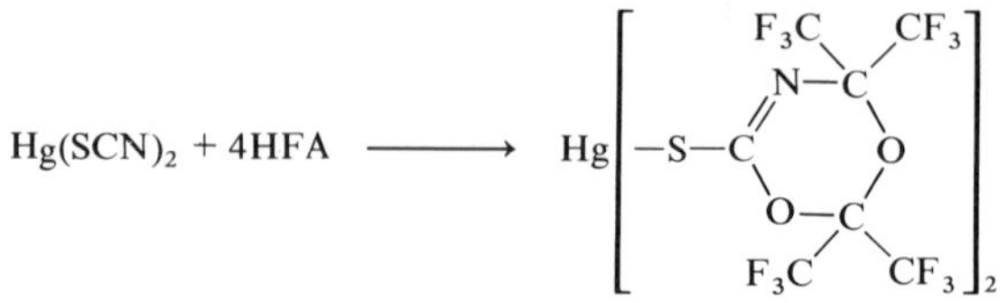

$$Hg(SCN)_2 + 4HFA \longrightarrow Hg\left[-S-C\underset{O-C}{\overset{N=C}{\diagup}}\begin{array}{c}F_3C \quad CF_3 \\ \diagdown / \\ O \\ \diagup \diagdown \\ F_3C \quad CF_3\end{array}\right]_2 \qquad (151)$$

(192)

Compound **192** is a useful precursor for the transfer of the ligand with halogen-containing molecules by precipitation of HgX_2 (*241*).

The reaction of mercury dicyanide with HFA yields a mixture of **193** and **194** [Eq. (152)]. Subsequent reaction of this mixture with diphenylchlorophosphane affords the bicyclic compound **80a**. This bicycle is structurally related to the product **80** from the reaction of HFA with phenyldicyanophosphane [Eq. (68)], which is also accessible from the cyanotrimethylsilane–HFA system [Eq. (12)] (*210*) (*vide supra*).

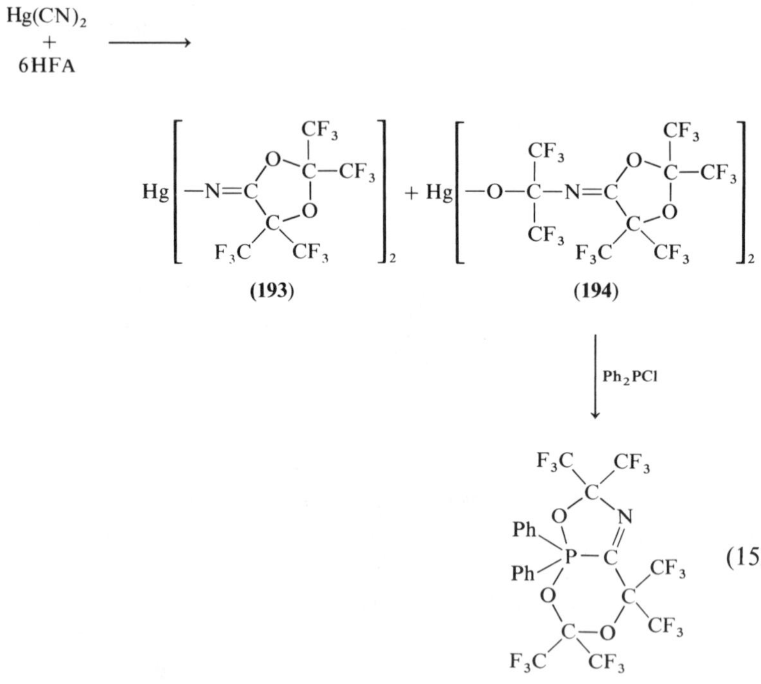

$$Hg(CN)_2$$
$$+ \longrightarrow$$
$$6\,HFA$$

(193)   **(194)**

$$\Big\downarrow Ph_2PCl$$

(152)

(80a)

G. GROUP VI AND VII ELEMENTS

Complexes of manganese, molybdenum, and tungsten containing dihydro- and tetrahydrofuranate ligands are obtained in analogy to the iron complexes **149** and **150** (Section V,A) (*174*).

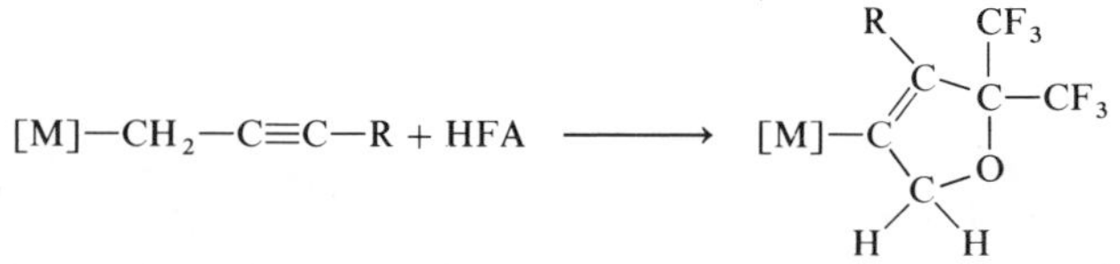

$$[M]-CH_2-C\equiv C-R + HFA \longrightarrow [M]-C \overset{R}{\underset{H}{\diagup}} \cdots \qquad (153)$$

(**195**), (**196**)

[**195**: [M] = CpMo(CO)$_3$, R = Ph; **196**: [M] = Mn(CO)$_5$, R = Me, Ph]

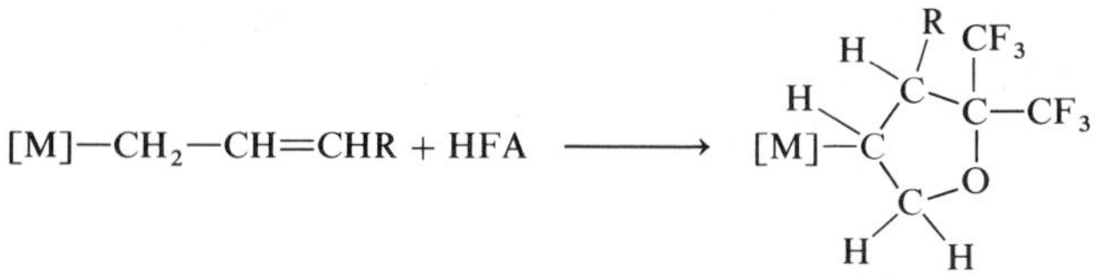

$$[M]-CH_2-CH=CHR + HFA \longrightarrow [M]-C \cdots \qquad (154)$$

(**197**)-(**199**)

[M]	R	
CpMo(CO)$_3$	Ph	**197**
CpW(CO)$_3$	Ph	**198**
Mn(CO)$_5$	H, Me	**199**

The pentacarbonyltungstenhydrogen sulfide anion has been reacted with excess HFA in acetone at ambient temperature. Equation (155) is one of the rare examples where fluorine abstraction, rather than insertion, occurs (*13*).

$$(CO)_5\bar{W}-S-H + HFA \longrightarrow (CO)_5\bar{W}-S-CF_2-\overset{O}{\overset{\|}{C}}-CF_3 + HF \quad (155)$$

(**200**)

The reaction of HFA and nitridotungsten trichloride followed by addition of tetraphenylarsonium chloride yields **201**. Single-crystal X-ray structure analysis proves the formation of this surprising compound (*236*).

$$2Cl_3W\equiv N + HFA + 2Ph_4As^+Cl^- \longrightarrow [Cl_5W\dot{-}N\dot{-}\overset{CF_3}{\underset{CF_3}{C}}\dot{-}N\dot{-}WCl_5]^{2-} \quad (156)$$

(**201**) 2Ph$_4$As$^+$

The W—N bond of 174.3(15) pm is a little longer than that in $[Cl_5WNC_2Cl_5]^-$, which was considered to be a triple bond. Since the geometry at nitrogen is almost linear [176.9(14)°], a triple-bonded resonance extreme with a positive formal charge on nitrogen may make a significant contribution.

H. Reactions with Polymeric Pyrazole Complexes

Several d and f transition metals have been found to yield polymeric complexes with pyrazoles (14, 178). Reactions of these polymers with HFA produce monomeric species, HFA adding to the free nitrogen atom and the metal. Substitution of one carbon atom adjacent to nitrogen also results in degradation of the polymers. However, no reaction has been found when both α carbon atoms are sterically hindered (14).

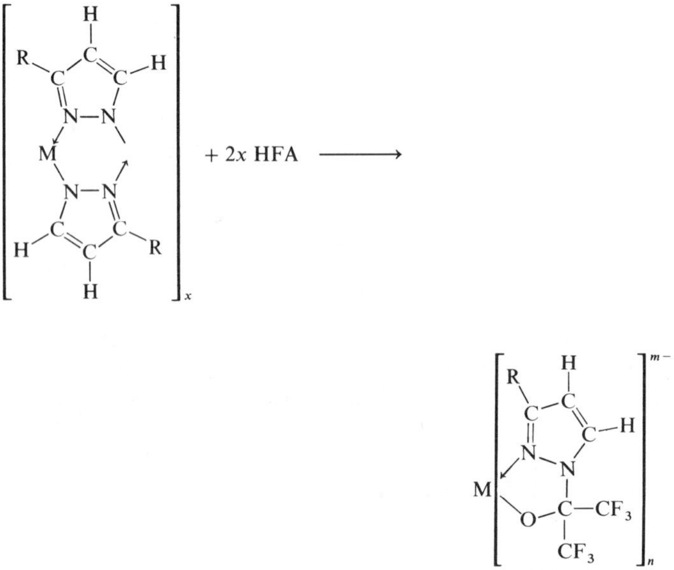

(157)

(202)–(213)

Most pyrazole complexes are inert to moist air, with only the silver and gold compounds **206** and **207** showing limited stability (178). The X-ray structures of the unsubstituted thorium and uranium derivatives **209a** and **210a**, which are isostructural, have been reported (280). The hexafluoro-acetonylpyrazole complexes are listed in Table IX.

TABLE IX

HEXAFLUOROACETONYLPYRAZOLE COMPLEXES

Compound	M	n	R	m	Reference
202	Fe	3	H	0	*178*
203	Ni	3	H	1	*178*
204	Cu	2, 3	H	0	*178*
205	Zn	3	H	1	*178*
206	Ag	1	H	0	*178*
207	Au	1	H	0	*178*
208	Eu	4	H	1	*178*
209	Th	4	H, Me	0	*14, 178*
210	U	4	H, Me	0	*14*
211	UO$_2$	2	H	0	*178*
212	Np	4	H, Me	0	*14*
213	Pu	4	H, Me	0	*14*

VI. Miscellaneous

A. REACTIONS WITH METALS AND METAL HALIDES

The reductive coupling of HFA with alkali metals to yield perfluoro-pinacolate (*7, 64*) has already been mentioned in earlier sections, together with metathetical reactions with a variety of dihalides. Similarly, free perfluoropinacol reacts with a series of metal halides in aqueous solution to yield anionic and neutral complexes of transition metals listed in Table X.

The moisture-sensitive complexes **215** and **217c** have been prepared from dilithium perfluoropinacolate and the appropriate dihalide in tetrahydro-furan. One molecule of THF is also coordinated to the metal (*64*).

Several "organic" reactions of HFA are catalyzed by Lewis acids (e.g., AlCl$_3$) (*168, 185*, and references cited therein).

Starting from the aldol condensation product of HFA with acetone (*255*), a series of nickel and copper di- and triamide complexes have been synthesized, which can undergo intramolecular condensation with elimination of water (*181*).

HFA forms 1:1 adducts with metal fluorides. The stability of the perfluoro-isopropoxides **226** increases with the size of the cation (*224*). The salts are the

TABLE X

ANIONIC AND NEUTRAL PERFLUOROPINACOLMETAL COMPLEXES

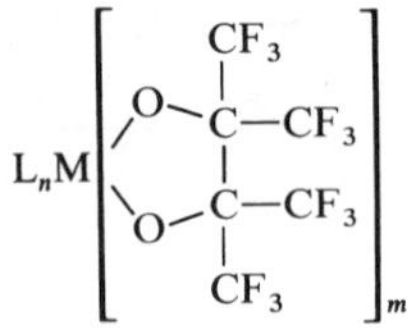

$$\left[L_nM \begin{array}{c} O-C \\ \\ O-C \end{array} \begin{array}{c} CF_3 \\ | \\ -CF_3 \\ | \\ -CF_3 \\ | \\ CF_3 \end{array} \right]_m$$

Compound	M	m	L	n	Reference
214	Al	3	K	3	8
215	Ti	1	Cl	2	64
216	VO	2	K	2	287
217a	CrO	2	K	1	287
217b	CrO	2	Cs	1	287
217c	Cr	1	O	2	64
218	Mn	2	K	2	8
219	Fe	3	K	3	8
220	Co·2H$_2$O	2	K	2	288
221a	Ni	2	K	2	8, 289
221b		1	L^a		289
222a	Cu	2	K	2	8
222b, c		1	(R$_2$NCH$_2$—)$_2$^{b}	1	289
223	Zn	2	K	2	8
224	Pd	1	PMe$_2$Ph	2	289
225	Pt	1	PMe$_2$Ph	2	289

^{a} L = Various neutral N- and P-containing ligands.
^{b} R = Me, Et.

reactive species, and a metal fluoride is necessary to promote reaction of HFA.

$$HFA + MF \longrightarrow \quad F-\overset{\overset{\displaystyle CF_3}{|}}{\underset{\underset{\displaystyle CF_3}{|}}{C}}-O^-M^+ \qquad (158)$$

$$(M = K, Rb, Cs, Ag, NEt_4) \qquad (226)$$

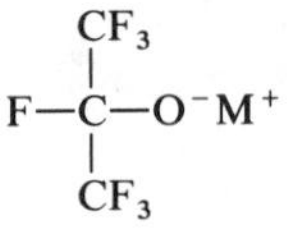

The lattice energy of the metal fluorides is the main factor in promoting the reaction. Though the lattice energy of NaF is smaller than that of AgF, no adduct formation has been observed. This is due to complex formation of AgF with the solvent acetonitrile prior to the reaction with HFA (99).

The potassium salt 226a has been found to react with pentafluorochloro-acetone. Subsequent treatment with KF and excess of pentafluorochloro-acetone yields perfluorinated polyethers 228 (167).

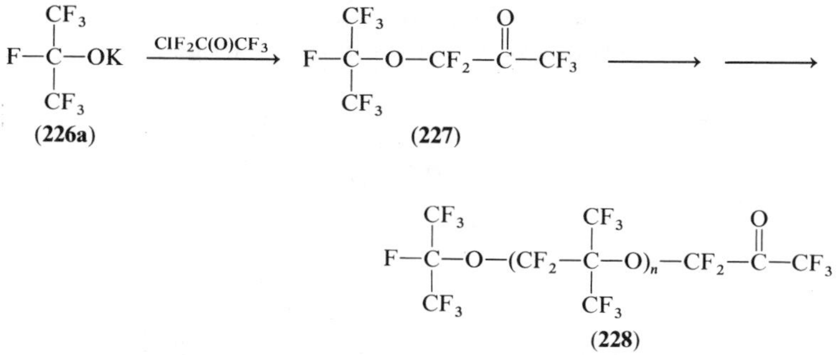

$$F-\underset{\underset{\text{CF}_3}{|}}{\overset{\overset{\text{CF}_3}{|}}{C}}-OK \xrightarrow{\text{ClF}_2\text{C(O)CF}_3} F-\underset{\underset{\text{CF}_3}{|}}{\overset{\overset{\text{CF}_3}{|}}{C}}-O-CF_2-\overset{O}{\overset{||}{C}}-CF_3 \longrightarrow \longrightarrow$$

(226a)   (227)

$$F-\underset{\underset{\text{CF}_3}{|}}{\overset{\overset{\text{CF}_3}{|}}{C}}-O-(CF_2-\underset{\underset{\text{CF}_3}{|}}{\overset{\overset{\text{CF}_3}{|}}{C}}-O)_n-CF_2-\overset{O}{\overset{||}{C}}-CF_3 \quad (159)$$

(228)

B. Boron

Only a few reports deal with the interaction of boron compounds with HFA. No addition products of HFA with monoborane have been detected in thermal (*101*) or in photochemical (*264*) reactions. However, co-photolysis of HFA with pentaborane(9) and 2,4-dicarbapentaborane(7) results in the insertion of HFA into B—H bonds (*16a*). Mechanisms are discussed in detail.

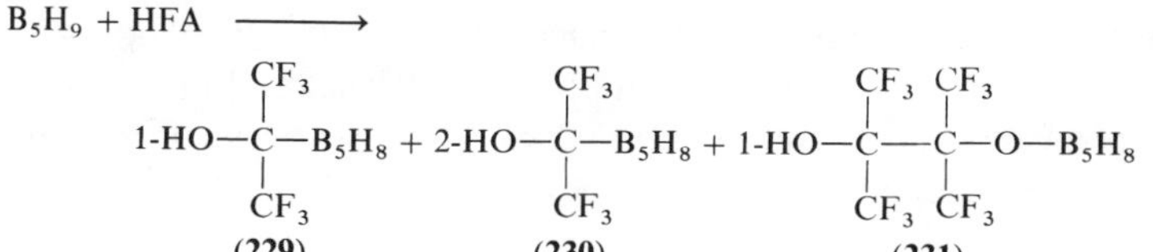

$$B_5H_9 + HFA \longrightarrow$$

$$1\text{-HO}-\underset{\underset{\text{CF}_3}{|}}{\overset{\overset{\text{CF}_3}{|}}{C}}-B_5H_8 + 2\text{-HO}-\underset{\underset{\text{CF}_3}{|}}{\overset{\overset{\text{CF}_3}{|}}{C}}-B_5H_8 + 1\text{-HO}-\underset{\underset{\text{CF}_3}{|}}{\overset{\overset{\text{CF}_3}{|}}{C}}-\underset{\underset{\text{CF}_3}{|}}{\overset{\overset{\text{CF}_3}{|}}{C}}-O-B_5H_8 \quad (160)$$

(229)   (230)   (231)

$$2,4\text{-C}_2B_5H_7 + HFA \longrightarrow 5\text{-HO}-\underset{\underset{\text{CF}_3}{|}}{\overset{\overset{\text{CF}_3}{|}}{C}}-C_2B_5H_6 + 5\text{-HO}-\underset{\underset{\text{CF}_3}{|}}{\overset{\overset{\text{CF}_3}{|}}{C}}-\underset{\underset{\text{CF}_3}{|}}{\overset{\overset{\text{CF}_3}{|}}{C}}-O-C_2B_5H_6$$

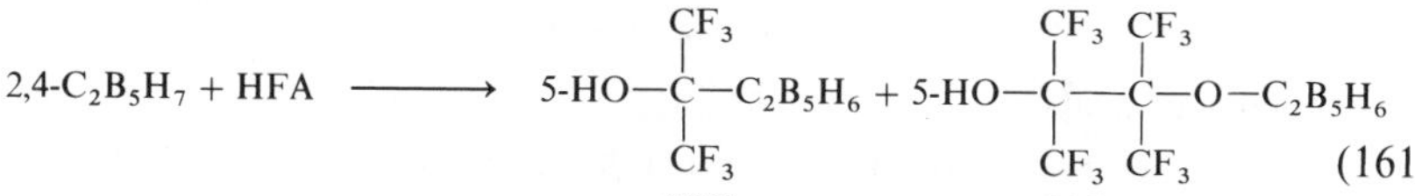

(232)   (233)   (161)

Photochemical insertion into a B—H bond occurs with borazine (*267*). The CF_3-substituted product **234b** arises from radical decomposition of the HFA molecule.

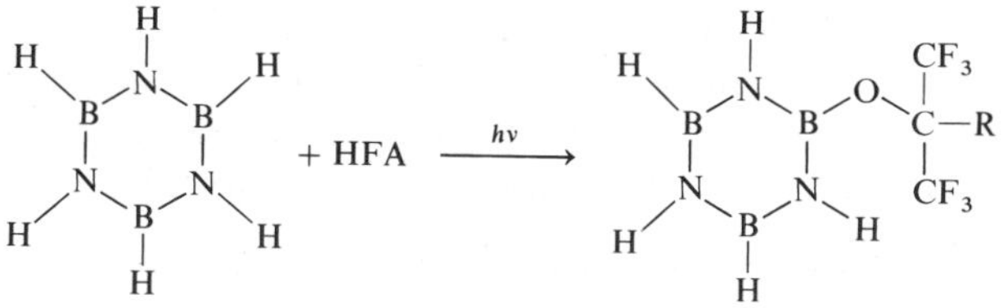

$$\text{(borazine)} + HFA \xrightarrow{h\nu} \text{(product)} \quad (162)$$

(234a,b)   (R = H, CF_3)

Phenyldichloroborane reacts with perfluoropinacolate to form a 1,3,2-dioxaborolane (7, 64). A spirobicyclic boranate **235** is generated from sodium borohydride and disodium pinacolate (7).

$$NaBH_4 + Na_2Pfp \longrightarrow Na^+ \quad \text{(235)} \qquad (163)$$

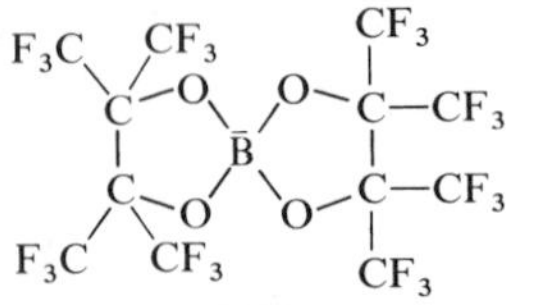

(235)

Allylboranes react with HFA in the presence of alcohols like nonanol or triethanolamine to yield partially fluorinated, unsaturated alcohols **236** (194).

$$R_2B-CH_2-\overset{R'}{\underset{}{C}}=CH_2 + HFA \xrightarrow{ROH} CH_2=\overset{R'}{\underset{}{C}} \quad (164)$$

(R = Pr, Bu; R' = H, Me)   (236)

Migration of the double bond and rearrangement of the skeleton have been observed with 2-butenylboranes (194).

Pentacoordinated (10-B-5) and hexacoordinated (12-B-6) boron compounds are available from the reaction of BCl_3 with the dilithio salt of the bis addition product of HFA and pyridine (265). These hypervalent compounds show signals in the ^{11}B NMR spectrum at very high field (173a).

C. HALOGENATION REACTIONS

Chlorine monofluoride reacts with HFA under the catalytic influence of cesium fluoride to form perfluoroisopropoxy hypochlorite (**237**) (292).

$$HFA + ClF \xrightarrow{CsF} \begin{matrix} CF_3 \\ | \\ FC-OCl \\ | \\ CF_3 \end{matrix} \qquad (165)$$

(237)

Fluorination of HFA with XeF_2 has not been observed (114).

Photochemical reaction of HFA with fluorine, also in the presence of oxygen, has been investigated by Aymonino (19). Depending on the stoichiometry, different product distributions have been observed.

$$HFA + F_2 \xrightarrow{h\nu} CF_3COF, COF_2, CF_3OF, CF_4 \qquad (166)$$

$$2HFA + O_2 + F_2 \xrightarrow{h\nu} CF_3COF, COF_2, CF_3OF, CF_3OOCF_3 \qquad (167)$$

Insertion of HFA into the C—F bonds of perfluorinated dicarboxylic acyl fluorides has been achieved thermally under the catalytic influence of potassium fluoride (70).

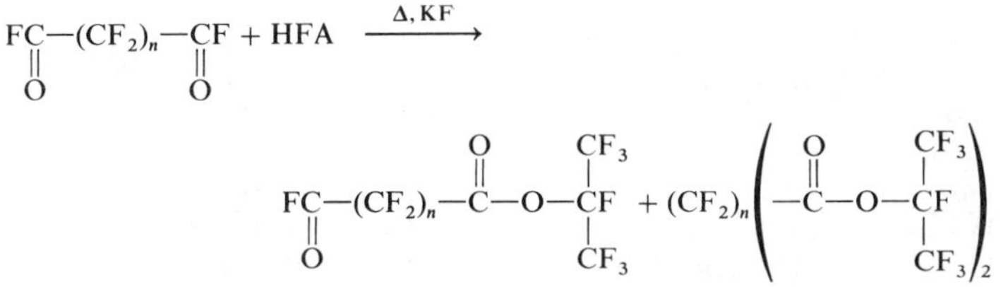

$$FC-(CF_2)_n-CF + HFA \xrightarrow{\Delta, KF}$$

$$FC-(CF_2)_n-\overset{O}{\underset{O}{C}}-O-\overset{CF_3}{\underset{CF_3}{CF}} + (CF_2)_n\left(-\overset{O}{C}-O-\overset{CF_3}{\underset{CF_3}{CF}}\right)_2 \qquad (168)$$

$$(238) \qquad\qquad (239)$$

A,ω-Diacyl-substituted perfluoro ethers react in a similar way (70).

VII. Summary

This article has summarized achievements in the synthesis of "inorganic" compounds using HFA as starting material.

Hexafluoroacetone has been found to be a very interesting synthon on the borderline between inorganic and organic chemistry, and has established its own chemistry with various nonmetallic and metallic substrates; only a few types of reactions resemble those of a regular ketone. The electron-withdrawing properties of the CF_3 groups are the stabilizing factor for structures that would otherwise be inaccessible. Only a few cases are known in which the electronic properties of HFA have inhibited any reaction.

Furthermore, the six fluorine atoms of HFA provide the chemist with a versatile tool for the investigation of dynamic processes by means of ^{19}F NMR spectroscopy.

Reactions of HFA with transition metals are known. Further studies in this direction will provide a better understanding of the chemistry of HFA.

REFERENCES

1. Abel, E. W., and Burton, C. A., *J. Fluorine Chem.* **14**, 105 (1979).
2. Abel, E. W., and Crow, J. P., *J. Chem. Soc. A*, p. 1361 (1968).
3. Abel, E. W., and Illingworth, S. M., *J. Organomet. Chem.* **17**, 161 (1969).
4. Abel, E. W., and Rowley, R. J., *J. Organomet. Chem.* **84**, 199 (1975).
5. Abel, E. W., and Sabberwal, I. H., *J. Chem. Soc. A*, p. 1105 (1968).
6. Abel, E. W., Walker, D. J., and Wingfield, J. N., *J. Chem. Soc. A*, p. 2642 (1968).
7. Allan, M., Janzen, A. F., and Willis, C. J., *Can. J. Chem.* **46**, 3671 (1968).
8. Allan, M., and Willis, C. J., *J. Am. Chem. Soc.* **90**, 5343 (1968).
9. Allwörden, U., Tseggai, I., and Röschenthaler, G.-V., *Phosphorus Sulfur* **21**, 177 (1984).

10. Aly, H. A. E., Barlow, J. H., Russell, D. R., Smith, D. J. H., Swindles, M., and Trippett, S., *J. Chem. Soc., Chem. Commun.*, p. 449 (1976).

11. Anderson, L. R., Ratcliffe, C. T., Young, D. E., and Fox, W. B., *J. Fluorine Chem.* **7**, 481 (1976).

12. Anderson, L. R., Young, D. E., and Fox, W. B., *J. Fluorine Chem.* **7**, 491 (1976).

13. Angelici, R. J., and Gingerich, R. G. W., *Organometallics* **2**, 89 (1983).

14. Andruchow, W., Jr., and Karraker, D. G., *Inorg. Chem.* **12**, 2194 (1973).

15. Ashley-Smith, J., Green, M., and Stone, F. G. A., *J. Chem. Soc. A*, p. 3019 (1969).

16. Ashley-Smith, J., Green, M., and Stone, F. G. A., *J. Chem. Soc. A*, p. 3161 (1970).

16a. Astheimer, R. J., and Sneddon, L. G., *Inorg. Chem.* **23**, 3207 (1984).

17. Astrologes, G. W., and Martin, J. C., *J. Am. Chem. Soc.* **97**, 6909 (1975).

18. Astrologes, G. W., and Martin, J. C., *J. Am. Chem. Soc.* **98**, 2895 (1976).

19. Aymonino, P. J., *An. Asoc. Quim. Argent.* **55**, 47 (1966).

19a. Baceiredo, A., Bertrand, G., Majoral, J.-P., Wermuth, U., and Schmutzler, R., *J. Am. Chem. Soc.* **106**, 7065 (1984).

20. Bagland, R. W., and Hartter, D. R., *J. Org. Chem.* **37**, 4136 (1972).

21. Bagnall, R. D., Coe, P. L., and Tatlow, J. C., *J. Chem. Soc., Perkin Trans. I*, p. 2277 (1972).

22. Barrau, J., Massol, M., Mesnard, D., and Satgé, J., *Recl. Trav. Chim. Pays-Bas* **92**, 321 (1973).

23. Beaulieu, W. B., Mercer, G. D., and Roundhill, D. M., *J. Am. Chem. Soc.* **100**, 1147 (1978).

24. Bell, T. N., and Johnson, B. B., *Aust. J. Chem.* **20**, 1545 (1967).

25. Bell, T. N., and Zucker, U. F., *Can. J. Chem.* **47**, 1701 (1969).

26. Birum, G. H., and Matthews, C. N., *J. Org. Chem.* **32**, 3554 (1967).

27. Blackmore, T., Bruce, M. I., and Stone, F. G. A., *J. Chem. Soc., Dalton Trans.*, p. 106 (1974).

28. Blackmore, T., Bruce, M. I., Stone, F. G. A., Davis, R. E., and Raghavan, N. V., *J. Organomet. Chem.* **49**, C-35 (1973).

29. Bland, W. J., Kemmitt, R. D. W., and Moore, R. D., *J. Chem. Soc., Dalton Trans.*, p. 1292 (1973).

30. Bockerman, G. N., and Parry, R. W., *J. Fluorine Chem.* **7**, 1 (1976).

31. Bond, A., and Green, M., *J. Chem. Soc., Chem. Commun.*, p. 12 (1974).

32. Bone, S., Trippett, S., and Whittle, P. J., *J. Chem. Soc., Perkin Trans. I*, p. 2125 (1974).

33. Bone, S., Trippett, S., and Whittle, P. J., *J. Chem. Soc., Perkin Trans. I*, p. 80 (1977).

34. Braun, R. A., *J. Am. Chem. Soc.* **87**, 5516 (1965).

35. Braun, R. A., *Inorg. Chem.* **5**, 1831 (1966).

36. Braun, R. A., *J. Org. Chem.* **31**, 1147 (1966).

37. Brierly, J., Dickstein, J. I., and Trippett, S., *Phosphorus Sulfur* **7**, 167 (1979).

38. Browning, J., Cook, D. J., Cundy, C. S., Green, M., and Stone, F. G. A., *J. Chem. Soc., Chem. Commun.*, p. 929 (1968).

39. Browning, J., Cundy, C. S., Green, M., and Stone, F. G. A., *J. Chem. Soc. A*, p. 20 (1969).

40. Browning, J., Empsall, H. D., Green, M., and Stone, F. G. A., *J. Chem. Soc., Dalton Trans.*, p. 381 (1973).

41. Browning, J., Green, M., and Stone, F. G. A., *J. Chem. Soc. A*, p. 453 (1971).

42. Bruce, M. I., Stone, F. G. A., and Thomson, B. J., *J. Org. Chem.* **77**, 77 (1974).

43. Bruker, A. B., Grinshtein, E. I., and Soborovskii, L. Z., *Zh. Obshch. Khim.* **36**, 1133 (1966); *J. Gen. Chem. USSR (Engl. Transl.)* **36**, 1146 (1966).

44. Burger, K., Albanbauer, J., and Eggersdorfer, M., *Angew. Chem.* **87**, 816 (1975); *Angew. Chem., Int. Ed. Engl.* **14**, 766 (1975).

45. Burger, K., Albanbauer, J., and Foag, W., *Angew. Chem.* **87**, 816 (1975); *Angew. Chem., Int. Ed. Engl.* **14**, 767 (1975).

46. Burger, K., Albanbauer, J., Kaefig, F., and Penninger, S., *Liebigs Ann. Chem.*, p. 624 (1977).
47. Burger, K., Fehn, J., and Thenn, H., *Angew. Chem.* **85**, 541 (1973); *Angew. Chem., Int. Ed. Engl.* **12**, 502 (1973).
48. Burger, K., and Ottlinger, R., *Tetrahedron Lett.*, p. 973 (1978).
49. Burger, K., Ottlinger, R., and Albanbauer, J., *Chem. Ber.* **110**, 2114 (1977).
50. Burger, K., Ottlinger, R., Goth, H., and Firl, J., *Chem. Ber.* **113**, 2699 (1980).
50a. Burger, K., Ottlinger, R., Goth, H., and Firl, J., *Chem. Ber.* **115**, 2494 (1982).
51. Burger, K., and Penninger, S., *Synthesis*, p. 524 (1978).
52. Burger, K., Penninger, S., and Greisel, M., *J. Fluorine Chem.* **15**, 1 (1980).
53. Burger, K., and Simmerl, R., *Synthesis* **83**, 237 (1983).
53a. Burger, K., and Simmerl, R., *Liebigs Ann. Chem.*, p. 982 (1984).
54. Burt, R., Cooke, M., and Green, M., *J. Chem. Soc. A*, p. 2975 (1970).
55. Bykhovskaya, E. G., Gontar', A. F., and Knunyants, I. L., *Izv. Akad. Nauk SSSR Ser. Khim.*, p. 436 (1984); *Bull. Acad. Sci. USSR, Div. Chem. Sci.* **33**, 399 (1984).
56. Caddy, P., Green, M., Howard, J. A. K., Squire, J. M., and White, N. J., *J. Chem. Soc., Dalton Trans.*, p. 400 (1981).
57. Chambers, R. D., and Clark, M., *Tetrahedron Lett.*, p. 2741 (1970).
58. Cherwinski, W. J., Johnson, B. F. G., and Lewis, J., *J. Org. Chem.* **52**, C61 (1973).
59. Cherwinski, W. J., Johnson, B. F. G., and Lewis, J., *J. Chem. Soc., Dalton Trans.*, p. 1405 (1974).
60. Chioccola, G., and Daly, J. J., *J. Chem. Soc. A*, p. 568 (1968).
61. Clark, B., Green, M., Osborn, R. B. L., and Stone, F. G. A., *J. Chem. Soc. A*, p. 167 (1968).
62. Clarke, D. A., Kemmitt, R. D. W., Mazid, M. A., Schilling, M. D., and Russell, D. R., *J. Chem. Soc., Chem. Commun.*, p. 744 (1978).
63. Clemens, J., Green, M., and Stone, F. G. A., *J. Chem. Soc., Dalton Trans.*, p. 375 (1973).
64. Conroy, A. P., and Dresdner, R. D., *Inorg. Chem.* **9**, 2739 (1970).
65. Cooke, M., and Green M., *J. Chem. Soc. A*, p. 651 (1969).
66. Cooksey, C. J., Dodd, D., Johnson, M. D., and Lockman, B. D., *J. Chem. Soc., Dalton Trans.*, p. 1814 (1978).
67. Corriu, R. J. P., Kpoton, A., Barrau, J., and Satgé, J., *J. Organomet. Chem.* **114**, 21 (1976).
68. Countryman, R., and Penfold, B. R., *J. Cryst. Mol. Struct.* **2**, 281 (1972).
69. Countryman, R., and Penfold, B. R., *J. Chem. Soc., Chem. Commun.*, p. 1598 (1971).
70. Croft, T. S., *J. Fluorine Chem.* **7**, 433 (1975).
71. Croft, T. S., and McBrady, J. J., *J. Fluorine Chem.* **6**, 213 (1975).
72. Cullen, W. R., and Styan, G. E., *J. Organomet. Chem.* **4**, 151 (1965).
73. Cullen, W. R., and Styan, G. E., *Inorg. Chem.* **4**, 1437 (1965).
74. Dahl, B. M., Dahl, O., and Trippett, S., *J. Chem. Soc., Perkin Trans.*, p. 2239 (1981).
75. Dakternieks, D., Röschenthaler, G.-V., Sauerbrey, K., and Schmutzler, R., *Chem. Ber.* **112**, 2380 (1979).
76. Davidson, J. L., Green, M., Stone, F. G. A., and Welch, A. J., *J. Chem. Soc., Dalton Trans.*, p. 2044 (1976).
77. Davies, C. H., Game, C. H., Green, M., and Stone, F. G. A., *J. Chem. Soc., Dalton Trans.*, p. 357 (1974).
78. Davison, J., and Solar, J. P., *J. Organomet. Chem.* **166**, C13 (1979).
79. Davydov, A. V., and Knunyants, I. L., *Zh. Vses. Khim. O-Va.* **22**, 358 (1977).
80. Davydov, A. V., Torgun, I. N., and Knunyants, I. L., *Zh. Obshch. Khim.* **50**, 936 (1980); *J. Gen. Chem. USSR (Engl. Transl.)* **50**, 759 (1980).
81. Del'tsova, D. P., Gambaryan, N. P., and Lur'e, E. P., *Izv. Akad. Nauk SSSR Ser. Khim.*, p. 1788 (1979); *Bull. Acad. Sci. USSR, Div. Chem. Sci.* **28**, 1648 (1979).
82. De Pasquale, R. J., *J. Fluorine Chem.* **8**, 311 (1976).

83. Dhathathreyan, K. S., and Roesky, H. W., unpublished results.
84. Dittmer, D. C., Lombardo, A., Batzold, F. H., and Greene, C. H., *J. Org. Chem.* **41**, 2976 (1976).
85. Dousse, G., Lavayssière, H., and Satgé, J., *Helv. Chim. Acta* **58**, 2610 (1975).
86. Dousse, G., Lavayssière, H., and Satgé, J., *Helv. Chim. Acta* **59**, 2961 (1976).
87. Dousse, G., and Satgé, J., *Recl. Trav. Chim. Pays-Bas* **90**, 221 (1971).
88. Drozd, G. I., and Ivin, S. Z., *Zh. Obshch. Khim.* **39**, 1179 (1969); *J. Gen. Chem. USSR (Engl. Transl.)* **39**, 1148 (1969).
89. Duff, R. E., Oram, R. K., and Trippett, S., *J. Chem. Soc., Chem. Commun.*, p. 1011 (1971).
90. Duff, E., Russell, D. R., and Trippett, S., *Phosphorus Relat. Group V Elem.* **4**, 203 (1974).
91. Duff, E., Trippett, S., and Whittle, P. J., *J. Chem. Soc., Perkin Trans. I*, p. 972 (1973).
92. Eikmeier, H.-B., Hodges, K. C., Stelzer, O., and Schmutzler, R., *Chem. Ber.* **111**, 2077 (1978).
93. Empsall, H. D., Green, M., Shakshooki, S. K., and Stone, F. G. A., *J. Chem. Soc. A*, p. 3472 (1971).
94. Empsall, H. D., Green, M., and Stone, F. G. A., *J. Chem. Soc., Dalton Trans.*, p. 96 (1972).
95. England, D. C. (to E. I. du Pont de Nemours), U.S. Patent 3,197,480 (1965); *Chem. Abstr.* **63**, 13216b (1965).
96. Evangelidou-Tsolis, E., and Ramirez, F., *Phosphorus Relat. Group V Elem.* **4**, 121 (1974).
97. Evangelidou-Tsolis, E., Ramirez, F., Pilot, J. F., and Smith, C. P., *Phosphorus Relat. Group V Elem.* **4**, 109 (1974).
98. Evans, D. A., Hurst, K. M., and Takacs, J. M., *J. Am. Chem. Soc.* **100**, 3467 (1978).
99. Evans, F. W., Litt, M. H., Weidler-Kubanek, A.-M., and Avonda, F. P., *J. Org. Chem.* **33**, 1837 (1968).
100. Farah, B. S., and Gilbert, E. E., *J. Org. Chem.* **30**, 1241 (1965).
101. Fehlner, T. P., *Inorg. Chem.* **12**, 98 (1973).
102. Fetyukhin, V. N., Voek, M. V., Dergunov, Y. I., and Samarai, L. I., *Zh. Obshch. Khim.* **51**, 1678 (1981); *J. Gen. Chem. USSR (Engl. Transl.)* **51**, 1431 (1981).
103. Field, L., and Sweetman, B. J., *J. Org. Chem.* **34**, 1799 (1969).
104. Fokin, A. V., Studnev, Y. N., Rapkin, A. I., Pasevina, K. I., Verenikin, O. V., and Kolomiets, A. F., *Izv. Akad. Nauk SSSR Ser. Khim.*, p. 1655 (1981); *Bull. Acad. Sci. USSR, Div. Chem. Sci.* **30**, 1344 (1981).
105. Forniés, J., Green, M., Laguna, A., Murray, M., Spencer, J. L., and Stone, F. G. A., *J. Chem. Soc., Dalton Trans.*, p. 1515 (1977).
106. Foss, V. L., Lukashev, N. V., Tsvetkov, Y. E., and Lutsenko, I. F., *Zh. Obshch. Khim.* **52**, 2183 (1982); *J. Gen. Chem. USSR (Engl. Transl.)* **52**, 1942 (1982).
106a. Francke, R., Dakternieks, D., Gable, R. W., Hoskins, B. F., and Röschenthaler, G.-V., *Chem. Ber.* **118**, 922 (1985).
107. Francke, R., Di Giacomo, R., Dakternieks, D., and Röschenthaler, G.-V., *Z. Anorg. Allg. Chem.* **519**, 141 (1984).
107a. Francke, R., Röschenthaler, G.-V., Di Giacomo, R., and Dakternieks, D., *Phosphorus Sulfur* **20**, 107 (1984).
108. Frye, C. L., Salinger, R. M., and Patin, T. J., *J. Am. Chem. Soc.* **88**, 2343 (1966).
109. Fukuhara, N., and Bigelow, L. A., *J. Am. Chem. Soc.* **63**, 788 (1941).
110. Gambaryan, N. P., Cheburkov, Y. A., and Knunyants, I. L., *Izv. Akad. Nauk. SSSR Ser. Khim.*, p. 1526 (1964); *Bull. Acad. Sci. USSR Div. Chem. Sci.*, p. 1433 (1964).
111. Gambaryan, N. P., Rokhlin, E. M., Zeifman, Y. V., Ching-Yun, C., and Knunyants, I. L., *Angew. Chem.* **78**, 1008 (1966); *Angew. Chem., Int. Ed. Engl.* **5**, 947 (1966).
112. Gambaryan, N. P., Rokhlin, E. M., Zeifman, Y. V., Simonyan, L. A., and Knunyants, I. L., *Dokl. Akad. Nauk SSSR Ser. Khim.* **166**, 864 (1966); *Dokl. Chem. (Engl. Transl.)* **166**, 161 (1966).

113. Germa, H., and Burgada, R., *Bull. Soc. Chim. Fr.*, p. 2007 (1975).

114. Gibson, J. A., Marat, R. K., and Janzen, A. F., *Can. J. Chem.* **53**, 3044 (1975).

115. Gibson, J. A., and Röschenthaler, G.-V., *J. Chem. Soc., Chem. Commun.*, p. 694 (1974).

116. Gibson, J. A., and Röschenthaler, G.-V., *J. Chem. Soc., Dalton Trans.*, p. 1440 (1976).

117. Gibson, J. A., Röschenthaler, G.-V., Sauerbrey, K., and Schmutzler, R., *Chem. Ber.* **110**, 3214 (1977).

118. Gibson, J. A., Röschenthaler, G.-V., Schomburg, D., and Sheldrick, W. S., *Chem. Ber.* **110**, 1887 (1977).

119. Gibson, J. A., Röschenthaler, G.-V., and Schmutzler, R., *J. Chem. Soc., Dalton Trans.*, p. 918 (1975).

120. Gibson, J. A., Röschenthaler, G.-V., and Schmutzler, R., *Z. Naturforsch. B: Anorg. Chem. Org. Chem.* **32**, 599 (1977).

121. Gibson, J. A., Röschenthaler, G.-V., Schmutzler, R., and Starke, R., *J. Chem. Soc., Dalton Trans.*, p. 450 (1977).

122. Gibson, J. A., Röschenthaler, G.-V., and Wray, V., *J. Chem. Soc., Dalton Trans.*, p. 1492 (1977).

123. Goggin, P. L., Goodfellow, R. J., Haddock, S.R., Taylor, B. F., and Marshall, I. R. H., *J. Chem. Soc., Dalton Trans.*, p. 459 (1976).

124. Greco, A., Green, M., Shakshooki, S. K., and Stone, F. G. A., *J. Chem. Soc., Chem. Commun.*, p. 1374 (1970).

125. Green, M., Cooke, M., and Kuc, T. A., *J. Chem. Soc. A*, p. 1200 (1971).

126. Green, M., Heathcock, S., and Wood, D. C., *J. Chem. Soc., Dalton Trans.*, p. 1564 (1973).

127. Green, M., Heathcock, S. M., Turney, T. W., and Mingos, D. M. P., *J. Chem. Soc., Dalton Trans.*, p. 204 (1977).

128. Green, M., Howard, J. A. K., Laguna, A., Murray, M., Spencer, J. L., and Stone, F. G. A., *J. Chem. Soc., Chem. Commun.*, p. 451 (1975).

129. Green, M., Howard, J. A. K., Laguna, A., Smart, L. E., Spencer, J. L., and Stone, F. G. A., *J. Chem. Soc., Dalton Trans.*, p. 278 (1977).

130. Green, M., Howard, J. A. K., Mitrprachachon, P., Pfeffer, M., Spencer, J. L., Stone, F. G. A., and Woodward, P., *J. Chem. Soc., Dalton Trans.*, p. 306 (1979).

131. Green, M., Howard, J. A. K., Spencer, J. L., and Stone, F. G. A., *J. Chem. Soc., Chem. Commun.*, p. 3 (1975).

132. Green, M., and Lewis, B., *J. Chem. Soc., Chem. Commun.*, p. 114 (1973).

133. Green, M., and Lewis, B., *J. Chem. Soc., Dalton Trans.*, p. 1137 (1975).

134. Green, M., Osborn, R. B. L., Rest, A. J., and Stone, F. G. A., *J. Chem. Soc. D*, p. 502 (1966).

135. Green, M., Shakshooki, S. K., and Stone, F. G. A., *J. Chem. Soc. A*, p. 2828 (1971).

136. Harris, J. F., *J. Org. Chem.* **30**, 2190 (1965).

136a. Hayashi, Y., Komiya, S., Yamamoto, T., and Yamamoto, A., *Chem. Lett.*, p. 1363 (1984).

137. Hayward, P. J., and Nyman, C. J., *J. Am. Chem. Soc.* **93**, 617 (1971).

138. Head, R. A., *J. Chem. Soc., Dalton Trans.*, p. 1637 (1982).

139. Hellwinkel, D., *Org. Phosphorus Compd.* **3**, 185 (1972).

140. Hermes, M. E., and Braun, R. A., *J. Org. Chem.* **31**, 2568 (1966).

141. Herz, J. E., and Cruz Montalvo, S., *J. Chem. Soc., Perkin Trans. I*, p. 1233 (1973).

142. Homsy, N. K., Ph.D. Thesis, University of Göttingen (1986).

143. Hoover, F. W., Stevenson, H. B., and Rothrock, H. S., *J. Org. Chem.* **28**, 1825 (1963).

143a. Howard, E. G., Sargeant, P. B., and Krespan, C. G., *J. Am. Chem. Soc.* **89**, 1422 (1967).

144. Howard, J. A., Russell, D. R., and Trippett, S., *J. Chem. Soc., Chem. Commun.*, p. 856 (1973).

145. Hunt, M. M., Kemmitt, D. W., Russell, D. R., and Tucker, P. A., *J. Chem. Soc., Dalton Trans.*, p. 287 (1979).

146. Igumnov, S. M., Sotnikov, N. V., and Sokol'skii, G. A., *Zh. Vses. Khim. O-va.* **26**, 98 (1980).

147. Igumnov, S. M., Sotnikov, N. V., Sokol'skii, G. A., and Knunyants, I. L., *Zh. Vses. Khim. O-va.* **26**, 97 (1980).

148. Ishihara, T., Shinjo, H., Inoue, Y., and Ando, T., *J. Fluorine Chem.* **22**, 1 (1982).

149. Ivin, S. Z., Promonenkova, V. K., and Fokin, E. A., *Zh. Obshch. Khim.* **37**, 2511 (1967); *J. Gen. Chem. USSR (Engl. Transl.)* **37**, 2388 (1967).

150. Janzen, A. F., Dalziel, J. R., Kay, S. N., and Galka, R., *J. Inorg. Nucl. Chem.* **43**, 629 (1981).

151. Janzen, A. F., Lemire, A. E., Marat, R. K., and Queen, A., *Can. J. Chem.* **61**, 2264 (1983).

152. Janzen, A. F., and Pollitt, R., *Can. J. Chem.* **48**, 1987 (1970).

153. Janzen, A. F., Rodesiler, P. F., and Willis, C. J., *J. Chem. Soc., Chem. Commun.*, p. 672 (1966).

154. Janzen, A. F., and Smyrl, T. G., *Can. J. Chem.* **50**, 1205 (1972).

155. Janzen, A. F., and Vaidya, O. M., *Can. J. Chem.* **51**, 1136 (1973).

156. Janzen, A. F., and Willis, C. J., *Can. J. Chem.* **43**, 3063 (1965).

157. Janzen, A. F., and Willis, C. J., *Can. J. Chem.* **44**, 745 (1966).

158. Janzen, A. F., and Willis, C. J., *Inorg. Chem.* **6**, 1900 (1967).

159. Johnson, M. P., and Trippett, S., *J. Chem. Soc., Perkin Trans. I*, p. 3074 (1981).

160. Johnson, M. P., and Trippett, S., *J. Chem. Soc., Perkin Trans. I*, p. 191 (1982).

161. Kibardin, A. M., Gazizov, T. K., and Pudovik, A. N., *Zh. Obshch. Khim.* **45**, 1982 (1975); *J. Gen. Chem. USSR (Engl. Transl.)* **45**, 1947 (1975).

162. Kibardin, A. M., Gazizov, T. K., and Pudovik, A. N., *Izv. Akad. Nauk SSSR, Ser. Khim.*, p. 1095 (1981); *Bull. Acad. Sci. USSR Div. Chem. Sci. (Engl. Transl.)* **30**, 855 (1981).

163. Knunyants, I. L., Ch'en, T.-Y., and Gambaryan, N. P., *Izv. Akad. Nauk. SSSR, Ser. Khim.*, p. 686 (1960); *Bull. Acad. Sci. USSR Div. Chem. Sci., (Engl. Transl.)*, p. 647 (1960).

164. Knunyants, I. L., Delyagina, I. N., and Igumnov, S. M., *Izv. Akad. Nauk. SSSR, Ser Khim.*, p. 860 (1981); *Bull. Acad. Sci. USSR Div. Chem. Sci., (Engl. Transl.)* **30**, 639 (1981).

165. Knunyants, I. L., Rokhlin, E. M., Gambaryan, N. P., Cheburkov, Y. A., and Ch'en T.-Y., *Khim. Nauka Promst.* **4**, 802 (1959).

165a. Kolodyazhnyi, O. I., *Zh. Obshch. Khim.* **54**, 966 (1984).

166. Konovalova, I. V., Burnaeva, L. A., Novikova, N. K., Kedrova, O. S., and Pudovik, A. N., *Zh. Obshch. Khim.* **51**, 995 (1981); *J. Gen. Chem. USSR (Engl. Transl.)* **51**, 831 (1981).

167. Krespan, C. G., *J. Org. Chem.* **43**, 637 (1978).

168. Krespan, C. G., and Middleton, W. J., *Fluorine Chem. Rev.* **1**, 145 (1967).

169. Kryukov, L. N., Isaev, V. L., Mal'kevich, L. Y., Truskanova, T. D., Sterlin, R. N., and Knunyants, I. L., *Zh. Vses. Khim. O-va.* **21**, 232 (1976).

170. Kryukov, L. N., Vitkovskii, V. S., Kryukova, L. Y., Isaev, V. L., Sterlin, R. N., and Knunyants, I. L., *Zh. Vses. Khim. O-va.* **22**, 355 (1977).

171. Kryukova, L. Y., Kryukov, L. N., Isaev, V. L., Sterlin, R. N., and Knunyants, I. L., *Zh. Vses. Khim. O-va.* **22**, 454 (1977).

172. Lavayssière, H., Dousse, G., and Satgé, J., *Helv. Chim. Acta* **59**, 1009 (1976).

173. Leader, G. R., *Anal. Chem.* **42**, 16 (1970).

173a. Lee, D. Y., and Martin, J. C., *J. Am. Chem. Soc.* **106**, 5745 (1984).

174. Lichtenberg, D. W., and Wojcicki, A., *Inorg. Chem.* **14**, 1295 (1975).

175. Lidy, W., and Sundermeyer, W., *Chem. Ber.* **106**, 587 (1973).

176. Lucas, J., Ph.D. Thesis, University of Frankfurt, 1984.

177. Lustig, M., and Hill, W. E., *Inorg. Chem.* **6**, 1448 (1967).

178. Mahler, W. (to E. I. du Pont de Nemours), U.S. Patent 3,265,705, 1966; *Chem. Abstr.* **65**, 16997 d (1966).

179. Martin, J. C., and Balthazor, T. M., *J. Am. Chem. Soc.* **99**, 152 (1977).

180. Martin, J. C., and Perozzi, E. F., *J. Am. Chem. Soc.* **96**, 3155 (1974).

181. Martin, J. W. L., and Willis, C. J., *Can. J. Chem.* **55**, 2459 (1977).
182. Massol, M., Barrau, J., and Satgé, J., *J. Organomet. Chem.* **25**, 81 (1970).
183. Mazhar-Ul-Haque, Caughlan, C. N., Ramirez, F., Pilot, J. F., and Smith, C. P., *J. Am. Chem. Soc.* **93**, 5229 (1971).
184. McGlinchey, M. J., and Stone, F. G. A., *J. Chem. Soc., Chem. Commun.*, p. 1265 (1970).
185. Middleton, W. J., *Kirk-Othmer Encycl. Chem. Technol. 3rd Ed.* **10**, 881 (1980).
186. Middleton, W. J., *J. Org. Chem.* **30**, 1402 (1965).
187. Middleton, W. J., and Carlson, H. D., *Org. Synth.* **50**, 81 (1970).
188. Middleton, W. J., England, D. C., and Krespan, C. G., *J. Org. Chem.* **32**, 948 (1967).
189. Middleton, W. J., and Krespan, C. G., *J. Org. Chem.* **30**, 1398 (1965).
190. Middleton, W. J., and Krespan, C. G., *J. Org. Chem.* **32**, 951 (1967).
191. Middleton, W. J., and Lindsey, R. V., *J. Am. Chem. Soc.* **86**, 4948 (1964).
192. Middleton, W. J., Metzger, D., Cunningham, K. B., and Krespan, C. G., *J. Heterocycl. Chem.* **7**, 1045 (1970).
193. Middleton, W. J., Metzger, D., and England, D. C., *J. Org. Chem.* **38**, 1751 (1973).
194. Mikhailov, B. M., Bubnov, Y. N., Tsyban', A. V., and Grigoryan, M. S., *J. Organomet. Chem.* **154**, 131 (1978).
195. Mir, Q. C., and Shreeve, J. M., *Inorg. Chem.* **19**, 1510 (1980).
196. Mitchell, C. M., and Stone, F. G. A., *J. Chem. Soc., Chem. Commun.*, p. 1263 (1970).
197. Modinos, A., and Woodward, P., *J. Chem. Soc., Dalton Trans.*, p. 2065 (1974).
198. Morton, D. W., and Neilson, R. H., *Organometallics* **1**, 289 (1982).
199. Mukhedkar, V. A., and Mukhedkar, A. J., *J. Inorg. Nucl. Chem.* **43**, 2801 (1981).
200. Mukhedkar, A. J., Mukhedkar, V. A., Green, M., and Stone, F. G. A., *J. Chem. Soc. A.*, p. 3166 (1970).
201. Neilson, R. H., and Goebel, D. W., *J. Chem. Soc., Chem. Commun.*, p. 769 (1979).
202. Newallis, P. E., and Rumanowski, E. I., *J. Org. Chem.* **29**, 3114 (1964).
203. Olah, G. A., and Pittman, C. U., *J. Am. Chem. Soc.* **88**, 3310 (1966).
204. Oram, R. K., and Trippett, S., *J. Chem. Soc., Perkin Trans. I*, p. 1300 (1973).
205. Oram, R. K., and Trippett, S., *J. Chem. Soc., Chem. Commun.*, p. 554 (1972).
206. Pellissier, N., *Org. Magn. Reson.* **9**, 563 (1977).
207. Perkins, C. W., and Martin, J. C., *J. Am. Chem. Soc.* **105**, 1377 (1983).
208. Pittman, A. G., and Sharp, D. L., *J. Org. Chem.* **31**, 2316 (1966).
209. Plakhova, V. F., and Gambaryan, N. P., *Izv. Akad. Nauk SSSR Ser. Khim.*, p. 681, (1962); *Bull. Acad. Sci. USSR, Div. Chem. Sci. (Engl. Transl.)*, p. 630 (1962).
210. Pogatzki, V. W., Ph.D. Thesis, University of Göttingen (1985).
211. Pudovik, A. N., Gazizov, T. K., and Kibardin, A. M., *Zh. Obshch. Khim.* **44**, 1210 (1974); *J. Gen. Chem. USSR (Engl. Transl.)* **44**, 1170 (1974).
212. Queen, A., Lemire, A. E., and Janzen, A. F., *Int. J. Chem. Kinet.* **13**, 411 (1981).
213. Raasch, M. S., *J. Org. Chem.* **37**, 1347 (1972).
214. Raasch, M. S., *J. Org. Chem.* **45**, 3517 (1980).
215. Raghavan, N. V., and Davis, R. E., *J. Cryst. Mol Struct.* **5**, 163 (1975).
216. Ramirez, F., *Bull. Chim. Soc. Fr.*, p. 3491 (1970).
217. Ramirez, F., Gulati, A. S., and Smith, C. P., *J. Am. Chem. Soc.* **89**, 6283 (1967).
218. Ramirez, F., Loewengart, G. V., Tsolis, E. A., and Tasaka, K., *J. Am. Chem. Soc.* **94**, 3531 (1972).
219. Ramirez, F., Pfohl, S., Tsolis, E. A., Pilot, J. F., Smith, C. P., Marquarding, D., Gillespie, P., and Hoffmann, P., *Phosphorus Relat. Group V Elem.* **1**, 1 (1971).
220. Ramirez, F., Smith, C. P., Gulati, A. S., and Patwardhan, A. V., *Tetrahedron Lett.*, p. 2151 (1966).
221. Ramirez, F., Smith, C. P., and Pilot, J. F., *J. Am. Chem. Soc.* **90**, 6726 (1968).

222. Ramirez, F., Smith, C. P., Pilot, J. F., and Gulati, A. S., *J. Org. Chem.* **33**, 3787 (1968).
223. Ramirez, F., Ugi, I., Lin, F., Pfohl, S., Hoffman, P., and Marquarding, D., *Tetrahedron* **30**, 371 (1974).
224. Redwood, M. E., and Willis, C. J., *Can. J. Chem.* **45**, 389 (1967).
225. Richter, R., Tucker, B., and Ulrich, H., *J. Org. Chem.* **48**, 1694 (1983).
226. Röschenthaler, G.-V., *Z. Naturforsch. B. Anorg. Chem. Org. Chem.* **33**, 131 (1978).
227. Röschenthaler, G.-V., *Z. Naturforsch. B. Anorg. Chem. Org. Chem.* **33**, 311 (1978).
228. Röschenthaler, G.-V., Gibson, J. A., Sauerbrey, K., and Schmutzler, R., *Z. Anorg. Allg. Chem.* **450**, 79 (1979).
229. Röschenthaler, G.-V., Gibson, J. A., and Schmutzler, R., *Chem. Ber.* **110**, 611 (1977).
230. Röschenthaler, G.-V., Sauerbrey, K., and Schmutzler, R., *Chem. Ber.* **111**, 3105 (1978).
231. Röschenthaler, G.-V., Sauerbrey, K., and Schmutzler, R., *Z. Naturforsch. B. Anorg. Chem. Org. Chem.* **34**, 107 (1979).
232. Röschenthaler, G.-V., Sauerbrey, K., and Schmutzler, R., *Isr. J. Chem.* **17**, 141 (1978).
233. Roesky, H. W., Djarrah, H., Lucas, J., Noltemeyer, M., and Sheldrick, G. M., *Angew. Chem.* **95**, 1029 (1983); *Angew. Chem., Int. Ed. Engl.* **22**, 1006 (1983).
234. Roesky, H. W., Hofmann, H., Noltemeyer, M., and Sheldrick G. M., *Z. Naturforsch. B. Anorg. Chem. Org. Chem.* **40**, 124 (1985).
235. Roesky, H. W., Homsy, N. K., Noltemeyer, M., and Sheldrick, G. M., *Angew. Chem.* **96**, 1002 (1984); *Angew. Chem., Int. Ed. Engl.* **23**, 1000 (1984).
236. Roesky, H. W., Homsy, N. K., Noltemeyer, M., and Sheldrick, G. M., *J. Chem. Soc., Dalton Trans.*, p. 2205 (1985).
237. Roesky, H. W., Lucas, J., Keller, K., Dhathathreyan, K. S., Noltemeyer, M., and Sheldrick, G. M., *Chem. Ber.* **118**, 2659 (1985).
238. Roesky, H. W., Lucas, J., Noltemeyer, M., and Sheldrick, G. M., to be published.
239. Roesky, H. W., Lucas, J., *et al.*, *Chem. Ber.* **118**, 2396 (1985).
240. Roesky, H. W., Lucas, J., Weber, K. L., Djarrah, H., Egert, E., Noltemeyer, M., and Sheldrick, G. M., *Chem. Ber.* **118**, 2396 (1985).
241. Roesky, H. W., Pogatzki, V.W., Dhathathreyan, K. S., Thiel, A., Schmidt, H.-G., Dyrbusch, M., Noltemeyer, M., and Sheldrick, G. M., *Chem. Ber.*, in press.
242. Pogatzki, V. W., and Roesky, H. W., *Chem. Ber.* **119**, 771 (1986).
243. Roesky, H. W., and Witt, M., *Rev. Inorg. Chem.* **4**, 45 (1982).
244. Rokhlin, E. M., Zeifman, Y. V., Cheburkov, Y. A., Gambaryan, N. P., and Knunyants, I. L., *Dokl. Akad. Nauk. SSSR* **161**, 1356 (1965); *Proc. Acad. Sci. USSR* **161**, 395 (1965).
245. Saunier, Y. M., Bougot, R. D., Danion, O., and Carrie, R., *Tetrahedron* **32**, 1995 (1976).
246. Schmidpeter, A., and v. Criegern, T., *Angew. Chem.* **90**, 64 (1978); *Angew. Chem., Int. Ed. Engl.* **17**, 55 (1978).
247. Schomburg, D., Weferling, N., and Schmutzler, R., *J. Chem. Soc., Chem. Commun.*, p. 810 (1981).
248. Seyferth, D., Tronich, W., Smith, W. E., and Hopper, S. P., *J. Organomet. Chem.* **67**, 341 (1974).
248a. Seyferth, D., Vick, S. D., and Shannon, M. L., *Organometallics* **3**, 1897 (1984).
249. Seyferth, D., Wursthorn, K. R., and Mammarella, R. E., *J. Org. Chem.* **42**, 3104 (1977).
250. Sheldrick, W. S., Schomburg, D., Schmidpeter, A., and v. Criegern, T., *Chem. Ber.* **113**, 55 (1980).
251. Shermolovich, Y. G., Kolesnik, N. P., Rozhkova, Z. Z., Kashkin, A. V., Bakhmutov, Y. L., and Markovskii, L. N., *Zh. Obshch. Khim.* **52**, 2526 (1982); *J. Gen. Chem. USSR (Engl. Transl.)* **52**, 2231 (1982).
252. Shimizu, N., and Bartlett, P. D., *J. Am. Chem. Soc.* **100**, 4261 (1978).
253. Simmons, H. E., and Wiley, D. W., *J. Am. Chem. Soc.* **82**, 2288 (1960).

254. Simonyan, L. A., Avetisyan, E. A., and Gambaryan, N. P., *Izv. Akad. Nauk SSSR, Ser. Khim.*, p. 2352 (1972); *Bull. Acad. Sci. USSR, Div. Chem. Sci. (Engl. Transl.)* **21**, 2294 (1972).
255. Simonyan, L. A., Gambaryan, N. P., and Knunyants, I. L., *Zh. Vses. Khim. O-va.* **11**, 467 (1966).
256. Sotnikov, N. V., Igumnov, S. M., Sokol'skii, G. A., and Knunyants, I. L., *Izv. Akad. Nauk SSSR Ser. Khim.*, p. 1671 (1980).
257. Sotnikov, N. V., Sokol'skii, G. A., and Knunyants, I. L., *Izv. Akad. Nauk SSSR Ser. Khim.*, p. 2168 (1977); *Bull. Acad. Sci. USSR Div. Chem. Sci. (Engl. Transl.)* **26**, 2009 (1979).
258. Steglich, W., Burger, K., Dürr, M., and Burgis, E., *Chem. Ber.* **107**, 1488 (1974).
259. Stockel, R. F., *Tetrahedron Lett.*, p. 2833 (1966).
260. Stockel, R. F., *J. Chem. Soc., Chem. Commun.*, p. 1594 (1968).
261. Stone, F. G. A., *Pure Appl. Chem.* **30**, 551 (1972).
262. Storzer, W., and Röschenthaler, G.-V., *Z. Naturforsch. B: Anorg. Chem. Org. Chem.* **33**, 305 (1978).
263. Storzer, W., Röschenthaler, G.-V., Schmutzler, R., and Sheldrick, W. S., *Chem. Ber.* **114**, 3609 (1981).
263a. Storzer, W., Schomburg, D., and Röschenthaler, G.-V., *Z. Naturforsch. B: Anorg. Chem. Org. Chem.* **36**, 1071 (1981).
264. Strong, R. L., Howard, W. M., and Tinklepaugh, R. L., *Ber. Bunsenges. Phys. Chem.* **72**, 200 (1968).
265. Taylor, S. L., Lee, D. Y., and Martin, J. C., *J. Org. Chem.* **48**, 4156 (1983).
266. Trippett, S., and Whittle, P. J., *J. Chem. Soc., Perkin Trans. I*, p. 2302 (1973).
267. Turbini, L. J., Golenwsky, G. M., and Porter, R. F., *Inorg. Chem.* **14**, 691 (1975).
268. Ugi, T., Marquarding, D., Klusacek, H., Gillespie, P., and Ramirez, F., *Acc. Chem. Res.* **4**, 288 (1971).
269. Ugi, I., and Ramirez, F., *Chem. Br.* **8**, 198 (1972).
270. Utebaev, U., Rokhlin, E. M., and Knunyants, I. L., *Izv. Akad. Nauk SSSR Ser. Khim.*, p. 2260 (1974); *Bull. Acad. Sci. USSR Div. Chem. Sci. (Engl. Transl.)* **23**, 2177 (1974).
271. Varetti, E. L., and Aymonino, P. J., *An. Asoc. Quim. Arg.* **58**, 17 (1970).
272. Varetti, E. L., and Aymonino, P. J., *J. Chem. Soc., Chem. Commun.*, p. 680 (1967).
273. Varwig, J., Ph.D. Thesis, University of Göttingen, 1976.
274. Varwig, J., and Mews, R., *Angew. Chem.* **89**, 675 (1977); *Angew. Chem. Int. Ed. Engl.* **16**, 646 (1977).
275. Vasil'ev, N. V., Kolomiets, A. F., and Sokol'skii, G. A., *Zh. Vses. Khim. O-va.* **25**, 703 (1980).
276. Vasil'ev, N. V., Kolomiets, A. F., and Sokol'skii, G. A., *Zh. Vses. Khim. O-va.* **26**, 350 (1981); *Mendeleev Chem. J.* **26** (2), 22 (1981).
276a. Vershinin, V. L., Vasil'ev, N. V., Kolomiets, A. F., and Sokol'skii, G.A., *Zh. Org. Khim.* **20**, 1806 (1984); *J. Org. Chem. USSR (Engl. Transl.)* **20**, 1646 (1984).
277. Volkholz, M., Stelzer, O., and Schmutzler, R., *Chem. Ber.* **111**, 890 (1978).
278. Volkovitskii, V. N., Knunyants, I. L., and Bykhovskaya, E. G., *Zh. Vses. Khim. O-va.* **18**, 112 (1973).
279. Volkovitskii, V. N., Zinov'eva, L. J., Bykhovskaya, E. G., and Knunyants, I. L., *Zh. Vses. Khim. O-va.* **19**, 470 (1974).
280. Volz, K., Zalkin, A. H., and Templeton, D. H., *Inorg. Chem.* **15**, 1827 (1976).
281. Weeks, P., and Gard, G. L., *J. Fluorine Chem.* **1**, 295 (1971/72).
282. Weferling, N., and Schmutzler, R., *Am. Chem. Soc. Symp. Ser.* **171**, 425 (1981).
283. Weferling, N., Schmutzler, R., and Sheldrick, W. S., *Liebigs Ann. Chem.*, p. 167 (1982).
284. Weidler-Kubanek, A.-M., and Litt, M. H., *J. Org. Chem.* **33**, 1844 (1968).

285. Weygand, F., and Burger, K., *Chem. Ber.* **99**, 2880 (1966).
286. Weygand, F., Burger, K., and Engelhardt, K., *Chem. Ber.* **99**, 1461 (1966).
287. Willis, C. J., *J. Chem. Soc., Chem. Commun.*, p. 944 (1972).
288. Willis, C. J., *J. Chem. Soc., Chem. Commun.*, p. 117 (1974).
289. Willis, C. J., and Cripps, W. S., *Can. J. Chem.* **53**, 809 (1975).
290. Witt, M. and Roesky, H. W., unpublished results.
291. Wittenbrook, L. S., *J. Heterocycl. Chem.* **12**, 37 (1975).
292. Young, D. E., Anderson, L. R., and Fox, W. B. (to Allied. Chem. Corp.) U.S. Patent 3,654,335 (1972); *Chem. Abstr.* **77**, 4893 (1972).
293. Zeifman, Y. V., Gambaryan, N. P., and Knunyants, I. L., *Dokl. Akad. Nauk SSSR, Ser. Khim.* **153**, 1334 (1963); *Dokl. Chem. Engl. Transl.* **153**, 1032 (1963).
294. Zeifman, Y. V., Gambaryan, N. P., and Knunyants, I. L., *Izv. Akad. Nauk SSSR, Ser. Khim.*, p. 450 (1965); *Bull. Acad. Sci. USSR Div. Chem. Sci. (Engl. Transl.)*, p. 435 (1965).
295. Zeifman, Y. V., Gambaryan, N. P., and Knunyants, I. L., *Izv. Akad. Nauk SSSR, Ser. Khim.*, p. 2046 (1965); *Bull. Acad. Sci. USSR Div. Chem. Sci. (Engl. Transl.)*, p. 2011 (1965).
296. Zeifman, Y. V., Ter-Gabriélyan, E. G., Del'tsova, D. P., and Gambaryan, N. P., *Izv. Akad. Nauk SSSR, Ser. Khim.*, p. 396 (1979); *Bull. Acad. Sci. USSR Div. Chem. Sci. (Engl. Transl.)* **28**, 366 (1979).

INDEX